ZEITSCHRIFT FÜR GEOMORPHOLOGIE

Annals of Geomorphology – Annales de Géomorphologie   Neue Folge

A journal recognized by the International Association of Geomorphologists (IAG)

Wiedergegründet von H. Mortensen, Göttingen – Herausgeber: W. Andres, Frankfurt/M. / V. R. Baker, Tucson / D. Barsch, Heidelberg / D. Busche, Würzburg / R. Dikau, Heidelberg / H. Hagedorn, Würzburg / E. Juvigné, Liège / S. Kozarski, Poznań / Y. Lageat, Clermont-Ferrand / E. Löffler, Saarbrücken / P. A. Pirazzoli, Meudon / Maria Sala, Barcelona / O. Slaymaker, Vancouver / L. Strömquist, Uppsala / M. A. Summerfield, Edinburgh / Heather A. Viles, Oxford / P. W. Williams, Auckland und K. H. Pfeffer, Tübingen, Schriftleiter.
Beirat: Hanna Bremer, Köln / K. M. Clayton, Norwich / R. W. Fairbridge, New York / A. Godard, Meudon / A. Gupta, Singapore / H. Heuberger, Salzburg / J. Hövermann, Göttingen / H. Mensching, Hamburg / K. Okunishi, Kyoto / A. Pissart, Liège / S. Rudberg, Västra Frölunda

Supplementband 106

Weathering – Erosion – Sedimentation

New studies in morphodynamics and processes

edited by K.-H. Pfeffer

with 154 figures and 33 tables

1996

GEBRÜDER BORNTRAEGER · BERLIN · STUTTGART

Die Deutsche Bibliothek – CIP-Einheitsaufnahme

[Zeitschrift für Geomorphologie / Supplementband]
Zeitschrift für Geomorphologie = Annals of geomorphology.
Supplementband. – Berlin ; Stuttgart : Borntraeger.

Früher Schriftenreihe
Reihe Supplementband zu: Zeitschrift für Geomorphologie
ISSN 0044-2798
NE: Annals of geomorphology / Supplementband

106. Weathering – erosion – sedimentation. – 1996

Weathering – erosion – sedimentation : new studies in morphodynamics and processes ; with 33 tables / ed. by K.-H. Pfeffer. – Berlin ; Stuttgart : Borntraeger, 1996
  (Annals of geomorphology : Supplementband ; 106)
  ISBN 3-443-21106-2
NE: Pfeffer, Karl-Heinz [Hrsg.]

ISBN 3-443-21106-2 / ISSN 0044-2798
© by Gebrüder Borntraeger, Berlin · Stuttgart, 1996
All rights reserved including translation into foreign languages. This journal or parts thereof may not be reproduced in any form without permission from the publishers.
Valid for users in USA: The appearance of the code at the bottom of the first page of an article in this journal indicates the copyright owner's consent that copies of the article may be made for personal or internal use, or for the personal or internal use of specific clients. This consent is given on the condition, however, that the copier pays the stated per-copy fee through the Copyright Clearance Center, Inc., P.O.B. 8891, Boston, Mass. 02114, for copying beyond that permitted by Sections 107 or 108 of the Copyright Law.
Printed in Germany by K. Triltsch, Würzburg

Contents

TWIDALE, C.R., J.R. VIDAL ROMANI, E.M. CAMPBELL & J.D. CENTENO: Sheet fractures: response to erosional offloading or to tectonic stress? (with 18 figures) 1– 24

VELDKAMP, A.: Late Cenozoic landform development in East Africa: The role of near base level planation within the dynamic etchplanation concept (with 3 figures) 25– 40

MURRAY-WALLACE, C.V., A.P. BELPERIO, J.H. CANN, D.J. HUNTLEY & J.R. PRESCOTT: Late Quaternary uplift history, Mount Gambier region, South Australia (with 7 figures and 2 tables) 41– 56

MATSUKURA, Y. & N. MATSUOKA: The effect of rock properties on rates of tafoni growth in coastal environments (with 6 figures and 4 tables) 57– 72

KIRCHNER, G.: Cavernous weathering in the Basin and Range area, southwestern USA and northwestern Mexico (with 10 figures and 2 tables) 73– 97

MALIK, J.N. & A.S. KHADKIKAR: Palaeoflood analysis of channel-fill deposits, central Tapti river basin, India (with 5 figures and 2 tables) 99–106

CONESA-GARCÍA, C., F. LÓPEZ-BERMÚDEZ & M.A. ROMERO-DÍAZ: Scale and morphometry interaction in the drainage network of badlands areas (with 5 figures and 9 tables) 107–124

EVANS, D.J.A.: A possible origin for a mega-fluting complex on the southern Alberta prairies, Canada (with 14 figures) 125–148

CALLES, B. & L. KULANDER: Likelihood of erosive rains in Lesotho (with 14 figures and 3 tables) 149–168

OGUCHI, T.: Late Quaternary hillslope erosion rates in Japanese mountains estimated from landform classification and morphometry (with 6 figures and 1 table) 169–181

WAINWRIGHT, J.: A comparison of the interrill infiltration, runoff and erosion characteristics of two contrasting 'badland' areas in southern France (with 7 figures and 2 tables) 183–198

ALLISON, R.J. & K.C. DAVIES: Ploughing blocks as evidence of down-slope sediment transport in the English Lake District (with 12 figures and 3 tables) 199–219

CHONGUIÇA, E.: Estimating sediment transport and reservoir sedimentation in an EIA framework (a case study applied to the Pequenos Limbobos Dam in Southern Mozambique) (with 7 figures and 2 tables) 221–237

XILIN LIU: Morphologic characteristics of debris flow fans in Xiaojiang Valley
of southwestern China (with 14 figures and 1 table) 239–254

WELLS, A.W. & M.R. BENNETT: A simple portable device for the measurement
of ground loss and surface changes (with 5 figures and 1 table) 255–265

ROMANESCU, G.: L'évolution hydrogéomorphologique du delta du Danube
Étape Pleistocène–Holocène inférieur (avec 22 figures et 1 tableau) 267–295

Preface

One of the aims of the Zeitschrift für Geomorphologie is to encourage papers from different geomorphological traditions, using different working methods. No schools of thought are favoured over others. In addition, the Zeitschrift für Geomorphologie tries to publish some rather innovative papers.

The liberal mentality of the editorial board and the international scope of the journal has led to an overwhelming number of high quality papers submitted recently to the Zeitschrift für Geomorphologie. The number and size of normal volumes of the journal are limited. To avoid unnecessary publication delays the editors have compiled this supplementary volume which mainly contains articles dealing with Weathering – Erosion – Sedimentation. The production of such a supplementary volume is considered to be a fair solution for the authors, readers and editors, allowing the rapid publication of papers originally destined for the normal issues.

K.-H. Pfeffer

Sheet fractures:
response to erosional offloading or to tectonic stress?

C. R. TWIDALE, Adelaide, J. R. VIDAL ROMANI, A. Coruña,
E. M. CAMPBELL, Adelaide and J. D. CENTENO, Madrid

with 18 figures

Summary. It has long been accepted that sheet fractures are due to pressure release consequent on erosional offloading, and the fractures are widely known as offloading joints, pressure release joints, or some similar term. All fractures are influenced by offloading for they close and disappear in depth as a result of lithostatic pressure. They find expression only as erosion brings them closer to the surface. But whether sheet fractures are entirely due to pressure release is questionable. Field evidence is cited that is inconsistent with pressure release but compatible with an origin involving compressional stress. For example: expansive tendencies consequent on pressure release would have been accommodated along pre-existing partings. There are discrepancies between the age of the sheet fractures and that of the land surfaces parallel to which, allegedly, they were formed. Sheet fractures and structures are characteristic of bornhardts, many of which evidently survive because they are in compression, with few open fractures. Minor forms associated with the release of compressive stress coexist with sheet fractures. Displacement along sheeting planes is evidenced. Some bornhardts display synformal structures, which is impossible in terms of offloading. The cross structures developed in sets of sheet fractures within the same bornhardt are also difficult to explain in terms of pressure release. The development of sheet partings parallel to land surfaces can be explained in terms of adjustment to planes of least principal stress. The development of sheet structure in two-stage (or multistage) bornhardts can be simulated experimentally.

Zusammenfassung. *Lagerklüfte als Konsequenz erosiver Entlastung oder tektonischen Stresses?* – Seit langem wird akzeptiert, daß Lagerklüfte im Zusammenhang mit Druckentlastung als Folge erosiver Entlastung entstehen. Diese Klüfte werden als erosive Entlastungs- oder Druckentlastungsklüfte bezeichnet. Alle Klüfte werden durch Entlastung beeinflußt, da sie mit zunehmender Tiefe als Folge des ansteigenden lithostatischen Drucks verschwinden. Sie treten erst in Erscheinung, wenn sie durch Erosion dichter an die Erdoberfläche gelangen. Ob jedoch Lagerklüfte zur Gänze dieses Ursprungs sind, ist fraglich. Geländebeobachtungen werden angeführt, die einer Genese durch Druckentlastung widersprechen, jedoch kompatibel mit der Entstehung in einem kompressiven Streßregime sind. So würden sich expansive Tendenzen als Folge der Druckentlastung entlang bereits bestehender Schwächezonen auswirken. Widersprüche bestehen zwischen dem Alter von Lagerklüften und dem der Landoberflächen, parallel zu denen sie angeblich gebildet worden sein sollen. Lagerklüfte und Strukturen sind charakteristisch für Inselberge, die ihre Existenz offensichtlich einem kompressiven Streßregime verdanken, die massiv sind und nur wenige offene Klüfte zeigen. Kleinformen, assoziiert mit der Freisetzung von kompressivem Streß, kommen gleichzeitig mit Lagerklüften vor. Versatz entlang von Lagerklüften läßt sich nachweisen. Manche Inselberge zeigen synformale Strukturen, die unmöglich durch Druckentlastung entstanden sein können. Kreuzungsstrukturen, die in Lagerkluftscharen derselben Inselberge entwickelt sind, können ebenfalls nur schwer mit der Druckentlastung erklärt werden. Die Entwicklung von Lagerklüften parallel zur Erdoberfläche kann mit der Anpassung an Flächen der geringsten Hauptnormalspannung erklärt werden. Die Entwicklung zwei- oder mehrphasiger Inselberge kann experimentell simuliert werden.

Résumé. *Les joints d'exfoliation: réponse à une décharge érosive ou à des efforts tectoniques?* – On a longtemps admis que l'exfoliation était due à la relaxation de contraintes par simple décharge érosive. C'est la raison pour laquelle ces diaclases courbes sont désignées, entre autres termes, sous les noms de joints de décharge ou de détente. Toutes les factures dépendent de la décharge érosive puisqu'elles se ferment et disparaissent en profondeur en liaison avec la pression lithostatique. Elles ne s'expriment qu'à seule condition que l'érosion les porte au voisinage de la surface, mais on peut se demander si tous les joints d'exfoliation sont produits par la seule décharge érosive. Les données de terrain suggèrent plutôt une origine par compression. Si l'expansion était due à la seule décharge, elle devrait se dissiper selon le système des discontinuités préalables, et il n'y aurait aucune raison qu'un nouveau système de fracturation se formât. Il existe des désaccords entre l'âge des joints d'exfoliation et celle des surfaces topographiques parallèlement auxquelles ils seraient supposés se former. De telles diaclases enveloppantes sont caractéristiques des bornhardts, dont beaucoup survivent d'évidence parce qu'ils sont en compression et montrent peu de fractures ouvertes. Des déplacements de long de ces plans courbes ont même été observés. Il est encore des bornhardts qui montrent des structures synformales qui sont incompatibles avec une décharge érosive. Le développement des différents systèmes de diaclasage qui recoupent ces mêmes joints d'exfoliation s'explique difficilement par des phénomènes de détente. Le fait que des diaclases courbes soient parallèles à la surface topographique peut s'expliquer par un ajustement aux plans de mondre tension. Les phénomènes d'exfoliation affectant des bornhardts bi- ou multi-phasés peuvent être expérimentalement reproduits.

Introduction

Sheet fractures are well and widely developed in massive rocks, i.e. rocks lacking other partings, and they have been discussed in the literature for more than a century. Yet there is no agreement as to their origin. Two contrasted interpretations hold sway. Almost without exception, earth scientists adhere to the pressure release or erosional offloading hypothesis. Engineers and engineering geologists, on the other hand, favour compressional stress.

After a review of nomenclature and a description of the characteristics of sheet fractures, a critique of previous explanations of their origin is presented. What are perceived to be critical lines of structural and morphological evidence bearing on the origin of sheet fractures are next reviewed, and this is followed by a discussion of the possible origins of the structures.

Nomenclature

Sheet fractures are antiformal, domical, synformal or basinal sets of partings that delineate more or less thick slabs or lenses of rock known as sheet structures (Photo 1). Unlike the thinner pseudobedding, spalling and lamination, sheet structure is at least 0.5 m, and frequently a few metres, thick. Sheet or sheeting fractures are also known as *Bankung, Lägerklufte, estructura en capas o descamacion*, exfoliation, and relief of load, offloading or pressure release joints (and several other, similar terms).

Exception can be taken to several of the names mentioned above either because they carry unwarranted implications or because they are used indiscriminately. "Relief of load" and similar terminologies pre-empt discussion of genesis. "Exfoliation" has been used of partings that define slivers 1 mm thick and slabs up to 10 m in thickness. The very term "joint" implies that no dislocation along the fracture has been demonstrated; which is

Fig. 1. Sheet fractures exposed in quarry near the Pindo, western (seaward) end of the Sierra Buxantes, some 70 km southwest of A Coruña, Galicia, Spain.

Fig. 2. The Kangaroo Tail, a sheet structure cutting across steeply dipping arkose, northwestern margin of Ayers Rock, central Australia.

Fig. 3. Sheet fractures cutting across columnar joints in dacite, Kolay Valley southern Gawler Ranges, South Australia.

manifestly incorrect in some instances. Here the term sheet fracture is applied to arcuate partings that most commonly run approximately parallel to the local land surface (Fig. 1) and is used in a genetically neutral sense.

Characteristics

Sheet fractures are found in a wide range of lithological environments, in various types of tectonic terranes (shields and cratons, orogens, platforms) and in many and varied climatic settings. They are characteristic of low, large-radius domes as well as of high, steep-sided bornhardts. They are especially well and widely developed in granite and other crystalline rocks, having been recorded, for example, in monzonite, granite gneiss, migmatite, syenite and gabbro (e.g. TWIDALE & BOURNE 1978a) but they are also developed in silicic volcanic rocks such as felsite, dacite and rhyolite, and in andesitic tuff, in sandstone terrains (Fig. 2), and, albeit rudimentarily, in metaquartzite, limestone and so on – sheet fractures are developed in many lithologies, though overwhelmingly in inherently strong massive rocks lacking open partings.

Where such features as orthogonal fracture systems, bedding, foliation, columnar joints due to cooling (Fig. 3), cross bedding, flow structures or rift and grain are present, sheet fractures cut across them (see e.g. DALE 1923). Many sheet fractures, in addition to being convex upward in hills are concave upward in valleys and merge laterally with synforms to produce a pattern of undulations (DALE 1923, p. 35), as for instance in the

Fig. 4. Multiple domes within bornhardt in the Kamiesberge, Namaqualand, South Africa.

Ozarks of Missouri, and in the Gawler Ranges, South Australia. They also commonly steepen markedly toward the vertical or near-vertical fracture zones that define the residuals (HILLS 1963, p. 368). Multiple structural domes are commonly developed within a single topographic form (Fig. 4). This is analogous to the "double sheet structure" described from parts of New England, U.S.A. (DALE 1923, p. 36).

Field evidence, as for instance at La Clarté, in Brittany, at Guitiriz in Galicia, northwestern Spain, and in the Rock of Ages Quarry, at Barre, Vermont, shows that though in general sheet structure thickens with depth (see e.g. DALE 1923), thinner sheets commonly occur between thicker ones and vice versa. Quarry and mine excavations suggest that at some sites sheet fractures extend to a few hundreds of metres below the land surface, though at other sites sheet fractures are evidently restricted to the near-surface zone.

At some sites, sheet structures increase in radius with depth (JAHNS 1943 – Fig. 5). In places, they thicken radially from the crestal zone, e.g. in the Leolooberge of southern Africa (LAGEAT 1989, Pl. IVB), and at Ucontitchie Hill, on northwestern Eyre Peninsula, South Australia.

Previous explanations and critique

Many explanations for sheet fractures have been proposed (for reviews see TWIDALE 1973, 1982, HOLZHAUSEN 1989) and two have received wide support. MERRILL (1897,

Fig. 5. Bank of sheet structures showing increased radius with depth, near the Pindo, Galicia, northwestern Spain.

p. 245), for example, considered that the partings are "...the result of torsional stress..." and BAIN (1931, p. 734) suggested that the (lateral) compression causing the (upwards) expansion and rupture was of tectonic origin. Lateral compression was the favoured explanation for sheet structure until early this century, and it is interesting to note that Gilbert, the originator of the rival offloading theory, later interpreted sheet structure in terms of compression (DALE 1923, p. 29).

The second, and most widely supported, hypothesis is that due to GILBERT (1904), who attributed sheet fractures to expansive or radial pressure release consequent on erosional offloading. The concept is simple and persuasive. Rocks like granite are emplaced at depth. Outcrops of granite imply erosional unloading of many hundreds of metres of superincumbent rock. In response to this unloading, the granite tends to expand radially, roughly normal to the land surface, and this stress is relieved by the development of the tangential fractures which are known as sheet fractures or offloading joints.

As CHAPMAN (1956) has pointed out, all fractures, whether faults or joints, are in a sense an expression of such erosional offloading, for all presumably disappear in depth under lithostatic loading. Strains find expression as partings when such loading has been reduced. But the strains were imposed by tectonic stresses, whereas proponents of the offloading hypothesis as outlined by GILBERT argue that sheeting is due wholly and solely to pressure release consequent on erosional offloading, and that it can occur even in otherwise unstressed rocks. Despite its being mechanically unsound (WOLTERS 1969), however, the offloading or pressure release hypothesis is widely accepted; so much so that many geologists habitually refer to sheet fractures as offloading joints.

Evidence and argument

(a) Some earlier evidence

Several lines of previously adduced evidence are especially relevant to the discussion. First, as was demonstrated over a century ago, slabs cut free in quarries expand. It was inferred that the rock mass was in substantial compression (NILES 1871, see also DALE 1923). Measurements of rock expansion in tunnels and direct measurements of crustal stress in many parts of the world show that horizontal stresses are considerable, that they are greater than vertical loading at the same sites, and that they are commonly greater than those theoretically anticipated from considerations of rock density and thickness (e.g. ISAACSON 1957, TALOBRÉ 1957, MOYE 1958). It has been shown that the horizontal stresses increase linearly with depth and that within some 10 m of the surface horizontal stresses are an order of magnitude greater than those in the vertical (HAST 1967, BROWN & HOEK 1978).

Such stresses are believed by some to result from strain energy locked into igneous rocks during solidification, in metamorphic rocks during recrystallisation, and in sedimentary strata during compaction or cementation (e.g. EMERY 1964). At the microscopic scale, an analogy is drawn between strain energy locked into a crystal and a compressed spring embedded in plastic. Others attribute horizontal stress to regional factors such as the tectonic loading at plate boundaries. Plate migration may induce horizontal stresses as a result of the differential movement of tectonic units within the continental crust.

Second, the evidence pointing to parts of the crust being in substantial compression is consistent with several observed features of sheet fractures. For example, the double sheet structures evidenced in many areas can be explained either in terms of cross folding or of shearing. Single superficial fractures such as are developed in well fractured silicic volcanics on Eyre Peninsula and in the Gawler Ranges, both in South Australia, are consistent with compressive stress being concentrated in the upper few metres of rock (HAST 1967, BROWN & HOEK 1978).

Third, sheet fractures cut across, are superimposed upon, and therefore postdate, a variety of structures and textures developed in the host rock, features ranging from orthogonal sets to bedding planes, columnar joints to foliation, intrusive contacts to rift and grain. Mechanically, it is unlikely that expansive tendencies consequent on offloading would not have been accommodated along pre-existing lines of weakness rather than induce a new set of fractures.

Fourth, there are inconsistencies between the known age of sheet fractures and the age of the surfaces to which they are supposedly related (TWIDALE 1971, p. 67). Thus on Dartmoor, southwestern England, the upland margins are scored by deep, narrow valleys that stand in contrast to the open shallow valleys of the high plain, and which are related to Quaternary changes of climate and sealevel (ORME 1963). Sheet fractures parallel the steep sides to these marginal valleys and in terms of pressure release are therefore also of Quaternary age. But radiometric dating shows that aplitic sills intruded along some sheet fractures are much older (essentially Mesozoic), so that the fractures must be still more ancient.

Persuasive as are these lines of evidence and argument, many aspects of sheet fractures are as well interpreted in terms of one hypothesis as of the other. The weight attached to evidence depends as much on training, experience and prejudice as on intrinsic merit (cf. READ 1957, p. xi). Here structural and morphological evidence that appears to be incompatible with the pressure release hypothesis is brought to bear on the problem.

(b) The dilemma presented by bornhardts

Bornhardts are domical hills characteristically developed in massive rocks, with well developed orthogonal fractures (which determine their plan form) and sheet fractures and sheet structures. They occur in multicyclic landscapes, in varied topographic and tectonic settings, and in a wide range of climatic conditions (e.g. TWIDALE & BOURNE 1978a, TWIDALE 1982). They occur in massifs, as in the dacitic Gawler and the granitic Everard ranges, both in the arid interior of South Australia, and the granitic and gneissic Kamiesberge of Namaqualand; and in isolated residuals or inselbergs.

Many geomorphologists interpret bornhardt inselbergs as the last, isolated, remnants resulting from long distance scarp retreat (e.g. KING 1949). In these terms the rounded form of the residuals is due to preferential weathering and erosion of the corners and edges of blocks or compartments, and sheet structure is due to pressure release. It is consequent on the morphology of the bornhardt.

The two stage hypothesis of bornhardt development, on the other hand (FALCONER 1911, but see also TWIDALE & VIDAL ROMANI 1994), is strongly supported by field evidence (TWIDALE 1982). Some residuals are associated with more resistant rock types, but many, perhaps most, bornhardts developed in granitic and other igneous or metamorphic materials apparently survive by virtue of the compartments on which they are developed being massive, i.e. lacking open fractures whereas the intervening compartments are well fractured (TWIDALE 1982). Those many residuals that are of the same rock type as underlies the adjacent valleys or plains have been explained in terms of their being developed on massive compartments related to horizontal compressional stress applied either directly (LAMEGO 1938) or in association with shear (TWIDALE 1980, 1982; also MERRILL 1897, see Fig. 6). Such compression would impede or even prevent the opening of any preexisting strains or fractures, thus making the compartment of rock compact or massive, and resistant to weathering and hence to erosion. It could also produce arcuate upward structures confined within blocks defined by orthogonal fractures, or antiforms and synforms which together form undulations underlying entire regions (DALE 1923, p. 35).

(c) Bornhardts in structurally anomalous settings

Sheet fractures developed in structurally anomalous contexts are frequently associated with bornhardts. Massive quartzites and sandstones most commonly give rise to plateau forms (where flat-lying) and to ridges and ranges (where tilted). Weathering along vertical fractures and preferential erosion of the weathered zones has in some areas of flat-lying arenaceous sequences given rise to fields of towers or domes (e.g YOUNG 1986), but domical residuals are developed on flat-lying sequences in the Colorado Plateau of Utah,

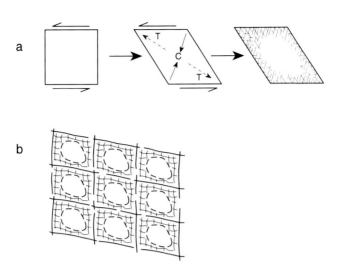

Fig. 6. Development of strain in sheared block (C – compression, T – tension).

U.S.A., for example the Kobon section of Zion National Park (YOUNG & YOUNG 1992, pp. 40–43). By no means all domical hills and towers developed in essentially undeformed sandstone or conglomerate display sheet fractures (e.g. the Bungle Bungle Ranges of northern Western Australia – see YOUNG 1986 – and the Mallos de Riglos, in the Pyrenees – see BARRÈRE 1968), and here any rounding that there is presumably results from preferential weathering. But some, at least, of the Kobon domes have well developed sets of sheet fractures. Here, the domical forms could result from the rounding of massive orthogonal blocks, with the sheet fractures having formed in parallel with the weathered outlines. Alternatively, shearing could cause the formation of arcuate fractures, the inner faces being concave inwards, at the corners of the fracture-defined blocks.

Domical forms are also well developed on deformed arenaceous sediments. Examples include Ayers Rock, central Australia (TWIDALE 1978), where the dip of the arkose approaches the vertical and where the sheet structure cuts across bedding (Fig. 2), and the adjacent conglomeratic Olgas complex (TWIDALE & BOURNE 1978b). Here only small compartments of the outcrop are upstanding. Thus, the ridges that usually would have formed along strike have been reduced to comparatively small domical massifs. They have been attributed to cross folding (TWIDALE 1978) though shearing may have played a role in their development (TWIDALE & CAMPBELL 1990). A small group of quartzitic residuals that includes Curtinye and Barna hills, is developed near Kimba, on northeastern Eyre Peninsula. In this area the Warrow Quartzite commonly gives rise to ridges, as in Darke Range and Caralue Bluff, but where rudimentary quaquaversally dipping sets of sheet fractures cut across the steeply dipping foliation, domical forms have developed (TWIDALE & CAMPBELL 1990). In these instances, it is difficult to avoid linking sheet fractures with domical forms, and concluding that it is structure that has determined surface morphology, and not the converse.

Fig. 7. A-tent (or pop-up) located on midslope of Enchanted Rock, in the Llano of central Texas, U.S.A.

(d) Minor morphological and structural features

If sheet fractures are associated with applied or tectonic stress, other forms and features may also be developed in response to it, and ought to be oriented consistently, and in geometrical relationship, with the direction of the stress.

Members of a suite of minor landforms construed as associated with the release of compressional stress (TWIDALE & SVED 1978) have been reported from many parts of the world. The assemblage comprises A-tents (or pop-ups), displaced slabs and horizontal wedges. Of these, A-tents (Fig. 7) are widely developed. They have been noted in many climatic environments (see TWIDALE 1982), and not only in granitic rocks but also in sandstone and limestone. The largest so far recorded in a natural setting occurs on Wudinna Hill, on northwestern Eyre Peninsula (JENNINGS & TWIDALE 1971). Large A-tents have been formed in recent times in the floors of quarries (e.g. EMERY 1964, COATES 1964) and in the scoured channels of rivers (e.g BOWLING & WOODWARD 1979), as well as on hillslopes (TWIDALE 1986).

A-tents are most plausibly explained in terms of release of compressive stress (TWIDALE & SVED 1978). During formation the slabs expand (by up to 4 % of their original length, but varying according to locality), because the deformation is non-elastic. Whatever its magnitude the expansion is irreversible and permanent.

A-tents have been interpreted as an expression of erosional offloading, despite the mechanism being found quantitatively inadequate (COATES 1964). In terms of this explanation, the features ought to be oriented consistently in relation to slope, whatever the inclination and aspect of the latter. Similarly, if they were due to insolation, they ought to be aligned along the contours of residuals and located on the sunny aspect, i.e. western and northern slopes in the southern hemisphere. Instead, on any given residual hill, and even in a given area, the crests of A-tents are roughly, and in some instances closely, aligned, and they are, moreover, distributed equally on all aspects. Thus, on Wudinna Hill (Fig. 8a), eight A-tents have been located, and their crests are aligned in an arc of 55°, and six in the range 215–240°, all magnetic (Fig. 8b). On nearby Polda Hill, the crests of all six A-tents lie on a large radius arc (Fig. 8c). Such consistent orientations are surely suggestive of a directionally applied stress, i.e. a tectonic origin. At some sites in Western Australia and in central Australia, the crests of the A-tents display two or more preferred orientations, presumably as a result of the application of compressive stress from different directions at different times, or reflecting the development of secondary shears and stresses (TWIDALE et al. 1993).

Ucontitchie Hill is notable for the development of wedges of rock of triangular cross section at the outer and lower edges of exposed sheet structure. Most are simple triangular wedges, but some are complex and consist of several minor constituents each of which can, however, be resolved as elements of a shattered single triangular wedge. The wedges vary in thickness between about 10 cm and 1 m, and are between 1 and 10 m long. Here, and at other sites, some remain in situ but others have been displaced laterally up to 2 m from their original position (Fig. 9), and moreover have expanded, for they do not fit the void in which they originated. Similar forms are reported from the Sierra Guadarrama of central Spain (see CENTENO 1989).

Some of these forms are explicable in terms of gravitational loading. Alternatively, they may be construed as due to differential movement along sheet fractures, induced by changing radii of curvature developed during antiform development (analogous to displacement along bedding planes during folding but in brittle granite resulting in fracture and dislodgement – Fig. 10). Expansion of the displaced mass on becoming unconfined could also cause such "bedding plane" dislocation.

Triangular wedges are also well developed in the Guitiriz area, some 60 km southeast of A Coruña, in Galicia, northwestern Spain. The granite in which they are found is a fine grained leucocratic rock, typically grey with black biotite, and of Hercynian age. The Mariz quarries exposures, though only 5–6 m deep maximum, consist of numerous slabs, most of them thin (10–15 cm), but with some up to 2 m thick, and thus qualifying as sheet structures (though the thinner slabs are of the same origin and are therefore relevant to the present discussion). They run parallel with the gentle slopes of the convex upward land surface. They cut across biotite lineation related to the emplacement of the granite. They form stacks within blocks defined by fractures which trend latitudinally and meridionally and which have been attributed to late Hercynian deformation. The sheet partings transect emplacement structures and textures. Some of the resultant structural domes are large radius, others small, and patterns of orthogonal cracking are developed on crests. In some instances the sheets plunge steeply toward the vertical fractures that

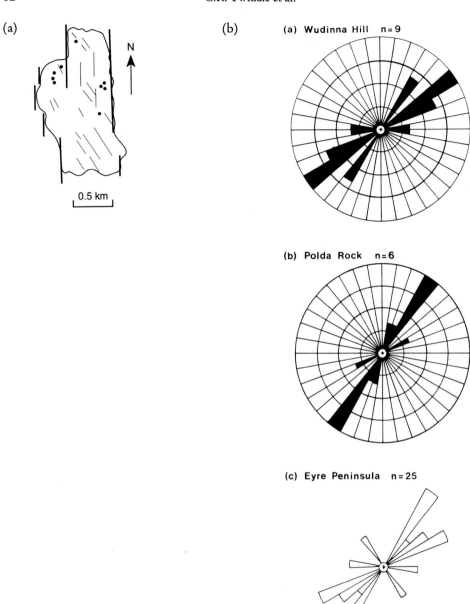

Fig. 8. (a) Location of A-tents on Wudinna Hill, Eyre Peninsula, South Australia. (b) Rose diagrams of orientation of crests of A-tents.

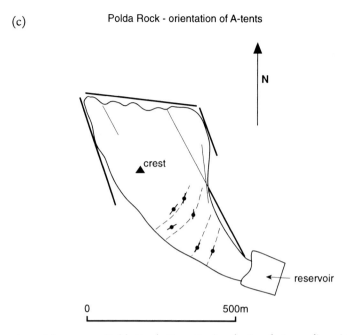

Fig. 8. (c) Location of A-tents on Polda Rock, Eyre Peninsula, South Australia, with suggested strain trajectories.

Fig. 9. Laterally displaced triangular wedge, eastern side of Ucontitchie Hill, northwestern Eyre Peninsula, South Australia.

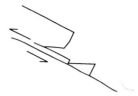

Fig. 10. Section showing suggested development of lateral wedge.

define the blocks or sectors. They have been interpreted as due to decompression consequent on erosional offloading.

Upward arching is, however, surely inconsistent with decompression or extension. In addition, overthrusting (Fig. 11a), imbricate structure and triangular wedges, most of them located between, rather than within, domical structures (Fig. 11b) are well developed. Unlike the wedges described from Ucontitchie Hill, these wedges have been squeezed upwards and are therefore referred to as vertical wedges. Such features cannot be related to tensional offloading and expansion but are best interpreted as due to recurrent compressional stress resulting first in arching, then rippling on fracture planes due to changes in radius of curvature. Following further shearing, the first-formed arches were distorted into domes to give an "egg box" effect (cf. RAMSAY & HUBER 1983, VIDAL ROMANI 1991). Such dislocations have developed after cooling, in the brittle phase. They predate some of the orthogonals and the foliation, because the latter transect secondary thrusts developed in the rippled sheet fractures. They could represent a late Hercynian event (e.g. ARANGUREN 1994). The assemblage of structures and forms at the Mariz quarries is inconsistent with the radial expansion implicit in the offloading hypothesis but is compatible with conditions of stress and strain within sheared orthogonal blocks (Fig. 6).

In the Sierra Guadarrama late Hercynian granites are subdivided by prominent meridional fractures, also of Hercynian age, and dated by the presence of Permian dykes. These fractures have been exploited by rivers so that many prominent valleys are linear. In one of the valleys of the Rio Tabalón, near Cadalso de los Vidrios, there is clear evidence of compressional stress (see CENTENO 1989). Displacement has taken place along a sheeting plane on the west side of the valley (Fig. 12) and imbrication occurs along another. An A-tent is developed on a nearby bornhardt, and what are construed as fault steps can be seen on the surface of the steeply dipping sheet structure that forms the opposite valley side. The orientations of these forms are consistent with E-W compression.

(d) Dislocation along sheet fractures

Though faulting can give rise to secondary partings which can be construed as sheet fractures (Fig. 13), the character of sheet fractures is critical to their interpretation. If offloading were responsible for sheet partings in the manner proposed by Gilbert and others, then, as only radial expansion is involved, there ought to be no differential movement along the fractures. In contrast, compression implies arching and changing radii of curvature with depth comparable to bedding plane dislocations in folded sequences. In this case there ought to be evidence of differential movement along the sheet fractures.

(a)

(b)

Fig. 11. (a) Overthrusting and (b) vertical triangular wedge in crest of dome, Mariz quarries, near Guitiriz, Galicia, northwestern Spain.

Fig. 12. Displacement along sheet fracture, Rio Tabalón valley, Sierra Guadarrama, central Spain.

Fig. 13. Part of the Rock of Ages Quarry, Barre, Vermont, in November 1965, showing general tendency for sheet structures to increase in thickness with depth, and faults running diagonally across the face from bottom left to top right.

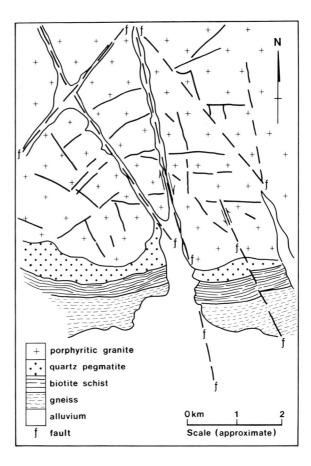

Fig. 14. Map of part of Kamiesberge, east of Garies, Namaqualand, South Africa, showing displacement along fractures delimiting blocks on which bornhardts developed.

If shearing has operated there ought to be evidence of faulting and of compression along one horizontal axis and extension along the other, normal to it.

Faults are difficult to locate in granitic terrains, and no doubt many of the fractures called joints are in fact faults (HILLS 1963, p. 365). In some bornhardt massifs with sheet fractures well developed, however, massifs such as the Kamiesberge of Namaqualand, South Africa, faulting is clearly demonstrated by displacement of veins and outcrops of distinctive rock types (Fig. 14), as well as by fault steps, slickensides and recrystallisation along the fracture planes. Similarly, in the Pindo, Galicia, sheet fractures close to those shown in Fig. 1 are displaced along a normal fault with slickensides, recrystallisation and drag.

The Ploumanac'h granite complex, of Hercynian age, is part of the Armorican Massif of northwestern France (BARRIÈRE 1976). Located on the north coast of Brittany around Ile Grande, between Brest and St Malo, several types of granite (as well as a basalt and mylonite) are exposed in a series of roughly concentric outcrops. The so-called 'red' granite, a coarse-grained biotitic rock, is extensively quarried. In an exposure near La Clarté, well-defined fault steps (Fig. 15) with associated slickensides and surficial recrystallisation are developed on a sheeting plane.

Fig. 15. Fault steps on sheet fracture exposed in quarry, La Clarté, Brittany, France.

On the other hand, there is evidence of tensional displacement between two sheets exposed on the eastern flank of the residual, with what appear to be fault steps pulled apart along a north-south or a NE–SW axis. Also at the northern and southern extremities of Ucontitchie Hill, sheeting fractures are gaping, with 15–20 cm between adjacent slabs. By contrast those exposed on the eastern and western flanks are tight, with partings only a centimetre wide maximum, and commonly much less. Thus there is suggestion of compression along one horizontal axis and of extension normal to it, consistent with the mass having been sheared (Fig. 6), and hence with the structural basis of the explanation of bornhardts suggested by TWIDALE (1980).

Lateral dislocation is also discernible on a recently formed sheet fracture at Quarry Hill, near Wudinna. The fracture was caused by blasting in the quarry in early 1993. As seen in the rock face, the fracture extends diagonally upwards from the point of detonation and approaches the convex upward slope at a moderate angle but as it extends outwards it comes into parallelism with the surface or plane of least principal stress (see MÜLLER 1964). The upper sheet moved south or downslope over the lower. Similarly at White Quarry, near Minnipa, northwestern Eyre Peninsula, the gritty grus found along some sheet partings has resulted from alteration of an already brecciated granite.

(e) Relation of sheet fractures to topography

The occurrence of bornhardts in multicyclic landscapes can be understood in terms of deep erosion of the rock mass, taking the land surface, including the bornhardts, into the deeper compressional zones of antiformal structures. But as was realised a century ago,

Fig. 16. Granite residual in the Joshua Tree National Monument, southern California, U.S.A., showing synformal sheet fracture intruded by sill and subsequently faulted.

some hill crests and synformal structures occur in association (HERRMANN 1895), and the same is true of some bornhardts. This is anomalous in terms of the offloading hypothesis, for if sheet fractures were related to erosional offloading then topography would have determined the geometry of the partings and in bornhardts, for example, the latter ought to be convex-upward, and in valleys, concave-upward; moreover, there ought to be a consistent positive relationship between structure and surface.

In many, perhaps most, exposures, whether natural or artificial, sheet structure and topographical surface are in close parallelism though there is no indication as to which precedes which, and cause and effect can readily be substituted, according to whether the sheet fractures are attributed to offloading or to compression. Elsewhere, there are obvious discordances between inclination of land surface and dip of sheeting fractures with the former more gentle than the latter, or with the surface steeper than the dip of the sheeting fractures. Weathering or erosion in well-jointed rocks subsequent to the formation of the sheet structure can reasonably be invoked.

At several sites, however, the discordances between surface and structure are much greater and more difficult of explanation, for the sheeting planes form synformal patterns within domical or other uplands. Thus at Tenaya Lake, in the Yosemite region of the Sierra Nevada, California, U.S.A., synformal traces are clearly visible in a monzonite dome (TWIDALE 1982a, p. 55). Little Shuteye Pass, also in the Sierra Nevada, is deservedly well known, if only through Hubbert's well known picture (see e.g. TWIDALE 1982a,

Fig. 17. Steeply dipping sheet fractures trending parallel to flanks of recent, deeply incised valley at the Pindo, northwestern Galicia, Spain. The geometry of these fractures contrasts markedly with the gently disposed fractures seen in the hillslope in the background.

p. 53) for its spectacular banks of convex-upward sheet structure exposed in the headwall of a cirque; but synformal fractures are visible near the top of the backwall. In the Joshua Tree National Monument, southern California, sheet structure is generally conformable with topography but there are exceptions which stand in inverse relationship to surface (Fig. 16). Similarly, synformal sheet fractures are exposed in a monzonite quarry at Central Tilba, southeastern New South Wales (GLYNN 1992) and in a granite quarry recently (1995) opened near Wudinna, northwestern Eyre Peninsula. Thus, in most instances sheet partings and surface are in parallel but at several sites, inverse relationships can be observed. Similarly, though conversely, valleys are frequently underlain by synformal fracture patterns, as for instance in the valley of Stout Creek, St Francis Mountains, Missouri, where sheeting is developed in rhyolite, and in the central Gawler Ranges. In some valley floors, however, as near Eucarro Dam, in the southern Gawler Ranges, convex-upward sheet partings are exposed.

Inverse relationships between structure and surface cannot be explained in terms of offloading. In bedrock that has been laterally compressed, they can be understood as a result of compression producing undulating structures underlying entire regions, as envisaged by DALE (1923, p. 35), followed by deep erosion into the tensional zones of antiforms and the development of associated drainage and topographic inversions.

Finally, the possibility of changes in the geometry of sheet fractures in response to changing bedrock topography undermines one of the most convincing arguments favouring offloading. At some sites, sheet fractures run in parallel with recently eroded surfaces, as for instances in glacially eroded regions (LEWIS 1954, KIERSCH 1964, GAGE 1966). They have been interpreted as consistent with offloading. But BAIN (1931, p. 734) pointed

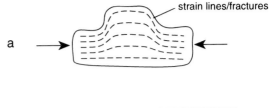

Fig. 18. (a) Stress trajectories in partly unconfined block subjected to horizontal compression (after HOLZHAUSEN 1989). (b) Application of HOLZHAUSEN model to development of bornhardt with sheet fractures in differentially weathered terrain.

out that in such cases the required pressure of ice causing compression of the underlying rock and its consequent expansion "greatly exceeds any conceivable stress on the stone due to glacial ice". Also, stress trajectories become aligned parallel to erosional or weathering surfaces (the surfaces of least principal stress: MÜLLER 1964), including the sidewalls of deeply incised valleys (Fig. 17). Such revised orientations may explain the spectacular sheeting fractures developed in the sides of canyons incised in flat-lying sandstone in the Colorado Plateau, U.S.A. (BRADLEY 1963), as well as overlapping slabs and otherwise curious local anomalies between fracture and surface; for some fractures may be relic from earlier land surfaces or previous weathering fronts, as for instance where sheet fractures change dip and overlap, as can be observed at various sites in Galicia, Texas, and in the Sierra Nevada of California.

Discussion

Clearly the field evidence discussed points to a relationship between sheet fractures and crustal stress. The evidence from the Mariz quarries seems incontrovertible on this point and much of the morphological evidence from Eyre Peninsula is consistent with such a conclusion. At the Mariz quarries, in the Wudinna district of Eyre Peninsula and in the Rio Tabalón, there is strong suggestion of tension along one horizontal axis and compression normal to this, indicating that the areas have been affected by shearing (Fig. 6), that the bornhardts are developed on compressed blocks and that sheet fractures as well as minor forms have evolved in response to such imposed stresses.

In these terms, the arcuate partings simply reflect the buckling of massive strong rocks. HOLZHAUSEN (1989) described arcuate-upward stress trajectories developed in a partly unconfined block subjected to uniaxial compression at its base (Fig. 18a). This situation can be translated to geomorphological reality. Consider for example bornhardts or domical hills developed in granitic terrains. There is clear evidence and argument that many such residuals are masses or columns of intrinsically fresh rock projecting into or through a mantle of weak, weathered material. According to HOLZHAUSEN's thesis, if the entire mass were compressed, the fresh cohesive rock would transmit the stresses, but the friable regolith would not. In this way the incipient bornhardts would acquire convex-upward sheet fractures (Fig. 18b).

Many bornhardts evolve in two main stages, one involving fracture-controlled subsurface weathering, and another the preferential stripping of some or all of the altered rock. As weathering reduced the size and changed the shape of any compact projections, so would strain trajectories accommodate to the changing outlines of the incipient bornhardt: as weathering by groundwaters proceeded, the flanks of the projections would be steepened, because groundwaters would run off them, and new, steep partings would develop in rough parallelism to the surface of the projecting mass of fresh rock. In this way, overlapping lenticular patterns would be formed. This is precisely the pattern of sheeting fractures found in many field localities.

Conclusion

Several lines of evidence and argument point to many sheet fractures being related to compressional stress. The evidence for substantial compression in many parts of the crust and in many granitic terranes is incontrovertible. Direct measurements show that the rocks are in substantial compression, possibly related to petrological processes superimposed on lithostatic loading, but more likely, in view of their geometry, due to tectonic forces. The practical experience of engineers confirms this, and there is in addition much geological and geomorphological argument and evidence pointing to the same conclusion. The comparison between stress trajectories developed in the unconfined section of a partly confined compressed block, and the convex upward sheeting fractures found in bornhardts that originated as projections standing above the general level of the weathering front, are especially telling. The geometrical relationships of sheeting fractures and A-tents are suggestive, as are dislocations along the sheet partings, the geometrical relationships of the fractures and topography, and, especially, the very survival of bornhardts.

Linking sheet fractures and structures with compressive stress offers an explanation of sheet fractures at least as worthy of consideration as that invoking pressure release. Certainly it is imprudent to beg the question and unthinkingly and uncritically refer to the fractures as offloading or pressure release joints.

Acknowledgements

The authors thank Professor David Stapledon (University of South Australia), Professor Jack Sharp (Austin, Texas), Dr Nicholas Gay (Johannesburg) and the late Dr George Sved (University of Adelaide), for their interest, advice and constructive comments.

References

ARANGUREN, A. (1994): Estructura cinematica del amplazamiento de los granitoides del Domo de Lugo y del Antiforme del Ollo de Sapo. – Serie Nova Terra **9**: 227 pp., Ediciòs do Castro.

BAIN, G.W. (1931): Spontaneous rock expansion. – J. Geol. **39**: 715–735.

BARRIÈRE, M. (1976): Architecture et dynamisme du complexe éruptif centre de Ploumanac'h (Bretagne). – Bur. Rech. Géol. Minières Bull. (Ser. 2) **13**: 247–295.

BARRÈRE, P. (1968): Le relief des Pyrénees centrales occidentales. – J. d'Études Pau-Biarritz **194**: 31–52.

BOWLING, A.J. & R.C. WOODWARD (1979): An investigation of near surface rock stresses at Copeton Damsite in New South Wales. – Aust. Geomech. J. **G9**: 5–13.

BRADLEY, W.C. (1963): Large scale exfoliation in massive sandstones of the Colorado Plateau. – Geol. Soc. Amer. Bull. **74**: 519–528.
BROWN, B.T. & E. HOEK (1978): Trends in relationships between measured rock in situ stresses and depth. – Inter. J. Rock Mech. Min. Sci. **16**: 211–215.
CENTENO, J.D. (1989): Quaternary evolution of relief in the southern slope of the Central Range of Spain. Residual forms as morphological indicators. – Cuad. Lab. Xeol. Laxe **13**: 79–82.
CHAPMAN, C.A. (1956): The control of jointing by topography. – J. Geol. **66**: 552–558.
COATES, D.F. (1964): Some cases of residual stress effects in engineering work. – In: JUDD, W.R. (ed.): The State of Stress in the Earth's Crust. – 679–688, Elsevier, New York.
DALE, T.N. (1923): The commercial granites of New England. – U.S. Geol. Surv. Bull. **738**.
EMERY, C.L. (1964): Strain energy in rocks. – In: JUDD, W.R. (ed.): The State of Stress in the Earth's Crust. – 235–260, Elsevier, New York.
FALCONER, J.D. (1911): The Geology and Geography of Northern Nigeria. – 295 pp., Macmillan, London.
GAGE, M. (1966): Franz Josef Glacier. – Ice **20**: 26–27.
GILBERT, G.K. (1904): Domes and dome structures of the High Sierra. – Geol. Soc. Amer. Bull. **15**: 29–36.
GLYNN, J-A.M. (1992): Geomorphological Evolution of the Mt Dromedary Igneous Complex, Far South Coast of N.S.W. – B. Sci. (Hons) thesis, Univ. Wollongong. (Unpubl.).
HAST, N. (1967): The state of stresses in the upper part of the Earth's crust. – Engng. Geol. **2** (1): 5–17.
HERRMANN, O. (1895): Technische Verwerthung der Lausitzer Granite. – Z. Prak. Geologie: 433–444.
HILLS, E.S. (1963): Elements of Structural Geology. – 483 pp., Methuen, London.
HOLZHAUSEN, G.R. (1989): Origin of sheet structure, 1. Morphology and boundary conditions. – Eng. Geol. **27**: 225–278.
ISAACSON, E. DE ST Q. (1957): Research into the rock burst problem on the Kolar Goldfield. – Mine Quarry Engng. **23**: 520–526.
JAHNS, R.H. (1943): Sheet structure in granites: its origin and use as a measure of glacial erosion in New England. – J. Geol. **51**: 71–98.
JENNINGS, J.N. & C.R. TWIDALE (1971): Origins and implications of the A-tent, a minor granite landform. – Aust. Geogr. Stud. **9**: 41–53.
KIERSCH, G.A. (1964): Vaiont Reservoir disaster. – Civ. Engng. **34**: 32–39.
KING, L.C. (1949): A theory of bornhardts. – Geogr. J. **112**: 83–87.
LAGEAT, Y. (1989): Le relief du Bushveld: une géomorphologie des roches basiques et ultrabasiques. – 596 pp., Faculté des Lettres et Sciences Humaines de l'Université Blaise-Pascal, Clermont-Ferrand.
LAMEGO, A.R. (1938): Escarpas do Rio de Janeiro. – Departamento Nacional da Produção Mineral (Brasil) Serie Geologia e Mineria Boletim **93**.
LEWIS, W.V. (1954): Pressure release and glacial erosion. – Jour. Glaciol. **2**: 417–422.
MERRILL, G.P. (1897): A Treatise on Rocks, Rock-weathering and Soils. – 411 pp., Macmillan, New York.
MOYE, D.G. (1958): Rock mechanics in the interpretation and construction of T.1 underground power station, Snowy Mountains, Australia. – 50 pp., A.G.M. Geol. Soc. Amer. Symp. Engng. Geol., St Louis, Nov. 1958.
MÜLLER, L. (1964): Application of rock mechanics in the design of rock slopes. – In: JUDD, W.R. (ed.): The State of Stress in the Earth's Crust. – 575–598, Elsevier, New York.
NILES, W.H. (1871): Peculiar phenomena observed in quarrying. – Proc. Boston Nat. Hist. Soc. **14**: 80–87.
ORME, A.R. (1963): The geomorphology of southern Dartmoor and the adjacent area. – In: SIMMONS, I.G. (ed.): Dartmoor Essays, pp. 31–72, Association for the Advancement of Science, Literature and Art, Torquay, Devonshire.

RAMSAY, J.G. & M.I. HUBER (1983): The Techniques of Modern Structural Geology. – 307 pp., Academic Press, London.
READ, H.H. (1957): The Granite Controversy. – 320 pp., Murby, London.
TALOBRÉ, J.A. (1957): La Mécanique des Roches Appliquée aux Travaux Publiques. – 333 pp., Dunod, Paris.
TWIDALE, C.R. (1971): Structural Landforms. – 247 pp. Australian University Press, Canberra.
– (1973): On the origin of sheet jointing. – Rock Mechanics **5**: 163–187.
– (1978): On the origin of Ayers Rock, central Australia. – Z. Geomorph. Suppl. **31**: 177–206.
– (1980): Origin of bornhardts. – J. Geol. Soc. Aust. **27**: 195–208.
– (1982): Granite Landforms. – 372 pp., Elsevier, Amsterdam.
– (1986): A recently formed A-tent on Mt Wudinna, Eyre Peninsula, South Australia. – Rev. Géomorph. Dynam. **35**: 21–24.
TWIDALE, C.R. & J.A. BOURNE (1978a): Bornhardts. – Z. Geomorph. Suppl. **31**: 111–137.
– – (1978b): Bornhardts developed in sedimentary rocks, central Australia. – S. Afr. Geogr. **60**: 34–50.
TWIDALE, C.R. & E.M. CAMPBELL (1990): Les Gawler Ranges, Australie du Sud: un massif de roches volcaniques silicieuses, la morphologie originale. – Rev. Géomorph. Dynam. **39**: 97–113.
TWIDALE, C.R. & G. SVED (1978): Minor granite landforms associated with the release of compressive stress. – Austr. Geogr. Stud. **16**: 161–174.
TWIDALE, C.R. & J.R. VIDAL ROMANI (1994): On the multistage development of etch forms. – Geomorph. **11**: 106–124.
TWIDALE, C.R., J.R. VIDAL ROMANI & E.M. CAMPBELL (1993): A-tents from The Granites, near Mt Magnet, Western Australia. – Rev. Géomorph. Dynam. **42**: 97–103.
VIDAL ROMANI, J.R. (1991): Kinds of plane fabric and their relation to the generation of granite forms. – Cuad. Lab. Xeol. Laxe **16**: 301–312.
WOLTERS, R. (1969): Zur Ursache der Entstehung oberflächenparalleler Klüfte. – Rock Mechanics **1**: 53–70.
YOUNG, R.W. (1986): Towerkarst in sandstone: Bungle Bungle massif, northwestern Australia. – Z. Geomorph. **30**: 189–202.
YOUNG, R.W. & A. YOUNG (1992): Sandstone Landforms. – 163 pp., Springer, Berlin.

Addresses of the authors: C.R. TWIDALE and E.M. CAMPBELL, Department of Geology and Geophysics, University of Adelaide, Adelaide, South Australia 5005, Australia. J.R. VIDAL ROMANI, Faculdad de Ciencias, Universidad de Coruña, Campus de Zapateira, 15071 A Coruña, Galicia, Spain. J.D. CENTENO, Departamento de Geodinamica, Faculdad de Ciencias Geologicas, Universidad Complutense de Madrid, 28040 Madrid, Spain.

Late Cenozoic landform development in East Africa:

The role of near base level planation within the dynamic etchplanation concept

A. VELDKAMP, Wageningen

with 3 figures

Summary. The Late Cenozoic landform development history of East Africa is explained insufficiently by the available landform genesis concepts. Although the 'dynamic etchplanation' concept is well able to explain the development of lithologically controlled relief, its explanation of planated areas does not fit reconstructed characteristics of the East African landscape history. Late Cenozoic plain formation in East Africa was an episodic event caused by contemporaneous erosion and deposition near base level. It is therefore proposed to elaborate the etchplanation concept with fluvial dynamics. Late Cenozoic planation events in East Africa were triggered by a major base level rise due to the combined effects of eustatic sea level rise and crustal movements associated with rift valley dynamics. The relative base level rise caused a sequence of active fluvial lateral aggradation and erosion within the landscape followed by lagoonal and sometimes even marine conditions shaping extensive plains. The interacting fluvial and etchplanation dynamics make a more comprehensive theory to explain the Late Cenozoic landscape development in East Africa.

Introduction

Theories on global long term landscape development have been relatively little discussed last decades as many regard them deficient in their treatment of exogenic geomorphic processes (SUMMERFIELD 1991). These theories have in common that they are applied globally, are conceptual in character and often static in their process approach. On the other hand, recent geomorphological studies are mostly locally or regionally valid and focused mainly on process dynamics. Ideally, a suitable geomorphological theory should bridge this scale gap by combining long term control effects of climate and tectonism with the short term dynamics of earth surface processes.

The East African landscape is well known for both its extensive lithological independent plains and its lithologically controlled relief above and in dissected parts of these plains (KING 1962, SAGGERRSON & BAKER 1965, SPÖNEMANN 1984, PYE et al. 1986, VELDKAMP & VISSER 1992). Recently, a reconstruction of the Late Cenozoic landscape development chronology in East Kenya was made by linking regional geomorphology and stratigraphy with radiometric dated volcanic and tectonic events (VELDKAMP & OOSTEROM 1994). This reconstruction demonstrated that the East African tectonic history has been rather eventful causing many changes in relative base level since the early Miocene. It also confirmed that existing long term landscape evolution theories, especially those focused on planation, are incorrect or too simplistic in their approach for East Africa. This result confirms the idea that more emphasis should be put on efforts to bridge the scale gap between the Quaternary process oriented studies and the global long term concepts. East Africa seems a very suitable study area to attempt a first step towards a more dynamic large scale landscape development concept for a longer time span (Late Cenozoic). The most plausible global landscape development concept for East Africa is the dynamic etchplanation theory (THOMAS 1994). This concept will be evaluated first followed by an attempt to elaborate it with fluvial system dynamics in order to give a more plausible explanation of the reconstructed characteristics of Late Cenozoic landscape development history of East Africa. The study area discussed will be limited to the region East of the Gregory rift valley (East Kenya and East Tanzania) because here a direct link with (paleo) base levels is possible and because Pleistocene rifting events complicated the landscape development further land inward (OLLIER 1981, EBINGER 1984).

Geological context of East Africa

The geology of East Africa (Fig. 1) is dominated by a rift system formed mainly within a Proterozoic orogenic belt surrounding the Archean cratons of central and eastern Africa. The Proterozoic Basement consists mainly of metamorphic rocks which originate from psammitic, (semi-) pelitic and calcareous sediments with thin intercalations of basic volcanics from sills or lava flows. These materials were deposited in geosynclinal troughs during the Middle Proterozoic (CHOUBERT & FAURE-MURET 1976). Deformation of the deposits took place in three major orogenic episodes (HACKMAN et al. 1990). The East African rift system lies atop a broad intra-continental swell, the East African Plateau, and consists of two branches, the Western and Eastern (Gregory) rift valleys which started to form since the Miocene. The sequence of active rifting began with doming and volcanism which was then followed by rifting (BAKER et al. 1971, HACKMAN et al. 1990, SUMMERFIELD 1991). In addition to upward vertical movements of the crust which gave rise to broad upwarps or swells, downward movements created basins. The intra-cratonic Zaire basin lies west of the western branch of the rift valley system, while the passive margin basins of Mombasa-Lamu and Tanzania, originating from Paleozoic/Mesozoic rift systems (CANNON et al. 1981), are found East of the Gregory rift valley along the actual coast. The succession of Mesozoic and Cenozoic marine and terrestrial deposits in the latter basins is known from oil-well logs and exposures (SAGGERSON & BAKER 1965, BISHOP 1966, WALTERS & LINTON 1973, CANNON et al. 1981, OOSTEROM 1988). The East African rift system is a true intra-plate rift and is characterized by prolonged volcanism and typically exhibit 1–2 km of down-faulting along the crest of substantial crustal upwarps. The volcanism and uplift result from the development of a hot region in the upper mantle, which itself probably originates from upwelling from the deeper mantle (KEARLEY & VINE 1990). The rift grabens are actually bordered by high angle normal fault systems. Along the length of the rift, approximately 100-km-long en echelon border-fault segments are linked by oblique slip transfer faults, ramps and monoclines within comparatively high strain accommodation zones. These transfer faults accommodate significant along axis variations in the elevation of basins and uplifted rift flanks but do not appear to extend outside the 40- to 70-km-wide rift basins cross the uplift flanks (BOS-

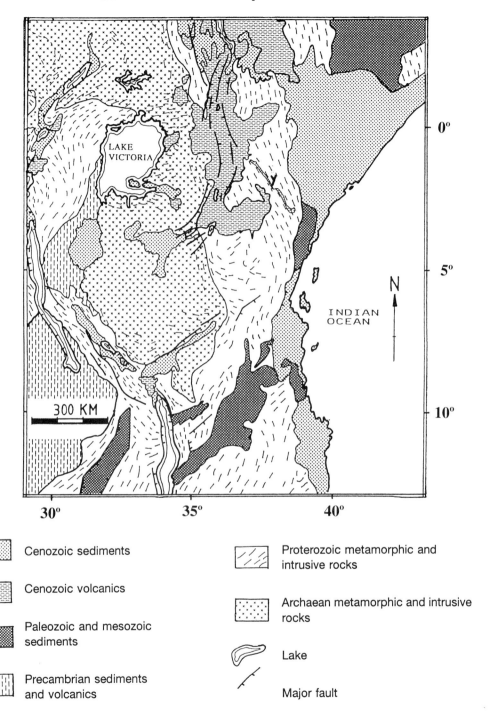

Fig. 1. Overview East African geology, based on information from BAKER et al. 1971, CHOUBERT et al. 1976, EBINGER 1989.

WORTH 1987, EBINGER 1989). The initial uplift of at least the eastern part of the plateau began in early Miocene time. During the Late Miocene, about 15 million years ago, great volumes of fluid phonolite erupted, filling up the crestal rift depression, and about 12 million years ago overflowed it east wards extending as the Yatta flow (Y, Fig. 1) to the Miocene coast line (EVERDEN & CURTIS 1965, BAKER et al. 1971). Associated with the rifts, but sometimes 100–200 km away from them, are the great volcanoes volcanic centres of Mt Kilimanjaro, Aberdares, Mt Kenya, Nyambeni range, Mt Elgon, Mt Marsabit. They all have complicated histories and have mostly been active since Late Tertiary to Quaternary times (BAKER et al. 1971, EBINGER 1984, BOSWORTH 1987, HACKMAN et al. 1990). Between the two rift valley branches the Lake Victoria basin developed during the Pleistocene as result of a drainage reversal caused by uplift of the western rift valley branch (OLLIER 1981).

Local Pliocene and early Pleistocene records are best known from the Gregory rift where large quantities volcanic ash and lake sediments have accumulated. The succession of the events in the Olduvia Gorge has been worked out by (HAY 1976). The oldest sediment dating from 2 to 1.75 Ma BP, were laid down in a perennial lake about 25 km wide. Other investigations have been carried out in the Turkana basin, were deltaic sediments laid down by the Omo form a long Pliocene-Pleistocene record (PATTERSON et al. 1970). The Turkana lake seems to have reached its greatest extent about 3.9 Ma BP indicating a similar time setting for high base level conditions as reconstructed in East Kenya (OOSTEROM 1988). Evidence for major Cenozoic sea level changes are found within the sediments of the Mombasa-Lamu and the coastal Tanzania basins (KARANJA 1983), while Quaternary sea level changes are witnessed by a flight of marine and fluvial terraces along the actual East African coast (COOKE 1974, OOSTEROM 1988).

East African geomorphology

Fig. 2 is a provisional (geo)morphological map of East Africa based on different physiographic, soil and geological maps and descriptions (KING 1962, SOMBROEK et al. 1982, SPÖNEMANN 1984, DE PAUW 1983, OOSTEROM 1988, VELDKAMP & OOSTEROM 1994). The East African geomorphology (Fig. 2) can be subdivided into three main units, *basins* in the coastal plains (e.g. Mombasa-Lamu and coastal Tanzanian basins), the Gregory *rift valley* with its associated volcanic region and *swells* of Basement and Karroo sediments containing both plains and dissected areas. This morphology can be more refined by discriminating the major volcanoes associated with rift valley genesis (BOSWORTH 1987), the different plains and dissected areas in between and above (inselberg complexes) these plains. Plain distributions and ages are only known for East Kenya to a certain extend, while the Tanzanian plains are hardly studied in detail making it even difficult to present a reliable map of their distribution. The unreliability of their distribution causes the provisional character of Fig. 2 and obstructed a more detailed subdivision which is almost certainly possible according a detailed study (see also OOSTEROM 1988). Due to the lack of accurate data the plain units are only grouped tentatively according to their approximate altitude and soil cover.

From the Indian Ocean land inward a coastal zone is found containing mainly marine and fluvial sediments locally covered by eolian sediments. The coastal zone shades grad-

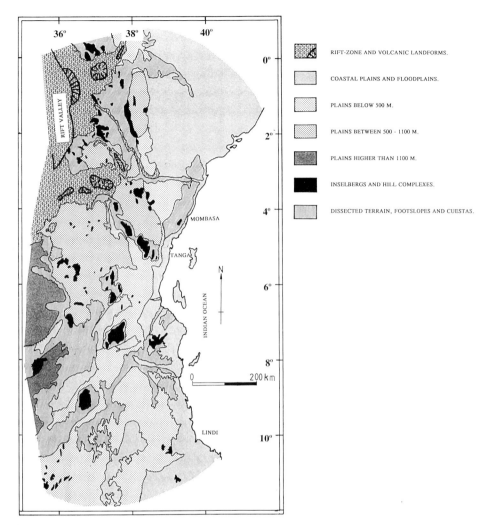

Fig. 2. Provisional Geomorphological map East Africa. Based on information from KING 1962, SOMBROEK et al. 1982, SPÖNEMANN 1984, OOSTEROM 1988, VELDKAMP & OOSTEROM 1994.

ually off into extensive sedimentary and erosional plains (plains below 500 m, Fig. 2) (SOMBROEK et al. 1982, OOSTEROM 1988). To the north and south of Tanga, such plains extend to Somalia and to coastal Mozambique. These plains are being backed by higher plains and inselberg complexes rising high above the plains. The higher plains are generally more tilted, weathered and dissected than the younger plains. In and along major incisions like Tana river, the typical lithologically controlled relief of a stripped etchplain is exposed (PYE et al. 1986, VELDKAMP & VISSER 1992). In northern Tanzania the lower plain zone is relatively narrow widening up north in Kenya and south in Tanzania. The plains rise up to the uplifted shoulders of the Gregory rift valley. Whether the central plateau of Tanzania (Plains higher than 1100 m, Fig. 2) has a similar genetical background as the other plains in Fig. 2 remains to investigated. The rift valley is generally 50–100 km wide with lakes occupying hollows in the floors of the troughs and escarpments rise as much as 1500 m to the uplifted shoulders. Summarizing, the East African landscape East of the Gregory rift valley and the Tanzanian central Plateau can be viewed as flight of plains with scarps and dissected terrain in between. The plains display various degrees of weathering and dissections depending on their geographical position.

Concepts of East African Landform genesis

Traditionally, the East African plains are explained by the often applied concept of cyclic and global pediplanation sensus L.C. KING (1962). Last decades, indications that KING's pediplanation concept is insufficient for the explanation of the extensive African plains have increased considerably (LE ROUX 1991, SUMMERFIELD 1991, VELDKAMP & OOSTEROM 1994). Other Kenyan investigations (SPÖNEMANN 1984, PYE et al. 1986, VELDKAMP & VISSER 1992) already demonstrated that the development of a lithological controlled morphology contradicts KING's concept and can be explained more conveniently by the processes of 'etching and stripping'. The comparable concepts of 'dopelten Einebnungsflächen' (BÜDEL 1957) or 'Dynamic Etchplanation' (THOMAS 1994) are essentially the same and are referred to in this paper as 'dynamic etchplanation'. Dynamic etchplanation has climatic and tectonic controls in time, determining whether etching or stripping is dominating the overall denudational activity. Within this theory the formation of extensive plains is caused by the prolonged operation of weathering with the gradual removal of the fine grained and solutional products. In Kenya this planation mechanism seems contradicted in three ways. 1) East Africa has known a general net uplift of hundreds of meters since Early Miocene (HACKMAN et al. 1990) which should have caused the etching and stripping processes to create more instead of less relief in time; 2) the periods during which extensive planation is known to have taken place, are, considered on geological time scale, limited (VELDKAMP & OOSTEROM 1994); 3) the reconstructed Late Cenozoic landscape development history of East Kenya shows that East African plains have been several times at near base level conditions (VELDKAMP & OOSTEROM 1994), a situation which must have hampered prolonged continuation of weathering processes considerably. These arguments suggest that the proposed planation mechanism within the dynamic etchplanation concept is incomplete for East Africa. When studying the original concepts it becomes obvious that they were originally designed for regional scales because much emphasis was put on slope processes, pediments and local streams. Effects of tectonic and climatic changes are treated in a generalistic way only, while the role of base level changes is hardly discussed. Other possible inconsistencies in the dynamic etchplanation theory are the lag effects between tectonic or base level changes and land inward responses, and the geographical control on climatic conditions. One of the most obvious dynamic links between any terrestrial environment and its base level is the fluvial system. Although it is argued that fluvial erosion is more dependent on the geomorphological position than on base level, investigations show that tropical fluvial dynamics can have considerable tectonic and base level control, as witnessed by fluvial terraces containing large amounts of coarse sediments and showing different longitudinal gradients (KROONENBERG & MELITZ 1983, OOSTEROM 1988, KLAMMER 1979).

The inactivity of tropical fluvial systems is also opposed by CRICKMAY (1975) and LOUIS (1964) who propose a major role of fluvial dynamics in the planation process. From above arguing it seems obvious that the dynamic etchplanation concept needs incorporation of fluvial dynamics, to make it more dynamic and applicable for larger scales. Process relationships with major *graded* fluvial systems are necessary. Previous research in Kenya revealed some characteristics of plain formation environment during the Late Cenozoic. These characteristics can be evaluated using known Quaternary fluvial dynamics and as such be incorporated within the dynamic etchplanation concept. By extending the concept in this way the landscape is viewed as a weathering mantle with a set of slopes within a dynamic fluvial system.

Reconstructed events of landscape evolution in East Africa

Based on dated volcanic rocks, tectonic displacements, sedimentary sequences, longitudinal profiles and landform morphology a tentative sequence of events in East Kenya could be reconstructed (VELDKAMP & OOSTEROM, 1994). The similarity in distribution, altitudes, morphology and soil covers of the various plains in both East Kenya and East Tanzania (Fig. 2) indicates that this reconstructed Late Cenozoic plain formation sequence is also valid for East Tanzania.

The Miocene doming of the initial Gregory rift valley caused erosional processes to dominate the landscape as evidenced by the preserved valley below the Yatta phonolite flow. The coast line reached, compared with the actual coast, far land inward to the end of the actual Yatta plateau. The coastal basins in Kenya and Tanzania were drowned as evidenced by lagoonal and marine sediments in these basins (KING 1962, SAGGERSON & BAKER 1965, BISHOP 1966, WALTERS & LINTON 1973, CANNON 1981, OOSTEROM 1988). During the Miocene eustatic sea level gradually lowered (HAQ et al. 1987) causing a shift in coastal zone to the East. During the Early Pliocene (ca. 5–2.9 Ma BP) an extensive planation took place. This planation was a near base level event as evidenced by reconstructed longitudinal profiles and unconformities in basin sediments, which all point to a relative coastal subsidence of East Kenya (VELDKAMP & OOSTEROM 1994). Evidences for a far land inward reaching transgression are also found in Tanzania were Pliocene marine sediments are found in the Tanzania basin (BISHOP 1966, KENT et al. 1971, COOKE 1974). The termination of the plain formation due to base level lowering seems to have happened relatively fast in East Kenya and was associated with rifting events in the Gregory rift valley and contemporaneous uplift of the eastern Gregory rift valley shoulder. The reconstructed eustatic sea level (HAQ et al. 1987) lowering of approximately 150 m during the same period is insufficient to explain the major change in base level alone. The relative base level lowering took place at the beginning of the Late Pliocene (around 2.9 Ma BP), initiating a period with dynamic etchplanation of the emerged plains. Etching was the dominating process on the plains and stripping along the plain edges and the incising fluvial systems. During the Early Pleistocene (ca. 2 Ma BP) a new episode of doming in Central Kenya/Tanzania and contemporaneous gradual coastal subsidence coincided with eustatic sea level rise (HAQ et al. 1987) occurred, starting a renewed near base level planation in East Africa. The accompanying transgression is still clearly evidenced by lagoonal and marine sediments in the Mombasa-Lamu basin in Kenya and in the coastal plain in Tanzania (SAGGERSON & BAKER 1965, WALTERS & LINTON 1973, SOMBROEK et al. 1982). Again rifting in the Gregory rift and relative quick uplift at ca. 1.7 Ma BP of East Kenya and Tanzania combined with 100 m eustatic sea level lowering (HAQ et al. 1987) seems to have been associated with the termination of plain formation. Since this last major rifting phase etching and stripping processes have shaped the East African landscape. During the Pleistocene and Holocene gradual uplift and relative minor base level changes have continued as evidenced by flights of fluvial and marine terraces (OOSTEROM 1988). Reconstructed tectonic movements during the Late Cenozoic along the northern Gregory rift valley corroborate the reconstructed crustal movements of this graben (HACKMAN et al. 1990). Recent numerical modelling exercises of tectonic evolution of continental rifts (VAN DER BEEK 1995) demonstrated the relevance of isostatic rebound of rift shoulders due to erosion and sedimentation. Although these coupled processes may partly account for the reconstructed vertical crustal movements in East Africa they are currently insufficient to explain the rapid and large tectonic movements.

During the Late Cenozoic as a whole etching and stripping shaped uninterruptedly the areas protruding above the changing base level. This accounts for the deeply weathered or strongly dissected character of the higher areas and plains situated above the lower plains and coastal zone. The reconstructed levelling seems to have been episodic and initiated by a relative rise in base level which must have had a major impact on fluvial system dynamics during the Late Cenozoic. The reconstructed base level changes of several hundreds of meters exceed the eustatic sea level changes reconstructed for the Cenozoic (HAQ et al. 1987). It is striking however, that the timing of major East African crustal movements appears to coincide with major reconstructed global sea level changes. Furthermore, recent research (DEMENOCAL 1995) has demonstrated that Plio-Pleistocene climate shows a stepwise increase in aridity with major changes at approximately 2.8, 1.7 and 1.0 Ma. The oldest two ages match surprisingly well with the reconstructed termination phases of major planation episodes in East Africa (2.9 and 1.7 Ma respectively).

Etchplanation and fluvial dynamics

In the original etchplanation concept alternating and contemporaneous etching and stripping are thought to cause dynamic denudation in time. Changes in process dominance are thought to be connected essentially in the periodicity of environmental changes (THOMAS 1991). Although flowing groundwater can effectively dissolve rock minerals, STALLARD (1988) envisages that weathering rates are ultimately limited by the formation of thick or impermeable soils restricting free access by water to unweathered material. An important implication of this view is that without continuous stripping, sediment transport by streams and steady groundwater flow which is also related to the fluvial system, the etching process will ultimately come to a stand still. On a regional scale the climatic conditions and fluvial system characteristics, determining storage time of sediments within a floodplain, can therefore be considered to be a main etch planation controlling factor. A domination of the stripping process, leading eventually to a relief increase due to the uncovering of the irregular basal weathering surface, is only possible when climatic conditions cause a less dense vegetation cover and when the fluvial system has sufficient erosion and transport capacity. The formation of high inselbergs (sometimes up to 300 m above the etchplain) which are thought to originate from repeated deep weathering and stripping, is only possible when the fluvial system has at least incised 300 m. This illustrates that the etching and stripping equilibrium is strongly related to fluvial system dynamics, which is directly and indirectly steered by climate, tectonism and base level.

East Africa has known considerably Quaternary net uplift causing most actual fluvial systems to incise into the uplifted areas (Fig. 2). This dissection and stripping of the landscape is still an actual process causing relatively high sediment loads in Kenyan and Tanzanian rivers (DUNNE et al. 1978, STROMQUIST 1981). To discuss possible past behaviour of the East African fluvial systems general descriptions of fluvial systems and their response to external and internal changes are sufficiently available (LEOPOLD et al. 1964, SCHUMM 1977, GREGORY et al. 1987). A numerical model study of long term fluvial dynamics based on the cited literature can give insight into possible effects of climatic (discharge and sediment supply), base level and tectonic changes on general fluvial dynamics (VELDKAMP & VERMEULEN 1989). Model simulations corroborate that the deeper and faster a fluvial incision takes place (due to regional uplift or base level lowering) the less lateral erosion a fluvial system will and can display. But at near base level conditions a fluvial system has a much higher lateral erosion capacity and shapes a wide effective flood plain (VELDKAMP & VAN DYKE 1994). Base level lowering or a regional uplift will often lead to a head ward incision of the fluvial system, eventually leading to an accelerated erosion of regoliths in the headwaters. Base level rise or crustal subsidence will lead to net sediment storage within the lower reaches of the system, while erosion in its head waters still continues. Under certain equilibrium conditions, as schematically explained by SUMMERFIELD (1991) no direct effects of changing base level can be expected. Possible climatic control on fluvial system behaviour is also well known from Quaternary reconstructions (GREGORY et al. 1987). When vegetation cover reduced dramatically with increasing aridity it caused an increase in (stripping) sediment supply during these arid episodes and a subsequently decrease in sediment supply was observed when vegetation cover restored. Glaciers on high mountain ranges and active volcanoes can supply extra sediment fluxes into a fluvial system complicating fluvial system response to other factors.

It was already mentioned that graded fluvial systems display a large lateral erosion capacity when they have a small longitudinal gradient as found near base level. Actual examples of dynamic lateral migrating rivers on cratons are known from the upper reaches of the Amazon who still receive large quantities of Andean sediments (RÄSÄNEN et al. 1987). Another example is found in the coastal plain of south-eastern United States where the fluvial systems are characterized by wide floodplains and valleys (MARKEWICH 1990). These observations and simulation results imply that the Late Cenozoic near base level conditions in East Africa must have considerably promoted the lateral erosion and migration capacity of the existing fluvial systems.

A proposed scheme of near base level planation

By combining the reconstructed events in the Late Cenozoic East African landscape with known fluvial system dynamics a more detailed reconstruction of the sequence of events leading to planation can be made. The role of climatic changes have been important in

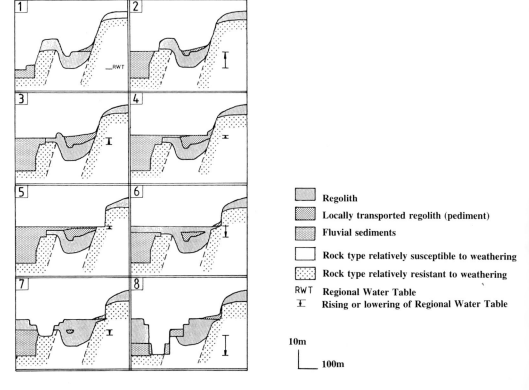

Fig. 3. Schematic presentation of eight stages during base level controlled planation sequence as reconstructed for East Africa.

[1] Weathering and stripping in an undulating landscape.

BASE LEVEL RISING

[2] Weathering continuous on higher part of landscape. Due to base level rise regional water table (RWT) rises and fluvial dynamics shifts from erosional to aggradation state limiting effectiveness regolith removal.
[3] Sediment from upstream is deposited filling in valleys and gradually hampering weathering processes. After valley infill lateral erosion and accretion of fluvial systems cause a effective levelling of large areas.
[4] The area becomes drowned gradually, stimulating laterite formation. Subsequent lagoonal conditions stimulate further in filling of depressions and eroding of remaining higher parts in the area.
[5] Eventually a lagoonal and marine environment with active wave erosion causes final levelling by cliff retreat.

BASE LEVEL LOWERING

[6] Base level lowering causes lowering RWT initiating renewed weathering without much erosion because the rising plain slope is too smooth and close to the equilibrium profile of a graded fluvial system.
[7] When the edge of the shaped plain is reached effective incision can start migrating land inward causing further drop in RWT and stimulating further etching and stripping processes.
[8] Both etching and stripping of the new raised plains starts and intensifies when the water table is further lowered in time.

determining the vegetation cover and related etching/stripping processes but their exact chronology is difficult to establish. East Africa has certainly known many oscillations in temperature and associated shifts in precipitation/evaporation ratios caused changes in vegetation cover as witnessed by changing lake levels, lake characteristics, deep sea sediment compositions (DeMENOCAL 1995). But since the exact impacts of these changes are still not well known it will be assumed that these climatic changes took place with a higher frequency than other major environmental changes linked to tectonism and sea level. The reconstruction will thus be focused on the effects of major base level changes caused by eustatic sea level and crustal movements.

Fig. 3 displays the discussed sequence in a very schematic way in a section parallel to the coast. It represents the vertical effects of changing process dominance due to base level changes at a fixed position in a planating area. As initial landscape an undulating area within a fluvial basin is assumed (Fig. 3, [1]), Both fluvial incision linked directly to a graded system and a local stream are considered in the reconstruction.

The reconstruction starts with a tectonic subsidence and/or sea level rise causing a major rise in base level. Such a rise will affect the lower reaches of the fluvial system first causing an upstream migration of the deposition zone and a rise in Regional Water Table (RWT) (Fig. 3, [2]). The fluvial system will fill in its own valleys and its lateral migration capacity will increase considerably. The local streams will gradually become chocked with stripped regoliths causing an increase in sediment residence time and an effective decrease in stripping (Fig. 3, [3]). In the head waters regolith erosion and weathering will continue (Fig. 3, [1]) and be related to the prevailing climate and relief. The upstream shifting deposition zone causes a further increase of the residence time of sediments in floodplains of lower and middle reaches, hampering local and regional stripping processes. When the base level continuous to rise the fluvial system will react by further valley infill and more lateral migration causing wider effective floodplains. Small tectonic tilts can potentially cause migration of such lateral sensitive system over extensive distances (RÄSÄNEN et al. 1987). Those part of the landscape not directly affected by the fluvial system will become locally waterlogged due to a gradual rise of the water table probably stimulating laterite formation. A continuing base level rise will gradually down the lower parts of the already fluvially flattened out landscape (Fig. 3, [4]) into a lagoonal environment. Finally wave erosion of standing water in a lagoonal or marine environment will give the finishing touch to the lower parts of the plain (Fig. 3, [5]).

The near equilibrium slope of the flat plain will prevent any direct erosional response at a subsequent base level lowering or stand still. Initially the lowering of the regional water table will stimulate weathering (Fig. 3, [6]). Only when the lowering base level reaches the steeper edge of the shaped plain a new effective land inward shifting incision can start (Fig. 3, [7]) causing a dissection of the formed plain. In the mean while effective weathering of the raised plain continuous, erasing most sedimentary evidence of past events, while plain incision continuous (Fig. 3, [8]). It depends on the geographical position when and how this incision takes place. The described sequence of events seems more than sufficient to explain the extensive planation as reconstructed in East Africa. The more landward parts of the plain will not know all eight described stages, the lagoonal and marine environments will mainly affect the lower parts. Stage 6, 7 and 8 are still traceable in the lowest plains while stages 7 & 8 dominate the higher plains. An appealing aspect of the proposed near base level planation scheme is that it happens by known processes and in known environments.

Discussion

The described planation mechanism contains characteristics of other proposed planation concepts. The dynamic etchplanation concept is not rejected but elaborated, while elements of CRICKMAY's (1975) hypothesis of unequal activity as well as a marine levelling (TOURAINE 1972) fit within the proposed scheme. This may suggest that the proposed planation sequence may be a step to a more general concept as it seems to unite different theories which probably contain all a nucleus of truth.

Another way to judge the proposed planation scheme on its merits is evaluating observations made in East Africa. Contrary to the lithological related dissected areas there is no obvious relationship between plain soils and underlying rock type, further has each plain its own relatively uniform soil type (same apparent age) (SOMBROEK et al. 1982, DE PAUW 1983). Within the proposed near base level planation concept a cover of transported regoliths, fluvial or lagoonal sediments is left on the plains serving as extensive and uniform parent material for the developing plain soils. The described planation also causes a rejuvenation of the plain as a whole in a relative sort time explaining the apparent uniform soil age. Evidence for a sedimentary cover of the Kenyan Plio-Pleistocene plain is relatively scare but still found (SAGGERSON & BAKER 1965). The occurrence of infilled valleys with lagoonal sediments including lignite in the Kenyan plain below 500 m with a Plio-Pleistocene age (SANDERS 1954, TOUBER 1979) near the transition zone to the higher situated Lower Pliocene plain gives clear evidence of the transition of fluvial to lagoonal environments (Fig. 3, [3] & [4]) contributing to plain formation. Other more circumstantial evidences of lagoonal and even marine environments are the extensive pothole zones and the cliff like morphology of the sea ward sites of inselberg complexes in East Kenya and East Tanzania (OOSTEROM 1988). The extensive pothole zones may also be attributed to the lateral migrating of fluvial systems over large areas. The latter is also suggested by extensive basal gravel layers consisting of rounded and subrounded clasts (OOSTEROM 1988). All these observations seem to confirm the overall picture of base level controlled planation and contradict the more commonly used etch planation and pediplanation concepts. Several cycles of major base level changes combined with a net tectonic uplift, resulted in a landscape with a series of plains. The oldest plains are highest elevated and have deeply weathered soils but are often strongly dissected. The younger and also lower situated plains are less dissected and contain locally unweathered sediments (transported regolith, fluvial, lagoonal or marine sediments) witnessing the various environments which contributed to the base level controlled levelling.

The proposed process sequence causing planation demonstrates that within the dynamic etchplanation concept plain formation on large scale can be satisfactory explained if the role of fluvial system dynamics and its controls are incorporated within this theory. The climate related etching and stripping processes effectively controlled the chemical denudation and regolith production, removal and supply to the fluvial system. The fluvial system acts as an important and dynamic link between these regional processes and more global factors such as tectonism and base level changes. Under certain conditions large plains can develop due to the interplay of rising base level, fluvial dynamics and etching/stripping processes. The proposed sequence of events has a very strong analogy with events on many continental shelves during the post glacial sea-level-rise. Taking into acount the magnitude and geographical position of the plains in East Africa and the fact that East African shelf is one of the narrowest in the world (KING 1962, p. 563), suggests that the extensive plains may be viewed as a kind of raised continental shelfs.

Examples of near base level planation outside East Africa

The reconstructed planation characteristics raise the question how regional these events have been. An attempt to correlate uplift, planation, continental margin sedimentation and sea level changes for the Late Mesozoic and Cenozoic of Africa was made by SUMMERFIELD (1991). His choice of time scale and the use of the pediplanation concept did not allow him a clear view of these correlations. His observation that along the margin of southern Africa major unconformities are found during the Late Pliocene-Early Pleistocene and Late Miocene-Early Pliocene correlate well with the reconstructed chronology for East Kenya. This apparent analogy is further corroborated by approximately 180 m uplifted Pliocene marine sediments in the coastal plain of South Africa (KING 1972) indicating a major shift of the deposition zone, which is base level related, and a tectonic control of this shift.

Comparable plain formation conditions, as found in East Africa, are proposed for South America by STALLARD (1988) who envisages sea level changes as major control on long term landscape development for cratons in South America. When sea level is high, sediment is deposited on the passive margins, continental platforms and intra-cratonic basins. At the same time near-sea-level planation surfaces are thought to develop on shield areas. His view is partly supported by ALEVA (1984) who notes that the development of major surfaces in the Amazon basin appears to coincide with major Plio-Pleistocene sea level high stands. He and KLAMMER (1978) also note that similar to East African plains, these extensive surfaces are also partially depositional in character.

Other base level related plains are found in the eastern Unites States as the "coastal plain" which is found up to 250 m and is described as both erosional and constructional terrain (DANIELS et al. 1978). The coastal plain is a seaward sloping plane about 160 km long underlain by Late Cretaceous to Holocene fluvial and marine sediments that dips gently to the Atlantic ocean. Within the 'coastal plain' several different plain levels divided by scarps can be divided (DANIELS et al. 1978). Within these plains the long term episodic character of erosion is suggested by clastic fluvial wedges. The plains have known a net uplift since Cenozoic and undergone several major changes in base level (MARKEWICH et al. 1990). Most plain levels are dated from Miocene to Pleistocene, a similar age range as for South and East Africa.

As suggested above a general tentative correlation between reconstructed East African plain formation episodes and sea level controlled landward shifts of coastal onlap (HAQ et al. 1987) seems evident. From the reconstructed East African landscape history, the analogue conditions and histories from South Africa, the Amazon basin and SE United States it can be proposed that in more than one region plain formation was *episodic* and related to both *near base level* deposition and erosion. As near base level planation seems not only confined to East Africa it becomes interesting to investigate whether the proposed role of fluvial system dynamics within the dynamic etchplanation concept is globally valid. The examples from the Amazon basin, south-east United States and South Africa, suggest that in those regions similar sequences of events took place during the Late Cenozoic. The reconstructed maximum changes in relative base level which occurred in South Africa (± 180 m), South America (± 180 m) and south-east United States (± 250 m) are less extreme than for the East African region (± 1100 m), and can be related mainly to eustatic sea level changes (HAQ et al. 1987) and/or geoid changes (MÖRNER 1987). The reconstructed East African base level changes exceed known sea level changes, indicating that significant tectonic movements, associated with the rift valley system development, played a dominant role in the landform development in East Africa during the Late Cenozoic.

Conclusions

The dynamic etchplanation concept combined with known long term fluvial dynamics allows a more plausible explanation of reconstructed East African landform development during the Late Cenozoic. Within this elaborated concept the dynamic equilibrium of etching and stripping processes is thought to be directly controlled by climate and indirectly by fluvial dynamics which are steered by climatic, tectonic and base level changes. Especially the role of base level seems to have been dominant within the Late Cenozoic landscape history of East Africa.

Further correlations and detailed descriptions combined with radiometric dating of plain sediments are needed to establish better insight in their rate of formation. The suggested analogy of the plain formation events in South Africa, the Amazon basin and south-eastern United States also needs further investigation. The existence of such a correlation implies a relationship between the chronology of fluctuating Cenozoic sea levels and planation episodes. The exact regional effects of these planation events seem both climatic and tectonic related. When more quantitative and accurate reconstructions are made, extensive numerical modelling exercises can shed more light on the plain formation history.

Acknowledgements

Prof. Dr Salomon Kroonenberg, Dr Piet Oosterom and Ir Philip Visser are gratefully acknowledged for their support, contributions and useful discussions. The International Soil Reference and Information Center (ISRIC) is thanked for their cooperation.

References

ALEVA, G.J.J. (1984): Lateritization, bauxitization and cyclic landscape development in the Guiana shield. – Proc. '84 Bauxite sympos., chapt. 13: 297–318.

BAKER, B.H., L.A.J. WILLIAMS, J.A. MILLER & F.J. FITCH (1971): Sequence and geochronology of the Kenya Rift volcanics. – Tectonophysics 11: 191–215.

BEEK, P. VAN DER (1995): Tectonic evolution of continental rifts. Inferences from numerical modelling and fission track thermochronology. – PhD thesis, 232 pp., Free University, Amsterdam.

BISHOP, W.W. (1966): Stratigraphical Geomorphology. A review of some East-African Landforms. – In: DURY, G.H. (ed.): Essays in Geomorphology.

BOSWORTH, W. (1987): Off-axis volcanism in the Gregory Rift, East Africa: implications for models of continental drifting. – Geology 15: 397–400.

BÜDEL, J. (1957): Die „Doppelten Einebenungsflächen" in den feuchten Tropen. – Z. Geomorph. N.F. 1: 201–228.

CANNON, R.T., W.M.N. SIMIYU SIAMBI & F.M. KARANJA (1981): The proto-Indian Ocean and a probable Paleozoic/Mesozoic triradial rift system in East Africa. – Earth and Planetary Science Letters 52: 419–426.

CHOUBERT, G. & A. FAURE-MURET (1976): Africa, explanatory note to sheet 6, 7 and 8. – In: Unesco & Commission for the Geological Map of the World. – Geological World Atlas, Unesco, Paris, France.

COOKE, H.J. (1974): Coastal geomorphology of Tanga. – Geogr. Rev. USA 64: 517–515.

CRICKMAY, C.H. (1975): The Hypothesis of unequal activity. – In: MELHORN, W.N. & R.C. FLEMAL (eds.): Theories of landform evolution: 103–109.

DANIELS, R.B., E.E. GAMBLE & W.H. WHEELER (1978): Age of Soil landscape in the Coastal plain of North Carolina. – Soil. Sci. Soc. Am. J. 42: 98–115.

deMENOCAL, P.B. (1995): Plio-Pleistocene Africa Climate. – Science 270: 53–59.

DUNNE, T., W.E. DIETRICH & M.J. BRUNENGO (1978): Recent and past erosion rates in semi-arid Kenya. – Z. Geomorph. N.F. Suppl. 29: 130–140.

EBINGER, C.J. (1989): Tectonic development of the western branch of the East African rift system. – Geol. Soc. Am. Bull. 101: 885–903.

EVERDEN, J.F. & G.H. CURTIS (1965): The potassium-argon dating of Late Cenozoic rocks in East Africa and Italy. – Current Anthropology 6: 343–385.

GREGORY, K.J., J. LEWIN & J.B. THORNES (eds.) (1987): Palaeohydrology in practice. – John Wiley & Sons Ltd., London.

HACKMAN, B.D., T.J. CHARSLEY, R.M. KEY & A.F. WILKINSON (1990): The development of the East African Rift System in north-central Kenya. – Tectonophysics 184: 189–211.

HAQ, B.U., J. HARDENBOL & P.R. VAIL (1987): Chronology of fluctuating sea levels since the Triassic. – Science 235: 1156–1166.

HAY, R.L. (1976): Geology of the Olduvia Gorge. A study of sedimentation in a semiarid basin. – University of California press Berkeley, Los Angeles, London.

KARANJA, F.M. (1983): The Carboniferous to Tertiary continental sediments of East Kenya. – 15 pp., Ministry of Kenya.

KEARLEY, P. & F.J. VINE (1990): Global tectonics. – Geoscience text Blackwell scientific publications, 302 pp., Oxford, London, Edinburgh, Boston, Melbourne.

KENT, P.E., J.A. HUNT & D.W. JOHNSTONE (1971): The geology and geophysics of coastal Tanzania. – Geoph. paper no. 6, 101 pp., Her Majesty's Stationary Office, London.

KING, L.C. (1962): The morphology of the Earth. A study and synthesis of world scenery. – Oliver and Boyd, Edinburgh, Great Britain.

– (1972): The coastal plain of Southeast Africa: its form, deposits and development. – Z. Geomorph. N.F. 16: 239–251.

KLAMMER, G. (1978): Reliefentwicklung im Amazonasbecken und pliopleistozäne Bewegungen des Meeresspiegels. – Z. Geomorph. N.F. 22: 390–416.

KROONENBERG, S.B. & P.J. MELITZ (1983): Summit level, bedrock control and the etchplain concept in the Basement of Suriname. – Geol. Mijnbouw 62: 389–399.

LEOPOLD, L.B., M.G. WOLMAN & J.P. MILLER (1964): Fluvial processes in geomorphology. – W.H. Freeman & Company, San Francisco.

LE ROUX, J.S. (1991): Is the pediplanation cycle a useful model? Evaluation in the Orange Free State (and elsewhere) in South Africa. – Z. Geomorph. N.F. 35: 175–185.

LOUIS, H. (1964): Über Rumpfflächen- und Talbildung in den wechselfeuchten Tropen, besonders nach Studien in Tanganjika. – Z. Geomorph. N.F. Sonderh. 8: 43–70.

MARKEWICH, H.W., M.J. PAVICH & G.R. BÜLL (1990): Contrasting soils and landscapes of the Piedmont and Coastal plain, eastern United States. – Geomorphology 3: 417–447.

MÖRNER, N.-A. (1987): Models of global sea level changes. – In: TOOLEY, M.J. & I. SHENNAN (eds.): Sea level changes. – 333–355, Blackwell 1986.
OLLIER, C.D. (1981): Tectonics and landforms (edited by K.M. CLAYTON). – Geomorphology texts 6, Longman, London, New York.
OOSTEROM, A.P. (1988): The geomorphology of Southeast Kenya. – PhD Thesis, 227 pp., Agricultural University, Wageningen.
PATTERSON, G., A.K. BEHRENSMEYER & W.D. SILL (1980): Geology and fauna of a new Pliocene locality in north western Kenya. – Nature 226: 918–921.
PAUW, E. DE (1983): Soils and physiography of Tanzania (Map 1:2,000,000). – FAO project crop monitoring and early warning systems. Netherlands Soil Survey Institute Wageningen.
PYE, K., A.S. GOUDIE & A. WATSON (1986): Petrological influence on differential weathering and inselberg development in the Kora area of Central Kenya. – Earth surface processes and landforms 11: 41–52.
RÄSÄNEN, M.E., J.S. SALO & R. KALLIOLA (1987): Fluvial perturbance in the Western Amazon Basin: regulation by long-term Sub-Andean Tectonic. – Science 238: 1398–1401.
SAGGERSON, E.P. & B.H. BAKER (1965): Post-Jurassic erosion-surfaces in eastern Kenya and their deformation in regulation to rift structure. – Quart. J. geol. Soc. Lond. 121: 51–72.
SANDERS, L.D. (1954): Geology of the Kitui area. – Report no 30, 37 pp., Geological Survey of Kenya, Government Printer, Nairobi.
SCHUMM, S.A. (1977): The fluvial system. – Wiley, New York.
SOMBROEK, W.G., H.M.H. BRAUN & B.J.A. VAN DER POUW (1982): Exploratory Soil map and Agro-climatic Zone map of Kenya. – Exploratory Soil Survey Report no E1, 56 pp., Kenya Soil Survey, Naurobi, Kenya.
SPÖNEMANN, J. (1984): Geomorphologie-Ostafrika (Kenya, Uganda, Tanzania) 2°N–2°S,32°–38°E. – Afrika-Kartenwerk, Serie E: Beiheft zu Blatt 2. Gebr. Borntraeger, Berlin, Stuttgart.
STALLARD, R.F. (1988): Weathering and erosion in the humid tropics. – In: LERMAN, A. & M. MEYBECK (eds.): Physical and chemical weathering in geochemical cycles. – 225–246, Kluwer Acad. Publ.
STROMQUIST, L. (1981): Recent studies on soil erosion, sediment transport and reservoir sedimentation in central Tanzania. – In: LAL, R. & E.W. RUSSELL (eds.): Tropical Hydrology. – 289–200, Wiley, Chichester.
SUMMERFIELD, M.A. (1991): Global geomorphology. An introduction to the study of landforms. – 537 pp., Longman Scientific Technical UK.
THOMAS, M.F. (1994): Geomorphology in the tropics. A study of weathering and denudation in low Latitudes. – 460 p., Wiley, Chichester.
TOURAINE, F. (1972): Erosion et planation. – Rev. Géogr. Alpine 60: 101–121.
VELDKAMP, A. & S.E.J.W. VERMEULEN (1989): River terrace formation, modelling, and 3-D graphical simulation. – Earth surf. proc. landforms 14: 641–654.
VELDKAMP, A. & P.W. VISSER (1992): Erosion surfaces in the Chuka-South Area, Central Kenya. – Z. Geomorph. N.F, Suppl. 84: 147–158.
VELDKAMP, A. & J.J. VAN DYKE (1994): Modelling of potential effects of long-term fluvial dynamics on possible geological storage facilities of nuclear waste in the Netherlands. – Geol. Mijnbouw 72: 237–249.
VELDKAMP, A. & A.P. OOSTEROM (1994): The role of episodic plain formation and continuous etching and stripping processes in the End-Tertiary landscape development of SE Kenya. – Z. Geomorph. N.F. 38, 75–90.
WALTERS, R. & R.E. LINTON (1973): The sedimentary basin of coastal Kenya. – In: BLANT, G. (ed.): Sedimentary basins of African Coasts. – Part 2, South and East Coast, 133–158, Ass. Afr. Geol. Surv. Paris.

Address of the author: A. VELDKAMP, Agricultural University Wageningen, Department of Soil Science and Geology, P.O. Box 37, 6700 AA Wageningen, The Netherlands.

Late Quaternary uplift history, Mount Gambier region, South Australia

C.V. MURRAY-WALLACE, Wollongong, A.P. BELPERIO, J.H. CANN, Adelaide,
D.J. HUNTLEY, Burnaby and J.R. PRESCOTT, Adelaide

with 7 figures and 2 tables

Summary. A series of Quaternary highstand shoreline deposits are preserved on the Coorong coastal plain of South Australia. One of these, the Woakwine Range, developed during the last interglacial maximum (oxygen isotope substage 5e), is preserved in pristine form for up to 250 km parallel to the modern shoreline. The barrier shoreline complexes consist of transgressive aeolian dunes of quartz-skeletal carbonate composition, that interfinger with estuarine-lagoon and lacustrine facies to the lee of the barrier structure. The estuarine-lagoon facies contain a peritidal fauna that permit confident estimation of past sea-level. A progressive change in the height of the back-barrier estuarine-lagoon facies along the coastal plain reveals significant differential uplift during the Late Quaternary. The last interglacial shoreline increases in elevation from 3 m at Salt Creek to 8 m near Robe, 100 km to the south, to 18 m at Mt. Gambier, a further 100 km south. Older shoreline deposits near Robe corresponding to stage 7e (Reedy Creek Range) and stage 9 (West Avenue Range) are preserved at 18 m and 24 m APSL respectively. Near Mt. Gambier, equivalent shoreline features mapped as Burleigh Range and Caveton Range are preserved at 34 and 38 m APSL respectively. Together, these data indicate spatial variation in rates of uplift with maxima centred on the Quaternary volcanic centres. Uplift rates range from 70 mm/ka near Robe to 130 mm/ka near the Holocene volcanic centres of Mounts Gambier and Schank.

Zusammenfassung. *Die jungquartäre Hebungsgeschichte der Mount Gambier-Region, Südaustralien.* – In der südaustralischen Coorong-Küstenebene ist eine Abfolge von Strandlinien eines quartären Meeresspiegelhochstands erhalten. Eine von ihnen, die Woakwine Range, entstand zum Höhepunkt des letzten Interglacials (Sauerstoffisotopensubstadium 5e) und ist auf fast 250 km Länge parallel zur heutigen Küstenlinie in ihrem ursprünglichen Zustand erhalten geblieben. Die Sandbarrierenkomplexe bestehen aus transgressiven Quarzsanddünen mit biogen-karbonatischem Bindemittel, die sich im Lee der Barriere mit estuarin-lagunärer und limnischer Fazies verzahnen. Die ästuarin-lagunäre Fazies enthält eine peritidale Fauna, aus der der ehemalige Meeresspiegel zuverlässig abgeleitet werden kann. Die zunehmende Höhenänderung der ästuarin-lagunären Fazies auf der Rückseite der Barriere weist auf bedeutende unterschiedliche Hebung während des Jungquartär hin. Die letztinterglaziale Strandlinie steigt von 3 m Höhe bei Salt Creek auf fast 8 m im 100 km südlich liegenden Robe an, und bei Mount Gambier, weitere 100 km südlich, auf 18 m. Ältere Strandliniensedimente bei Robe, die den Stadien 7e (Reedy Creek Range) und 9 (West Avenue Range) entsprechen, sind in 18 bzw. 24 m Meereshöhe erhalten. Äquivalente Küstenlinien bei Mt. Gambier, die als Burleigh und Caveton Range kartiert worden sind, sind in 34 und 38 m Höhe erhalten geblieben. Im Zusammenhang belegen diese Daten räumliche Unterschiede der Hebungsraten, wobei die Maxima in den Zentren des quartären Vulkanismus liegen. Die Hebungsbeträge reichen von 70 mm/ka bei Robe bis 130 mm/ka nahe der holozänen Vulkangebiete von Mt. Gambier und Mt. Schank.

Résumé. *L'histoire du soulèvement du Quaternaire supérieur dans la région du Mont Gambier, Australie méridionale.* – Une série de dépôts littoraux quaternaires est préservée au-dessus du niveau marin actuel sur la plaine côtière de Coorong dans le sud de l'Australie. L'un d'entre eux, la Woakwine Range, date du maximum de la dernière période interglaciaire (appartenant au cinquième stade isotopique). Remarquablement préservée, ces dépôts se suivent sur quelque 250 km parallèlement au littoral actuel. Ils consistent en dunes de quartz et de carbonate d'origine biogénique qui passent latéralement, en position d'abri en arrière d'une barre, à des faciès lagunaires et lacustres. Ces derniers contiennent une faune péritidale qui autorise une estimation satisfaisante de l'ancien niveau marin. Le changement progressif d'altitude de ces dépôts suggère un soulèvement différencié durant le Quaternaire supérieur. Le dernier niveau marin interglaciaire passe de + 3 m à Salt Creek à +8 m près de Robe, à 100 km au sud, et à + 18 m au Mont Gambier, situé 100 km encore plus au sud. Des dépôts côtiers plus anciens, près de Robe, correspondant au stade isotopique 7e (Reedy Creek Range) et au stade 9 (West Avenue Range) sont préservés à + 18 m et + 24 m au-dessus du niveau marin actuel. Près du Mont Gambier, des dépôts équivalents sont respectivement conservés à +34 m (Burleigh Range) et + 38 m (Caveton Range). L'ensemble de ces données indique des variations spatiales dans les vitesses de soulèvement, les chiffres les plus élevés correspondant aux centres volcaniques quaternaires. Ils sont compris entre 70 mm/ka près de Robe à 130 mm/ka près des volcans holocènes des Monts Gambier et Schank.

1 Introduction

Since the early observations of WOODS (1862), a number of studies have been undertaken in the southeast of South Australia, examining aspects of the region's Quaternary coastal history (e.g. SPRIGG 1952, COOK et al. 1977, VON DER BORCH et al. 1980, SCHWEBEL 1984). These investigations have highlighted the geological significance of the Coorong coastal plain in a global context, for it preserves a detailed record of barrier shoreline sedimentation during interglacial high sea-level stands through much of the Quaternary. The "staircase" of emergent shoreline barriers represents one of the world's most visible and comprehensive Quaternary coastal records. The combination of a moderate rate of tectonic uplift, surficial calcrete development, restricted surface drainage and minimal erosion has helped preserve this remarkable record.

Although SPRIGG (1952) assigned preliminary ages to some of the coastal barriers based on the timescale of the earth's orbital parameters as outlined in the Milankovitch hypothesis, it was not until recent advances in Quaternary dating methods, that the age and structure of the individual barriers could be confidently determined. Collectively, the following geochronological elements have been identified in the morphostratigraphy of the Coorong coastal plain: (a) the Brunhes-Matuyama boundary (780 ka), between the East and West Naracoorte Ranges (COOK et al. 1977, IDNURM & COOK 1980); (b) a last interglacial age (oxygen isotope substage 5e, ~125 ka) for the Woakwine Range (VON DER BORCH et al. 1980, SCHWEBEL 1984, MURRAY-WALLACE et al. 1991), and (c) a sequential progression of thermoluminescence ages for the sea level highstands from stage 5 to stage 21 (HUNTLEY et al. 1993a, b, 1994).

SPRIGG (1952) first examined the relationship of palaeo shoreline altitude to the nature of tectonic uplift along the coastal plain, based on the height above present sea level of flats in front of the stranded beaches, as well as the elevation of the top of the Naracoorte Range. He noted that the uplift rate varied spatially being higher in the south. The greatest site density in SPRIGG's study was between Kingston and Robe (Fig. 1). Here, we examine the spatial variability in Late Quaternary neotectonic uplift of the Coorong coastal plain and based on the elevation of the three barrier shoreline successions, extend the spatial coverage of Sprigg's study to Mount Gambier. Analyses and dating of emergent intertidal coastal facies preserved at the rear of the barriers, which we consider provides a more accurate measure of past sea level, forms the basis of this study. In particular, the last interglacial, penultimate interglacial and pre-penultimate interglacial shorelines are used as datums across the coastal plain to document spatial variability of uplift. We have used amino acid racemisation, luminescence and radiocarbon dating as a chronological framework to constrain the morphostratigraphic interpretation of the barrier shorelines.

Geomorphologic and climatic setting

The Coorong coastal plain is mantled by a series of stranded barrier shoreline deposits that represent well-defined but subdued features in the regional landscape. The relict barriers have been locally referred to as "dune ranges" after local place names. Their maximum height with respect to their back-barrier depressions seldom exceeds 35 m. The Holocene dune barrier (Younghusband Peninsula) and the Coorong lagoon provide a modern analogue for the stranded Pleistocene structures. The barriers consist predominantly of biogenic, skeletal carbonate sands with rare conglomerates and thin shell beds, and interfinger in their lee with fossil mollusc-bearing intertidal, estuarine lagoon and lacustrine sediments. The back-barrier facies are critical both for calculating past sea level and for providing fossil molluscs for amino acid racemisation dating. The barriers range in age from Early Pleistocene to the present and collectively represent one of the world's longest emergent Quaternary sea-level records (Sprigg 1952, Cook et al. 1977, Schwebel 1984, Belperio & Cann 1990). The barriers are shoreline sand bodies associated with successive Quaternary sea-level high stands and have been preserved by progressive uplift of the coastal plain and pervasive calcrete development. The driving force for coastal plain uplift is believed to be Quaternary volcanism centred about Mt. Burr and Mt. Gambier (Sheard 1990). Between Robe and Naracoorte a clear physical separation of the barriers is evident, but they coalesce in both northerly and southerly directions due to decreasing uplift away from the locii of volcanic activity (Fig. 1).

Quaternary volcanism on the coastal plain occurs in two distinctive associations (Fig. 1). The Early Pleistocene Mount Burr volcanic province includes 15 major volcanic centres (Sheard 1990). The volcanic features pre-date the earliest Pleistocene shoreline deposits (Naracoorte Range; ~800 ka) as evidenced by tombolo-like recurvature of shoreline bodies in the lee of what would have been a volcanic archipeligo (Fig. 1). Mount Gambier and nearby Mount Schank represent a more recent (Holocene) phase of volcanism. These volcanic features clearly erupted through and now blanket some of the Pleistocene shoreline deposits. Thermoluminescence dating of quartz sand buried beneath a lava flow adjacent to Mount Schank yielded an age of 4930 ± 540 yr (Smith & Prescott 1987) and palaeomagnetic data suggest that these two centres are broadly coeval (Barbetti & Sheard 1981).

The regional climate is warm temperate Mediterranean with cool wet winters and dry hot summers. Annual rainfall generally exceeds 450 mm with the highest rainfall recorded at Mount Gambier (Penney 1983). Mean annual temperature, a relevant consideration for amino acid racemisation studies, ranges from approximately 15.5 °C at Keith to 13.2 °C at Mount Gambier (Penney 1983). The present coastline experiences microtidal and winter storm-dominated high wave energy regimes.

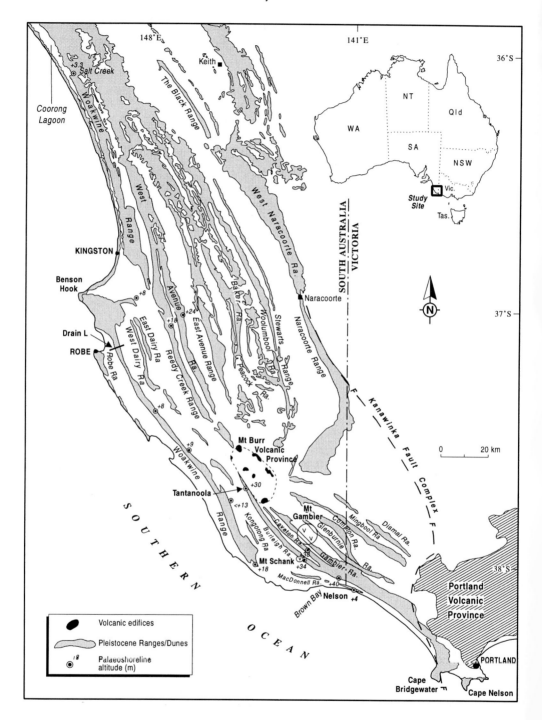

Fig. 1. The Coorong coastal plain showing the location of Pleistocene coastal barriers, sample sites and palaeoshoreline altitudes.

Table 1. Extent of amino acid racemisation (total acid hydrolysate) in molluscs from the Robe region, South Australia.

Locality/ Height APSL (m)**	Species & number of analyses (n)	Depth of burial (m)	Amino acid D/L ratio*					Age ka BP
			ALA	VAL	ASP	PHE	GLU	
Fresh Dip Lake, Robe Beachport 1:50,000 Map 6823-III 54HUD949759 (n.d.)	*Katelysia rhytiphora* (3)	1.4	0.18±0.01 (5.6)	0.05±0.002 (4.0)	0.31±0.06 (19.4)	0.12±0.002 (1.7)	0.10±0.002 (2.0)	3.68±0.11 (SUA-3028)
Lagoon facies behind Woakwine Range (Woakwine Drain) Hatherleigh 1:50,000 Map 6923-III 54HVD150618 (8 m)	*K. rhytiphora* (2)	1	0.59±0.08 (13.6)	0.22±0.02 (9.1)	0.56±0.001 (0.2)	0.43±0.04 (9.3)	0.34±0.02 (5.9)	~125
Lagoon facies behind Woakwine Range Robe 1:50,000 Map 6823-I 54HUD996867 (7 m)	*Anadara trapezia* (2)	1	–	0.20±0.01 (5.0)	0.47±0.003 (0.6)	0.42±0.01 (2.4)	0.32±0.01 (3.1)	~125
Lagoon facies behind Woakwine Range Robe 1:50,000 Map 6823-I 54HUD996867 (7 m)	*K. rhytiphora* (1)	1	0.43±0.02 (4.7)	0.21±0.03 (14.3)	–	0.45±0.007 (1.6)	0.32±0.003 (0.9)	~125
Lagoon facies behind Woakwine Range Robe 1:50,000 Map 6823-I 54HVD084898 (5 m)	*K. rhytiphora* (1)	1	0.59±0.01 (1.7)	0.20±0.01 (5.0)	0.54±0.03 (5.6)	–	0.31±0.01 (3.2)	~125
Lagoon facies behind Woakwine Range Robe 1:50,000 Map 6823-I 54HVD084898 (5 m)	*A. trapezia* (1)	1	0.60±0.01 (1.7)	0.24±0.01 (4.2)	0.67±0.01 (1.5)	0.47±0.01 (2.1)	0.37±0.01 (2.7)	~125
Lagoon facies behind Woakwine and East Dairy Ranges Konetta 1:50,000 Map 6923-IV 54HVD139909 (5 m)	*K. rhytiphora* (1)	1	0.52±0.03 (5.8)	0.17±0.002 (1.2)	0.47±0.01 (2.1)	0.30±0.01 (3.3)	0.20±0.03 (15.0)	~125
Lagoon facies behind Woakwine Range, on old Penola Road, Beachport 1:50,000 Map 6823-II 54HVD050755 (9 m)	*K. scalarina* (2)	2	0.56±0.02 (3.6)	0.19±0.01 (5.3)	0.53±0.02 (3.8)	–	0.31±0.03 (9.7)	~125
Lagoon facies, Hopkins River Estuary, Warrnambool Victoria (2 m)	*A. trapezia* (4)	2	0.70	0.34±0.03 (8.8)	0.62±0.11 (17.7)	0.55±0.11 (20.0)	0.47±0.06 (12.8)	212±7.3 (^{234}U/^{230}Th)
Beach facies, Burleigh Range, Gambier, 1:50,000 Map 7022-II 54HVD795008 (34 m)	*K. rhytiphora* (6)	5	0.82±0.08 (9.8)	0.39±0.07 (17.9)	0.65±0.07 (10.8)	0.69±0.06 (8.7)	0.52±0.04 (7.7)	237±16 (TL)
Lagoon facies, Caveton Range, Gambier 1:50,000 Map 7022-II 54HVD799049 (38 m)	*K. rhytiphora* (6)	2.5	0.78±0.01 (1.3)	0.36±0.02 (5.6)	0.64±0.03 (4.7)	0.64±0.10 (15.6)	0.47±0.03 (6.4)	320±22 (TL)

* Amino acids: ALA – alanine; VAL – valine; ASP – aspartic acid; PHE – phenylalanine and GLU – glutamic acid. Error terms indicate inter-shell D/L ratio variation (1σ). Coefficients of variation (%) are indicated in parentheses.

** Height above present sea level (APSL) in metres indicated in parentheses; Government of South Australia topographic map sheet and grid co-ordinates indicated; n.d. not determined.

2 Dating methods

Amino acid racemisation

The principles on which the amino acid racemisation dating method is based have been summarized by numerous workers with recent overviews by WEHMILLER (1993) and MURRAY-WALLACE (1993). Amino acid racemisation analyses were undertaken on aragonitic bivalve molluscs following established methods (MURRAY-WALLACE 1993). Analyses of the N-pentafluoropropionyl D, L-amino acid 2-propyl esters were performed using a HEWLETT-PACKARD 5890A Series II gas chromatograph with a flame ionisation detector and a 25 m fused silica capillary column coated with the stationary phase Chirasil-L-Val. Analyses generally used 1 g of shell calcium carbonate and, where possible, from the hinge region to avoid intra-shell amino acid D/L ratio variation. All samples analysed were well-buried (>> 1 m), thus reducing the influence of diurnal and seasonal temperature fluctuations, such that longer-term temperature variations associated with climate change represented the dominant influence on diagenetic racemisation. Amino acid racemisation data are presented in Table 1.

Thermoluminescence

Thermoluminescence analyses followed the regeneration method described in detail elsewhere (HUNTLEY et al. 1993a, 1994). Planchets holding approximately 20 mg of 90–125 µm quartz grains were used for equivalent dose determination. Thermoluminescence was measured using an EMI9635Q photomultiplier tube with the optical filters: two Kopp 7–59's, one Schott KG1 and one ND2. Quartz from a modern dune at Brown Bay, on the nearby coast was used as a modern analogue and used to corect for the non-zero thermoluminescence at deposition. Equivalent doses and dose rates were determined by three different methods as desribed by HUNTLEY et al. (1993a) are given in Table 2.

3 Regional variation in age and elevation of palaeoshorelines

The last interglacial, Woakwine Range

The Woakwine Range can be traced essentially uninterrupted for up to 250 km, from Salt Creek in the north to near Mt. Schank in the southeast of the coastal plain (Fig. 1).

Table 2. Thermoluminescence dating results.

Dune range	Laboratory identification	$D_{eq}^{\#}$ Gy	Present total dose rate (Gy·ka^{-1})			TL age ka	δ^{18}O stage
			A	B	C		
Brown Bay	BB1S SESA-98	0*	–	–	–	0*	modern
at Nelson	NT1S SESA-97	87±5	0.474±0.017	0.497±0.024	0.500±0.023	178±13	7a
Burleigh	BU1S SESA-94	178±10	0.809±0.023	0.722±0.027	0.752±0.034	237±16	7e
Caveton	CN1S SESA-95	262±16	0.778±0.023	0.840±0.032	0.853±0.028	320±22	9

* by definition, # used SESA-98 for "zero", method A: from γ-spectrometry. B: K from atomic absorption, Th from neutron activation analysis and U from delayed neutron analysis. C: K from XRS, and Th and U from thick-source alpha counting.

The barrier complex is ascribed a last interglacial age (Oxygen isotope substage 5e) based on uranium-series disequilibrium (SCHWEBEL 1984), luminescence (HUNTLEY et al. 1993a, b, 1994) and amino acid racemisation dating (VON DER BORCH et al. 1980, MURRAY-WALLACE & BELPERIO 1991) with numeric ages ranging between 100 and 132 ka. Sediments of equivalent age and lithology, referred to the Glanville Formation occur extensively along the coast of South Australia in shallow subcrop and outcrop (MURRAY-WALLACE & BELPERIO 1991, BELPERIO et al. 1995). The sediments comprise free-flowing to semi-indurated, medium to coarse-grained skeletal carbonate sands with variable proportions of quartz. They are capped by a laminar or pisolitic calcrete carapace (30 to 50 cm thick), a ubiquitous element of the regional landscape. Calcretes mantle many Quaternary landforms and preserve much of the former coastal morphology of South Australia.

As revealed in large drainage channels, the Woakwine Range consists internally of littoral and transgressive dune facies that interfinger in their lee with back-barrier lagoon sediments. Intertidal fossil mollusc assemblages in the lagoon facies permit confident estimation of past sea-level. Intertidal fossils include the bivalves *Katelysia* spp., *Anapella cycladea*, *Tellina deltoidalis* and the gastropod *Batillaria diemenensis*, which grazes on the surfaces of sand flats. In places, dolomitised marls of lacustrine origin, which formed as the last phase of lagoon infilling and isolation from the sea, overlie the intertidal facies. These sediments contain fossil ostracods and the gastropod *Coxiella confusa*, indicative of highly saline, restricted coastal and lacustrine environments. These settings are analagous to the contemporary Coorong Lagoon and associated dolomitic saline lakes. Where present, the intertidal and saline lagoon facies provide a powerful basis for calculating past sea level to within ± 1 m relative to present.

Fossil molluscs from the lagoon facies at the rear of the Woakwine Range are generally well-preserved, articulated and represent in situ death assemblages. In places they include articulated *Katelysia* sp., that were asphyxiated by the landward migration of the coastal dune-barrier facies (Figs. 2 and 3). Equivalent processes and patterns of sedimentation may be observed today along the modern, Holocene coastal barrier and the Coorong Lagoon. On the ocean (west) side of the Woakwine Range, rocky shoreline deposits developed on local exposures of the underlying Miocene Gambier Limestone. These deposits include fossil shells of the gastropods *Haliotis* sp. and *Turbo* sp., as well as the opercula and columella of *Turbo*, set within semi-consolidated, medium to very coarse-grained skeletal carbonate sands.

Amino acid racemisation analyses were performed on fossil molluscs from five sites along the Woakwine Range (Table 1). The relative extent of racemisation for the different amino acids for samples from the Woakwine Range is consistent with previously published results for the last interglacial Glanville Formation from numerous locations in South Australia (MURRAY-WALLACE & BELPERIO 1991; Table 1). An age of 125 ka may be confidently assigned to the fossil molluscs of the Woakwine Range on the basis of amino acid racemisation, by analogy with amino acid data for other last interglacial coastal deposits in southern Australia (MURRAY-WALLACE & BELPERIO 1991). Coefficients of variation [CV%=sd/(D/L) × 100] for intershell and between-site amino acid D/L ratio variation are generally less than 10 % for multiple samples. These data are also consistent with thermoluminescence and optical dating which yielded numeric ages at three sites in

Fig. 2. Articulated examples of the cockle, *Katelysia rhytiphora* that have moved up profile from the underlying estuarine lagoon facies and have been asphyxiated by landward migrating coastal dunes within the last interglacial Woakwine Range. The sample site is located within Drain L cutting, a flood mitigation channel and is in the lee of the barrier structure. See figure 1 for location. Specimens of *K. rhytiphora* were sampled from this locality for amino acid racemisation analyses.

Fig. 3. Detail from Fig. 1. Two in situ articulated *Katelysia rhytiphora*.

the range 132 to 114 ka (HUNTLEY et al. 1993a, b, 1994). The amino acid data and palaeogeographic mapping indicates that at that time, a major lagoonal waterway existed in the lee of the Woakwine Range with the sea onlapping against and around the pre-existing East and West Dairy Ranges (the "*Anadara* high sea-level deposits" of SPRIGG 1952).

Middle Pleistocene deposits between Robe and Mt. Gambier

Near Robe, successively older shorelines, formed during Stage 7e (Reedy Creek Range) and Stage 9 (West Avenue Range), have been delineated (SCHWEBEL 1984, BELPERIO & CANN 1990). Thermoluminescence ages (HUNTLEY et al. 1993a) are consistent with these identifications. In the Mt. Gambier region, shoreline structures of presumed equivalent ages are represented by the Burleigh and the Caveton Ranges (Fig. 1).

Burleigh Range

The Burleigh Range, located southwest of Mt. Gambier, is a prominent shoreline barrier structure that trends in a NW-SE direction (Fig. 1). The Holocene volcano, Mt. Schank erupted on the seaward side of the Burleigh Range, and basaltic pyroclastics from the volcanic centre are draped over this dune range. The Burleigh Range can be traced for 50 km from the Mt. Burr Volcanic Province to the Glenelg River at the South Australian – Victorian border where it merges with a number of older ranges. Palaeosea-level was determined at 30 m above present sea level (APSL) at Tantanoola and 34 m APSL near Mt. Schank.

Samples for amino acid racemisation and thernmoluminescence dating were collected from a beach facies within the best exposure, a road cutting through the Burleigh Range (Figs. 1 and 4, Table 1). Here, three lensoid-shaped, tightly-packed seaward dipping gravel beds occur within laminar and trough cross-bedded calcarenites. They consist predominantly of rounded, oblate to equant clasts of flint, interpreted as high energy shoreface cobble deposits. Similar flint-dominated cobble deposits occur today on the modern beaches near Pt. MacDonnell. Fragments of bivalves, including the cockle *K. rhytiphora*, and small gastropods are present within the sandy matrix of the gravels. Despite karstification and development of solution pipes the dune range appears to have retained much of its original morphology.

Thermoluminescence dating of quartz grains extracted from the sandy matrix of the shoreface gravels of the Burleigh Range yielded on age of 237 ± 16 ka (Table 2). This age is supported by amino acid racemisation data derived from fossil shells of *K. rhytiphora*, taken from the same gravels. The extent of racemisation of amino acids in these shells is consistently greater than for any of the last interglacial fossil molluscs (Table 1), and is similar to that determined for *Anadara trapezia* from another southern Australian site (Warrnambool, Victoria), shown to have a uranium-thorium age of 212 ± 7.3 ka (SHERWOOD et al. 1994). Thus we conclude that the sediments of the Burleigh Range were deposited during stage 7 of the marine oxygen isotope record (MARTINSON et al. 1987, Table 2). The Burleigh Range is thus correlated with Reedy Creek Range further north (HUNTLEY et al. 1993a). At Nelson, thermoluminescence dating of coastal dune facies in

Fig. 4. Shingle beach facies revealed in a road cutting through the Burleigh Range, approximately 2 km due east of Mount Schank.

front of the Burleigh Range yielded an age of 178 ± 13 ka (Table 2), suggesting that this small shoreline structure is correlative with the West Dairy Range, the Kongorong Dune and the MacDonnell Range and corresponds with an interstadial (stage 7a) rather than interglacial sea level highstand (Fig. 1).

Caveton Range

The Caveton Range is situated approximately 2 km inland from the Burleigh Range (Fig. 1). Samples were collected for amino acid racemisation and thermoluminescence analyses from "back-barrier" sediments at the rear of the structure. Pale yellowish grey, fine to medium grained skeletal carbonate-quartz sands are exposed in all small excavation. The shoaling upward, subtidal-intertidal succession contains the fossil molluscs *Anapella cycladea*, *Tellina deltoidalis*, *K. rhytiphora* and *Fulvia tenuicostata*. Juvenile *A. cycladea* are particularly abundant in the upper 30 cm calcreted portion of this exposure. Pristine *Ammonia beccarii* is the only species of foraminifer present within these sediments (n = 276) indicating a restricted lagoonal depositional environment equivalent to the modern Coorong Lagoon. Palaeosea-level was determined to be 38 m APSL from the intertidal-lagoonal indicators.

Results of amino acid racemisation and thermoluminescence analyses differ for this deposit. Although fossil molluscs from the Caveton Range yielded higher D/L ratios than those derived from the last interglacial fossils, these ratios are not significantly different from those determined for the molluscs of the Burleigh Range, of oxygen isotope stage 7 age. In contrast, thermoluminescence dating of quartz sand associated with the fossil molluscs yielded an age of 320 ± 22 ka (Table 2), indicating an equivalence with stage 9 of the marine oxygen isotope record (MARTINSON et al. 1987). As the Burleigh and Caveton Ranges represent major geomorphological features of the regional landscape, suggestive that they formed during different interglacials, we place greater credence in the TL data from the Caveton Range. On this basis, the Caveton Range is correlated with the West Avenue Range to the north of the Mt. Burr volcanic province.

The lower extent of racemisation of amino acids in the total acid hydrolysate for the Caveton Range samples may result from in situ leaching, involving the preferential loss of the more extensively racemised free amino acids and low molecular weight peptides. This results in a lower extent of racemisation for the total acid hydrolysate than would otherwise be expected on the basis of the "true" age of the fossils. This phenomenon was noted by MURRAY-WALLACE & BELPERIO (1994) in last interglacial specimens of the foraminifer *Marginopora vertebralis*. The low coefficients of variation (Table 1) would suggest, however, that the six *Katelysia* specimens were uniformly affected by leaching, which may be unusual in natural systems. As it is theoretically possible to distinguish age on the basis of racemisation in the temperate climate settings from which the fossils were obtained, further work is required to assess the basis for the low D/L values from the Caveton Range fossils. In this context, the amino acid results from the Caveton deposit represent only a minimum age estimate.

With the exception of the results for the Caveton Range fossils, all the amino acid racemisation results are in accord with an empirical model previously reported, that relates latitude and current mean annual temperature (and as a corollary, diagenetic temperatures) for Late Quaternary fossil-rich coastal deposits in southern Australia (MURRAY-WALLACE et al. 1991). Similar schemes have been used extensively in the northern hemisphere (WEHMILLER 1993) to assign ages to coastal deposits, particularly for sites where there are no corals suitable for uranium-series dating.

Holocene lagoon sediments, Robe – Woakwine corridor

Holocene back-barrier lagoon sediments occur within a corridor between the Robe and Woakwine Ranges, approximately 12 km SE of Robe (Figs. 1 and 5). The estuarine lagoon infill comprises an intertidal shoaling-upward fossil mollusc assemblage that includes *Katelysia* sp. *Brachidontes* sp., *Tellina* sp., *Pecten* sp. and *Ostrea* sp. Articulated specimens of *K. scalarina* and *K. rhytiphora* were selected for radiocarbon dating to calibrate the amino acid racemisation data (Table 1). The radiocarbon age of 3680 ± 110 yr cal BP (SUA-3028) is corrected for the marine reservoir-effect for southern Australia ocean surface waters (GILLESPIE & POLACH 1979) and converted to sidereal years (STUIVER et al. 1986).

4 Neotectonics

The last interglacial maximum spans the interval from 135 to 115 ka (LAMBECK & NAKADA 1992, CHEN et al. 1991, ZHU et al. 1993). MURRAY-WALLACE & BELPERIO (1991)

Fig. 5. Pit exposure through a Holocene coquina, part of the back-barrier lagoon facies of Robe Range. The majority of bivalve molluscs are disarticulated and indicate post mortem transport. The pen is 135 mm long.

document an Australian datum of + 2 m APSL for eustatic sea level during the last interglacial. An assumption implicit in this study is that the sea surface attained similar positions during earlier Quaternary interglacial maxima as indicated by the oxygen isotope record (MARTINSON et al. 1987). This local datum has been used to quantify differential coastal neotectonics around the Australian continent. At Salt Creek on the northern extremity of the Coorong Coastal Plain, the height of the back-barrier intertidal lagoon facies of the Woakwine Range is 3 m APSL, indicating minimal post-depositional uplift at this site. Coalescing of the numerous coastal barriers on the plain intuitively supports this and we confidently calculate that there has been minimal uplift in this northernmost region (i.e. 1 ± 1 m).

The height of the back-barrier lagoon facies of the Woakwine Range rises progressively in a south-southeasterly direction, from 3 m APSL at Salt Creek to 18 m APSL at the southeastern extremity of this dune range (Figs. 1 and 6). Similar systematic variations in palaeoshoreline altitude occur for the older Pleistocene shorelines. Local datums for stage 7 and stage 9 palaeo sea levels are documented by BELPERIO (1995). Near Robe, the lagoon facies of the Woakwine Range occurs 8 m APSL indicating an uplift rate of 50 mm/ka for this portion of the coastal plain. Nearby inland, the Reedy Creek Range (oxygen isotope stage 7 at 18 ± 1 m APSL) and West Avenue Range (oxygen isotope stage 9 at 24 ± 1 m APSL) indicate slightly higher, longer term rates of uplift of 80 mm/ka. However, the uncertainties inherent in measuring palaeoshoreline elevation and inferring the height of sea level at the time of deposition of the older barriers, may imply that the longer-term uplift rate in the Robe region is not significantly different to that calculated for the last interglacial Woakwine Range in the Robe area. Thus, an uplift rate

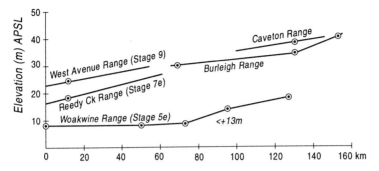

Fig. 6. Height above present sea-level of intertidal back-barrier estuarine lagoon and intertidal facies of the last interglacial Woakwine Range (oxygen isotope substage 5e), Reedy Creek-Burleigh Range (substage 7e) and West Avenue-Caveton Range (stage 9). Locations of data are indicated in Fig. 1. Note that the slopes of the lines are based on the elevations of the last interglacial back-barrier lagoon facies as determined from elevations on the 1:50,000 topographic maps listed in Table 1.

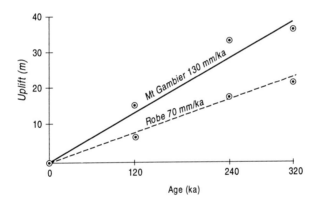

Fig. 7. Uplift curves (lines of best fit) for Mount Gambier and Robe, South Australia from elevation of dated palaeo shoreline indicators. A local datum of + 2 m APSL for the last interglacial sea surface has been adopted in calculation of uplift rates.

of 70 mm/ka, based on all these data, as previously determined by BELPERIO & CANN (1990) is considered representative for the Robe region. At the southernmost extremity of the Woakwine Range near Mt. Gambier, shoreline facies occur 18 m APSL, indicating 16 m of uplift since the last interglacial maximum and an uplift rate of 130 mm/ka. This calculation is based on a thermoluminescence age of 132 ± 9 ka for the last interglacial Woakwine Range (HUNTLEY et al. 1994). Longer-term uplift rates for the nearby Burleigh and Caveton Ranges are similar, being 140 and 120 mm/ka respectively; thus the uplift rates since the late Middle Pleistocene at least, have always been greater near Mt. Gambier than near Robe. These data thus indicate significant spatial differences in the rates of neotectonic uplift along the coastal plain (Figs. 6 and 7). In a global context, uplift of this magnitude is regarded as slow to moderate (BOWDEN & COLHOUN 1984).

A crucial question remains, however, regarding the relationship between rates of uplift and distinct episodes of volcanic eruption. Overall, the separation of Quaternary barrier shorelines implies consistency of uplift throughout the Pleistocene and an increase in rate of uplift towards Mt. Gambier. COOK et al. (1977) identified significant localised crustal doming, trending NW-SE with the centre of uplift situated to the northwest of Mt. Gambier. This was based on the elevation of the upper surface of the Miocene Gambier Limestone which represents the local basement to the Quaternary sequences. The location of doming coincides with the mapped distribution of Quaternary volcanoes (SHEARD 1990). Although long-term uplift of the Coorong coastal plain has been related to Quatrnary volcanism (SPRIGG 1952, COOK et al. 1977, SCHWEBEL 1984), the exact nature of the mechanisms responsible for uplift remains unclear. Crustal doming and continuing uplift, as attested to by the Woakwine Range, is still apparent, which suggests the presence of a magma chamber at depth within the crust and the possibility of further volcanism in this region.

5   Conclusions

Amino acid racemisation and thermoluminescence dating support an age framework of sequential Quaternary sea level highstand coastal deposition on the Coorong Coastal Plain. The Woakwine, Burleigh and Caveton Ranges are associated with last interglacial, penultimate and pre-penultimate highstand events that correspond to oxygen isotope stages 5e, 7e and 9 respectively. Fossiliferous intertidal, back-barrier lagoon facies represent precise indicators of past sea level and reveal spatial variation in the uplift of the coastal plain during the Late Quaternary. The greatest amount of uplift coincides with the Holocene volcanic centres of Mt. Gambier and Mt. Schank, where 16 m of uplift has occurred since the last interglacial (oxygen isotope substage 5e). A similar pattern of spatial variation in uplift rates is also apparent from oxygen isotope stages 7 and 9 shorelines. This method of defining coastal neotectonism from palaeo-shoreline elevation can be further refined as more data are obtained and interpreted in the spatial analysis.

Acknowledgements

We thank Maxine Matatko for assistance with amino acid racemisation sample preparation. Gillian Robertson and the late John Hutton participated in the collection of the samples for thermoluminescence dating and some of the analyses. Most of the laboratory work for the thermoluminescence dates was performed by G.O. Morariu. Mike Barbetti and Gillian Taylor undertook radiocarbon dating. This research was financially supported by the Australian Research Council small grant scheme and the Natural Sciences and Engineering Research Council of Canada. The assistance of Frank and Sarah Glatz and Iradj Yassini with German and French translations of the abstract is gratefully acknowledged. APB publishes with the permission of the Director-General of the South Australian Department of Mines and Energy.

References

BARBETTI, M. & M.J. SHEARD (1981): Palaeomagnetic results from Mounts Gambier and Schank, South Australia. – J. Geol. Soc. Aust. **28**: 385–394.

BELPERIO, A.P. (1995): The Quaternary. – In: DREXEL, J.F. & W.V. PREISS (eds.): The Geology of South Australia, Volume 2: The Phanerozoic. – Bulletin 54, Geological Survey of South Australia.

BELPERIO, A.P. & J.H. CANN (1990): Quaternary evolution of the Robe – Naracoorte Coastal Plain: An excursion Guide. – Department of Mines and Energy, South Australia, Report Book 90/27, 29 pp.

BELPERIO, A.P., C.V. MURRAY-WALLACE & J.H. CANN (1995): The last interglacial shoreline in southern Australia: morphostratigraphic variations in a temperate carbonate setting. – Quat. Int. **26**: 7–19.

BOWDEN, A.R. & E.A. COLHOUN (1984): Quaternary emergent shorelines of Tasmania. – In: THOM, B.G. (ed.): Coastal geomorphology in Australia. – 313–342, Academic Press, Sydney.

CANN, J.H. & J.D.A. CLARKE (1993): The significance of *Marginopora vertebralis* (foraminifera) in surficial sediments at Esperance, Western Australia, and in last interglacial sediments in northern Spencer Gulf, South Australia. – Mar. Geol. **111**: 171–187.

CHEN, J., H.A. CURRAN, B. WHITE & G.J. WASSERBURG (1991): Precise chronology of the last interglacial period: ^{234}U-^{230}Th data from fossil coral reefs in the Bahamas. – Geol. Soc. Am. Bull. **103**: 82–97.

COOK, P.J., J.B. COLWELL, J.B. FIRMAN, J.M. LINDSAY, D.A. SCHWEBEL & C.C. VON DER BORCH (1977): Late Cainozoic sequence of the South East of South Australia and Pleistocene sea level changes. – Bur. Min. Res. J. Geol. Geophys. **2**: 81–88.

GILLESPIE, R. & H.A. POLACH (1979): The suitability of marine shells for radiocarbon dating of Australian Prehistory. – In: BERGER, R. & H. SUESS (eds.): Proceedings of the Ninth International Conference on Radiocarbon Dating. – 404–421, University of California Press.

HUNTLEY, D.J., J.T. HUTTON & J.R. PRESCOTT (1993a): The stranded beach-dune sequence of south-east South Australia: A test of thermoluminescence dating, 0–800 ka. – Quat. Sci. Rev. **12**: 1–20.

– – – (1993b): Optical dating using inclusions within quartz grains. – Geology **21**: 1087–1090.

– – – (1994): Further thermoluminescence dates from the dune sequence in the south-east of south Australia. – Quat. Sci. Rev. **13**: 201–207.

IDNURM, M. & P.J. COOK (1980): Palaeomagnetism of beach ridges in south Australia and the Milankovitch theory of ice ages. – Nature **286**: 699–702.

LAMBECK, K. & M. NAKADA (1992): Constraints on the age and duration of the last interglacial period and on sea-level variations. – Nature **357**: 125–128.

MARTINSON, D.G., N.G. PISIAS, J.D. HAYS, J. IMBRIE, T.C. MOORE & N.J. SHACKLETON (1987): Age dating and the orbital theory of the ice ages: Development of a high-resolution 0 to 300 000-year chronostratigraphy. – Quat. Res. **27**: 1–29.

MURRAY-WALLACE, C.V. (1993): A review of the application of the amino acid racemisation reaction to archaeological dating. – The Artefact **16**: 19–26.

MURRAY-WALLACE, C.V. & A.P. BELPERIO (1991): The last interglacial shoreline in Australia – A review. – Quat. Sci. Rev. **10**: 441–461.

– – (1994): Identification of remanié fossils using amino acid racemisation. – Alcheringa **18**: 219–227.

MURRAY-WALLACE, C.V., A.P. BELPERIO, K. PICKER & R.W.L. KIMBER (1991): Coastal aminostratigraphy of the last interglaciation in southern Australia. – Quat. Res. **35**: 63–71.

PENNY, C.L. (1983): Climate. – In: TYLER, M.J., C.R. TWIDALE, J.K. LING & HOLMES, J.W. (eds.): Natural history of the South East. – 85–93, Royal Society of South Australia, Adelaide.

SCHWEBEL, D.A. (1984): Quaternary stratigraphy and sea-level variation in the southeast of South Australia. – In: THOM, B.G. (ed.): Coastal geomorphology in Australia. – 291–311, Academic Press, Sydney.

SHEARD, M.J. (1990): A guide to the Quaternary volcanoes in the lower south-east of South Australia. – Mines and Energy Review, South Australia 157: 40–50.
SHERWOOD, J., M. BARBETTI, R. DITCHBURN, R.W.L. KIMBER, W. MCCABE, C.V. MURRAY-WALLACE, J.R. PRESCOTT & N. WHITEHEAD (1994): A comparative study of Quaternary dating techniques applied to sedimentary deposits in southwest Victoria, Australia. – Quat. Geochron. (Quat. Sci. Rev.) 13: 95–110.
SMITH, B.W. & J.R. PRESCOTT (1987): Thermoluminescence dating of the eruption at Mt. Schank, South Australia. – Aust. J. Earth Sci. 34: 335–342.
SPRIGG, R.C. (1952): The geology of the south-east province, South Australia, with special reference to Quatrenay coast-line migrations and modern beach developments. – Geol. Surv. S. Aust. Bulletin 29: 120 pp.
STUIVER, M., G.W. PEARSON & T. BRAZIUNAS (1986): Radiocarbon age calibration of marine samples back to 9000 cal yr BP. – Radiocarbon 28: 980–1021.
VON DER BORCH, C.C., J.L. BADA & D.A. SCHWEBEL (1980): Amino acid racemization dating of Late Quaternary strandline events of the coastal plain sequence, southern South Australia. – Trans. R. Soc. S. Aust. 104: 167–170.
WEHMILLER, J.F. (1993): Applications of organic geochemistry for Quaternary research: Aminostratigraphy and aminochronology. – In: ENGEL, M.H. & S.A. MACKO (eds.): Organic Geochemistry. – 755–783, Plenum Press, New York.
WOODS, J.E.T. (1862): Geological observations in South Australia – principally in the district southeast of Adelaide. – Longman, London.
ZHU, Z.R., K.-H. WYRWOLL, L.B. COLLINS, J.H. CHEN, G.J. WASSERBURG & A. EISENHAUER (1993): High precision U-series dating of last interglacial events by mass spectrometry: Houtman Abrolhos Islands, Western Australia. – Earth and Planet Sci. Lett. 118: 281–293.

Addresses of the authors: Dr. C.V. MURRAY-WALLACE, Quaternary Environments Research Centre, School of Geosciences, University of Wollongong, New South Wales, Australia 2522. Dr. A.P. BELPERIO, Regional Geology Branch, South Australian Department of Mines and Energy, P.O. Box 151 Eastwood, South Australia, 5063. Assoc. Prof. J.H. CANN, School of Pure and Applied Sciences, University of South Australia (Salisbury Campus), Salisbury East, South Australia, 5109. Prof. D.J. HUNTLEY, Department of Physics, Simon Fraser University, Burnaby, British Columbia, V5A 1S6, Canada. Prof. J.R. PRESCOTT, Department of Physics and Mathematical Physics, University of Adelaide, South Australia, 5505, Australia.

The effect of rock properties on rates of tafoni growth in coastal environments

Y. Matsukura and N. Matsuoka, Tsukuba, Japan

with 6 figures and 4 tables

Summary. Relationships between rates of tafoni growth and rock properties were investigated at twenty four study sites in Japan where many tafoni of a range of sizes occur on the face of marine cliffs on uplifted shore platforms composed of various rock types. The rate of tafoni growth was defined as the ratio of (1) the mean value of the ten largest depths of thirty tafoni for each site to (2) the period of their growth; that is, from the time of emergence of the shore platform to the present. The salt crystallisation hypothesis indicates that rock properties such as pore-size distribution and rock strength affect the rate of tafoni growth. These two rock properties were measured in the laboratory using intact rock samples collected from the studied shore platforms. A comparison between the rate of tafoni growth and the measured rock properties shows that tafoni growth is more rapid where rocks have a larger proportion of micro-pores and/or lower tensile strength.

Zusammenfassung. *Die Auswirkung von Gesteinseigenschaften auf die Wachstumsrate von Tafoni in Küsten-Milieus.* – An 24 Lokalitäten in Japan, an denen viele Tafoni unterschiedlicher Größe auf den Flächen herausgehobener Brandungsterrassen mariner Kliffe, bestehend aus verschiedenen Gesteinstypen, vorkommen, wurde das Verhältnis zwischen der Wachstumsrate von Tafoni und den Gesteinseigenschaften untersucht. Die Wachstumsrate der Tafoni wurde als das Verhältnis vom (1) Mittelwert der zehn größten Tiefen von dreißig Tafoni jeder Lokalität zur (2) Zeitspanne ihres Wachstums, d.h. von der Zeit des Auftauchens der Brandungsterrasse bis zur Gegenwart, bestimmt. Die Salzkristallisationshypothese deutet darauf hin, daß die Porengrößenverteilung und die Gesteinswiderstandsfähigkeit die Wachstumsrate der Tafoni beeinflussen. Zwei Gesteinseigenschaften Porengrößenverteilung und Gesteinswiderstandsfähigkeit, wurden im Labor unter Verwendung unbeschädigter Gesteinsproben, die von den untersuchten Brandungsterrassen gesammelt wurden, analysiert. Ein Vergleich zwischen den Wachstumsraten der Tafoni und den bestimmten Gesteinseigenschaften zeigt, daß das Tafoniwachstum schneller ist, wo die Gesteine eine größere Anzahl an Mikroporen und/oder eine geringere Zugfestigkeit aufweisen.

Résumé. *Conséquences des propriétés de la roche sur la vitesse d'agrandissement des taffonis dans les environnements côtiers.* – Les relations entre la vitesse d'agrandissement des taffonis et les propriétés de la roche ont été précisément examinées en vingt-quatre sites du Japon où des taffonis se développent au front des falaises marines et sur les plateformes littorales soulevées. La vitesse de croissance des taffonis a été définie à partir du rapport entre la moyenne de dix valeurs des trente plus grandes profondeurs en chacun des sites et la longueur de leur évolution depuis l'émersion des sites choisis. L'hypothèse de l'haloclastisme implique que la vitesse de croissance des taffonis dépende des propriétés des roches, à savoir la distribution des tailles des pores et la résistance des matériaux. Deux propriétés ont été mesurées en laboratoire sur des échantillons sains prélevés sur les plateformes littorales: la plus ou moins grande proportion de micro-pores et/ou la plus ou moins grande résistance à une faible traction.

Introduction

Although coastal tafoni and honeycombs with a variety of sizes have been reported from many locations in the world (e.g. as compiled by SUNAMURA 1992, pp. 204–207, table 8-1), only sporadic information about the rate of their development is available, as summarised by MOTTERSHEAD (1994, table 10.7). This is due mainly to difficulties in finding the time of initiation, and duration, of their development. MATSUKURA & MATSUOKA (1991) assumed that the formation of tafoni, occurring on the marine cliffs of uplifted shore platforms, started just after their emergence. They therefore expressed the duration of tafoni growth by the time from the emergence to the present. A large proportion of rocky coasts of Japan are accompanied by uplifted shore platforms which have been formed by intermittent crustal uplifts associated with earthquakes during the Late Quaternary. Recent progress in tectonic geomorphology enables us to date the emergence of these uplifted shore platforms, and accordingly to estimate the rate of growth of tafoni and honeycombs. Since no explicit distinction has been made between small tafoni and honeycombs (TRENHAILE 1987, p. 49), both features are collectively called "tafoni" in this study.

SUNAMURA's (1992) table shows that coastal tafoni develop on a variety of rocks. However, no studies have previously been undertaken to elucidate the influence of rock properties on tafoni development. This study tackles this problem on several coasts in Japan based on (1) field investigations of tafoni formed on uplifted shore platforms whose time of emergence can be estimated, and (2) laboratory tests of the physical and mechanical properties of the rocks forming the shore platforms.

The study sites

Total twenty four study sites were selected in three areas – Kanto, Sado Island and Kii Peninsula (Fig. 1). Table 1 summarises meteorological data (Japan Meteorological Agency 1982) from representative stations (Fig. 1). The data show that all three areas have a humid temperate climate, as indicated by the mean annual air temperature of 13–17 °C, mean annual precipitation of 1,800–2,300 mm, and the mean relative humidity of 72–75 %. Kii Peninsula has a larger mean annual precipitation and a slightly greater average of sunshine hours than the other two areas.

Table 1. Meteorological data for three decades (1951–1980).

Area	Kanto	Sado Island	Kii Peninsula
Location of meteorological station	Choshi (N 35° 43' E 140° 51')	Aikawa (N 38° 01' E 138° 15')	Shio-no-misaki (N 33° 27' E 135° 46')
Mean annual air temperature (°C)	15.2	13.1	16.9
Mean humidity (%)	75	75	72
Annual precipitation (mm)	1,692	1,640	2,766
Duration of sunshine (hours/year)	2,013	1,854	2,299

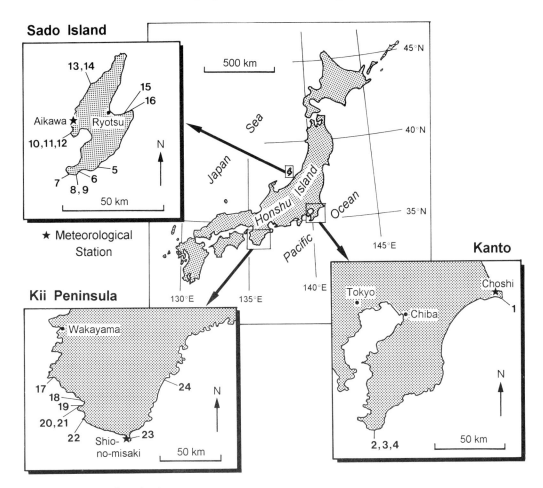

Fig. 1. Location of study sites.

Coasts in Kanto and Kii Peninsula, both facing to the Pacific Ocean, frequently experience large waves caused by typhoons through summer to autumn. The mean tidal range is about 1.0 m in Kanto and 1.3 m in Kii Peninsula. Sado Island is surrounded by the Japan Sea where the tidal range is as small as 0.2 m and the sea condition is calm in summer. In the three areas, strong northwesterly winds, which prevail in winter, supply abundant sea spray to the coasts. The strong winds accompanying typhoons similarly supply sea spray to the coasts.

Field investigations

1   *Method*

Field investigations were carried out from 1989 to 1990. These included measurements of the location of tafoni (altitude and distance from the shoreline), size of tafoni and

Table 2. Physiographical setting of study areas and results of field measurements.

Study site	Landform[1]	Altitude H (m)	Distance from shore line, L (m)	Lithology[2]	Schmidt hammer rebound value, R (%)	Tafoni size		
						D_l (mm)	D_s (mm)	D_d (mm)
Kanto								
1. Inubou-saki	P	0.5-1.0	5-10	Ss	48.4	42.5	35.8	40.0
2. Nojima-zaki	P	1.0-2.0	10	Tc	no data	68.2	54.8	73.9
3. Nojima-zaki	P	4.0-6.0	40	Tc	29.3	257	192	149
4. Nojima-zaki	P	8.5-9.0	60	Tc	21.5	385	374	202
Sado Island								
5. Sugi-no-ura	P	1.0-1.5	9	Tf	50.2	74.0	43.0	98.2
6. Ogi-kou	P	0.5-2.0	0-7	Bt	60.1	18.5	12.9	10.0
7. Shukunegi	P	0.5-2.0	0-1	Tf	45.3	121	97.0	109
8. Keijima-yajima	P	0.2-0.5	0-7	Tf	26.4	186	125	93.0
9. Keijima-yajima	P	1.5-3.0	10	Tf	42.5	689	489	372
10. Nagate-misaki	C	0.25-0.75	0-0.5	Tf	50.3	54.0	43.5	29.7
11. Nagate-misaki	C	0.75-2.1	0.5-1.0	Tf	56.1	235	182	137
12. Nagate-misaki	C	2.1-5.5	1.0-3.0	Tf	55.1	610	439	326
13. Iri-saki	C	1.2-2.0	0.5-2.0	Tf	47.4	184	142	115
14. Iri-saki	C	0.8-1.2	0.5-1.0	An	75.7	30.4	17.3	8.3
15. Habutao	C	1.3-3.0	2-3	An	70.4	291	225	229
16. Tsugami-jima	P	0.8-3.0	0.5-2	Tf	58.3	130	77.0	65.5
Kii Peninsula								
17. Shofuki-iwa	P	2.0-3.3	0-10	Ss	67.2	no tafoni		
18. Tenjin-zaki	P	0.8-2.4	23-25	Cg	22.2	197	157	124
19. Koga-no-ura	C	1.5-3.5	3	Ss	44.1	405	250	275
20. Shirahama-rinkai	P	1.1-1.8	0.5-3	Ss	38.0	126	102	109
21. Shirahama-rinkai	P	4.2-5.2	6-7	Ss	38.3	191	129	175
22. Bansho-zaki	P	3.0-3.5	0-5	Ss	49.6	79.7	58.6	64.5
23. Shio-no-misaki	P	2.0-2.5	5-10	Gr	52.4	41.2	33.8	23.6
24. Shishi-iwa	C	3.0-5.0	10	Tf	50.1	422	306	296

[1] P: Uplifted shore platform C: Sea cliff
[2] Ss: Sandstone Tc: Tuffaceous conglomerate Tf: Tuff Bt: Basalt, An: Andesite Cg: Conglomerate Gr: Granite

bedrock hardness, identification of lithology, and the collection of intact block samples for laboratory tests of rock properties.

The height of the location of tafoni above sea level was measured using a hand level and it was converted to the value above mean sea level with the help of tide tables; the measurement error appears to be less than ± 25 cm. The horizontal distance from the shoreline to the position of the tafoni was measured using a tape. The size of tafoni was determined using the method of MATSUKURA & MATSUOKA (1991). Thirty tafoni were randomly selected at each site and three dimensions (long axis, short axis, and depth) of tafoni were measured using a ruler and/or vernier calipers. The ten largest values for each dimension were averaged to give representative values of tafoni size which are denoted here as D_l, D_s, and D_d, respectively. Bedrock hardness was measured with a Schmidt rock hammer. Thirty rebound values were obtained at each site, and their mean value was used for this study.

2   Results

Results of the field investigations are summarised in Table 2. Some features of the tafoni at these study sites are shown in Fig. 2. Tafoni occur mainly on the face of marine cliffs of uplifted shore platforms (cf. Fig. 3). In rare cases, they occur on flat surfaces or on the landward cliffs of uplifted shore platforms. They are also occasionally found on the face of vertical cliffs such as plunging cliffs (denoted as "sea cliffs" in Table 2). Almost all the tafoni are formed in the altitude, H, ranging from 0.2 m to 5 m, and with a horizontal distance from the shoreline, L, being less than 10 m. This observation indicates that tafoni prevail in the spray zone. Whereas the spatial variation in tafoni size is small within an individual study site, the intersite variation is quite large. The maximum data indicate that D_l = 689 mm, D_s = 489 mm, and D_d = 372 mm in Site 8. The minimum D_l and D_d are 18.5 mm and 8.3 mm in Sites 6 and 14, respectively. The depth of tafoni is proportional to the long and short axis dimensions; for example, the ratio of depth to short axis length (D_d/D_s) lies mostly in 0.7–1.0. Tafoni develop on various rocks such as tuff, sandstone, conglomerate, granite, andesite and basalt. No tafoni were, however, found on sandstone at Site 17. Larger tafoni develop on bedrock with smaller Schmidt hammer rebound values, R, such as tuff and conglomerate, whereas smaller ones occur on bedrock with larger R-values such as basalt, andesite, granite and sandstone. The size of tafoni increases with increasing H-value, where multiple uplifted platforms or high sea cliffs have been developed in the same lithology. At Nojima-zaki (Sites 2, 3, & 4), for example, the D_d-value increases from 73.9 mm on the lowest (H = 1.0–2.0 m), to 149 mm on the middle (H = 4.0–6.0 m), and ultimately to 202 mm on the highest (H = 8.8–9.0 m) platforms as reported previously (MATSUKURA & MATSUOKA 1991). A similar relationship was found at all other locations such as Keijima-yajima (D_d = 93 mm for H = 0.5–2.0 m at Sites 8 and D_d = 372 mm for H = 1.5–3.0 m at Site 9). Nagate-misaki (D_d = 29.7 mm for H = 0.25–0.75 m at Sites 10, D_d = 137 mm for H = 0.75–2.1 m at Site 11, and D_d = 326 mm for H = 2.1–5.5 m at Site 12), and Shirahama-rinkai (D_d = 109 mm for H = 1.1–1.8 m at Site 20, and D_d = 175 mm for H = 3.0–3.5 m at Site 21).

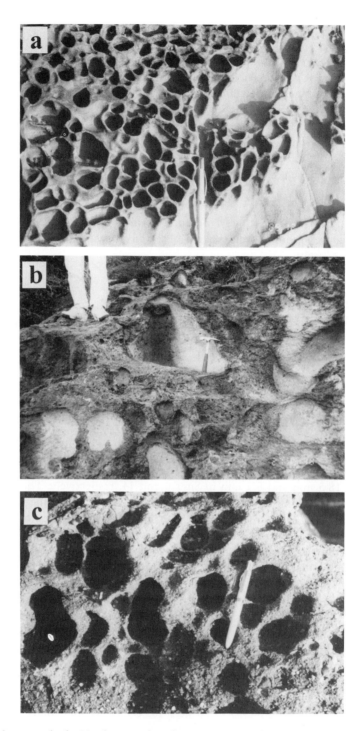

Fig. 2. Some features of tafoni in the several study sites: (a) Site 1 (Inubo-saki in Kanto) where tafoni having a depth, D_d, of 40 mm develop on the surface of the sandstone shore-platform; (b) Site 11 (Nagate-misaki in Sado Island) where tafoni (D_d = 137 mm) develop on the vertical cliff made of tuff; and (c) Site 18 (Tenjin-zaki in Kii Peninsula) where tafoni (D_d = 124 mm) develop on the cliff behind the conglomerate shore-platform.

Fig. 3. Uplifted shore platform in Shukunegi, which emerged on the 1802 Ogi earthquake. Tafoni are found on the marine cliff of the platform (in front of the photo).

Rates of tafoni growth

1 Estimation of time of emergence

The period of tafoni growth was estimated by the same approach as described by MATSUKURA & MATSUOKA (1991). For example, Sites 2 and 3 at Nojima-zaki, Kanto, are located on marine cliffs of uplifted shore platforms which emerged, respectively, during the Kanto (Taisho) earthquake (M 7.9) in 1923 and the Genroku earthquake (M 8.2) in 1703. This indicates that the period for tafoni growth, denoted as T, is 66 y (to the time of investigation in 1989) for Site 2 and 286 y for Site 3. Both Sites 7 and 8 in Sado Island emerged during the Ogi earthquake (M 6.6) in 1802 when the maximum uplift was about 2 m (OTA et al. 1976). The value of T on these sites is 188 years betweem 1802 and the time of invetigation in 1990.

Where the exact age of emergence is unknown, the age was estimated from the value of the mean rate of surface uplift or the radiocarbon and tephrostratigraphic ages of uplifted shore platforms reported in the literature. For the case of Kanto and Sado Island, the mean rate of surface uplift was estimated to be 0.4 m/1,000 y for Inubo-saki (OTA et al. 1985), 3.0 m/1,000 y for Nojima-zaki (NAKATA et al. 1980), and 0.9 m/1,000 y for southwest part of Sado Island (OTA et al. 1976, TAMURA 1979). Using these data, the height of tafoni can be converted into the time of emergence, and thereby yield the period of tafoni growth. For the case of Kii Peninsula, MAEMOKU & TSUBONO (1990) have classified six former sea levels (I through VI) using notches and shore platforms. They also estimated emergence age of platforms based on radiocarbon dating of fossil shells: 200–600 y for level-VI platform, 800–1,800 y for level V, 2,000–2,400 y for level IV,

Table 3. Estimation of rates of tafoni growth.

Site No	Altitude H (m)	Rate of uplift or level of uplifted shore platforms	Reference[2]	Period T (years)	Rate of tafoni growth D_d / T (mm/year)
1	0.5-1.0	0.4 m/1000 y	1	1,250-2,500	0.024 (0.032-0.016)
2	1.0-2.0	Taisho Earthquake (A.D. 1923)	2	66	1.12
3	4.0-6.0	Genroku Earthquake (A.D. 1703)	2	286	0.521
4	8.5-9.0	3.0 m/1000 y [1)]	2	1,300-1,500	0.145 (0.157-0.135)
5	1.0-1.5	0.75 m/1000 y	3	1,333-2,000	0.0641 (0.0736-0.0491)
6	0.2-1.5	0.9 m/1000 y	3	277-1,666	0.021 (0.036-0.0060)
7	0.5-2.0	Ogi Earthquake (A.D. 1802)	4	188	0.581
8	0.2-0.5	Ogi Earthquake (A.D. 1802)	4	188	0.495
9	1.5-3.0	0.9 m/1000 y	3	1,700-3,300	0.166 (0.291-0.113)
10	0.25-0.75	1.1 m/1000 y	3	227-682	0.0872 (0.131-0.0435)
11	0.75-2.1	1.1 m/1000 y	3	682-1,909	0.137 (0.202-0.072)
12	2.1-5.5	1.1 m/1000 y	3	1,909-5,000	0.118 (0.171-0.0652)
13	1.2-2.0	1.3 m/1000 y	3	923-1,538	0.0992 (0.124-0.0744)
14	0.8-1.2	1.3 m/1000 y	3	615-923	0.0112 (0.0135-0.00899)
15	1.3-3.0	0.6 m/1000 y	3	2,200-5,000	0.0750 (0.104-0.0458)
16	0.8-3.0	0.6 m/1000 y	3	1,300-5,000	0.0317 (0.0503-0.0131)
17	2.0-3.3	level III	5	2,600-3,800	0
18	0.8-2.4	level VI	5	200-600	0.415 (0.622-0.207)
19	1.5-3.5	level III	5	2,600-3,800	0.0892 (0.106-0.0723)
20	1.1-1.8	level IV	5	2,000-2,400	0.0503 (0.055-0.0455)
21	4.2-5.2	level II	5	4,000-5,000	0.0393 (0.0437-0.0349)
22	3.0-3.5	level III	5	2,600-3,800	0.0209 (0.0248-0.0170)
23	2.0-2.5	level IV	5	2,000-2,400	0.0108 (0.0118-0.0098)
24	3.0-5.0	level I	5	5,500-6,000	0.0516 (0.0538-0.0493)

[1)] The relationship between the altitude of shore platform, H (m), and the age of emergence, T^* (1,000 y), is exactly expressed by: $H = 3.0\ T^* + 4.5$

[2)] 1: OTA et al. (1985) 2: NAKATA et al. (1980) 3: TAMURA (1979) 4: OTA et al. (1976) 5: MAEMOKU & TSUBONO (1990)

2,600–3,800 y for level III, 4,000–5,000 y for level II, and 5,500–6,000 y for level I. The time of emergence of Sites 17 through 24 was estimated using the diagram presented by MAEMOKU & TSUBONO (1990, fig. 6) showing the distribution and altitude of these platforms around Kii Peninsula.

The calculated values of the time of emergence, that is the periods of tafoni growth (T-value), are shown in Table 3. These indicate that the formation of the studied tafoni took mostly 10^2–10^3 y with the maximum period of 5,500–6,000 y at Site 24.

2 Calculation of rates of tafoni growth

Tafoni grow with time after emergence. When a growing tafone merges with adjacent tafoni, it suddenly increases its diameter due to coalescence. On the other hand, the depth of tafoni is likely to increase gradually with time. Depth is, therefore, a more reasonable measure for the rate of tafoni growth.

The mean rate of tafoni growth at each site was calculated by D_d/T: the results are shown in the last row of Table 3. Where tafoni were found on different altitudes within the same lithology at the same location, two or three rate measurements were obtained (Table 3): Sites 2 (1.12 mm/y), 3 (0.521 mm/y), and 4 (0.145 mm/y) at Nojima-zaki, Sites 8 (0.495 mm/y) and 9 (0.166 mm/y) at Keijima-yajima, and Sites 10 (0.0872 mm/y), 11 (0.137 mm/y), and 12 (0.118 mm/y) at Nagate-misaki, and Sites 20 (0.0503 mm/y) and 21 (0.0393 mm/y) at Shirahama-rinkai. Except for the data of Nagate-misaki which show relatively constant rates, the other data reveal that the higher platforms with a longer period of tafoni development have a lower rate of tafoni growth. As suggested by MATSU-KURA & MATSUOKA (1991), the rate of tafoni growth is not linear but an exponential decay function of time, with the highest rates at the initial stage. The degree of decay with tafoni development is not, however, identical in the four locations: the reason for this discrepancy is not known at present but may be dependent on environmental conditions.

The present study assumes simply that the rate of tafoni growth is constant for the duration of their development. Mean rates of tafoni development in the above mentioned "four specific locations" were, therefore, calculated by averaging the data for two or three altitudes, which resulted in 0.595 mm/y for Nojima-zaki, 0.33 mm/y for Keijima-yajima, 0.114 mm/y for Nagate-misaki, and 0.045 mm/y for Shirahama-rinkai.

The calculated results show that higher rates of tafoni growth are found at Sites 2, 3, & 4 (0.595 mm/y). Site 7 (0.581 mm/y), Sites 8 & 9 (0.33 mm/y), Site 18 (0.415 mm/y), and lower rates at Site 14 (0.0112 mm/y). Site 17 (0 mm/y, i.e. no tafoni formation), and Site 23 (0.0108 mm/y).

Rock properties

Using intact rock samples taken at each site, physical and mechanical properties were determined in the laboratory. The properties included dry unit weight, porosity, pore size distribution (PSD), and compressive and tensile strength. Dry unit weight, porosity and PSD were measured with a mercury intrusion porosimeter. AUTO-PORE #9200 manufactured by Micrometrics Co., USA, applying the same method as that used by SUZUKI & MATSUKURA (1992). The "significant range" of PSD reading is from $10^{1.5}$ µm to 4.6

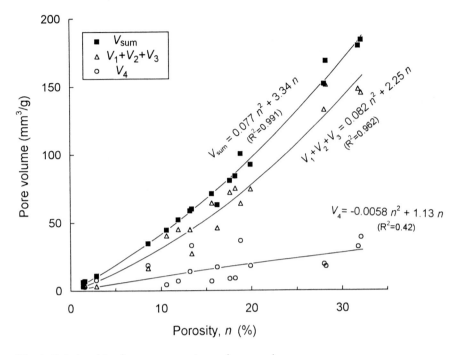

Fig. 4. Relationships between porosity and pore volume.

× 10^{-3} μm. Defining the pore diameter as d, pores are tentatively divided into four grades, d_1 through d_4: large ($10^{1.5}$ μm ≥ d_1 > $10^{0.5}$ μm); medium ($10^{0.5}$ μm ≥ d_2 > $10^{-0.5}$ μm); small ($10^{-0.5}$ μm ≥ d_3 > $10^{-1.5}$ μm); and very small ($10^{-1.5}$ μm ≥ d_4 ≥ 4.6 × 10^{-3} μm). Pore volumes per unit weight of rock specimens are denoted as V_1, V_2, V_3, and V_4 (mm³/g) for the four grades, respectively. The sum of V_1, V_2, V_3, and V_4 is called a total pore volume, V_{sum} (mm³/g).

Four or five dry specimens for each location were subjected to both uniaxial and radial compression tests, the former determining compressive strength, S_c, and the latter tensile strength, S_t. The specimens for the S_c-tests were made into cylinders with a diameter of 25 or 35 mm and a length of 50 or 70 mm; and those for the S_t-tests were formed into disks with the same diameters as above and a thickness of 25 or 35 mm.

Table 4 shows the results obtained. These show that total pore volume, V_{sum}, is highly correlated with porosity, n, that rocks with larger n have larger values of $V_1 + V_2 + V_3$, but that a weak correlation exists between n and V_4-values (Fig. 4). Rocks with larger R-values (Table 2) generally have larger S_c and S_t values while the values of S_c or S_t are rather weakly correlated with n (Fig. 5) because of the variety of rock types present.

Relationship between rates of tafoni growth and rock properties

Although various hypotheses have been proposed for the origin and growth of coastal tafoni (e.g. TRENHAILE 1987, pp. 44–51), the salt weathering hypothesis (e.g. BARTRUM

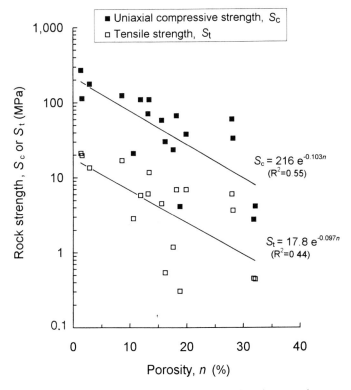

Fig. 5. Relationships between porosity and rock strength.

1936, CAILLEUX 1953) has become particularly popular. This hypothesis is supported by recent studies where salts have been detected in pitted rock surfaces by X-ray diffraction analysis and SEM observations (e.g. GILL et al. 1981, MCGREEVY 1985, MATSUKURA & MATSUOKA 1991) and where salts have been shown to be concentrated on the outer layers of tafoni walls (MOTTERSHEAD & PYE 1994). In the present study sites white efflorescence was frequently observed on the surface of the inner wall of tafoni. X-ray diffraction analysis indicates that the efflorescence consists of several kinds of salts such as halite (NaCl), epsomite ($MgSO_4 \cdot 7H_2O$), and gypsum ($CaSO_4 \cdot 2H_2O$). The flat bases of large tafoni are frequently littered with fine rock fragments which have been produced by flaking and have fallen from their ceilings and backwalls. The presence of salts and littered rock fragments suggest that salt weathering plays an important role in tafoni development. These salts are undoubtedly supplied by sea spray at the sites studied here.

The salt weathering hypothesis demonstrates that the primary agent producing rock fracture and/or exfoliation is the crystallisation pressure generated upon the evaporation of salt water absorbed in rock pores. Pore structure, therefore, can be an important factor controlling crystallisation pressure.

Assuming that the stress due to crystallisation is based upon the chemical potential of a crystal growing from a solution (EVERETT 1961, WELLMAN & WILSON 1965, FITZ-

Table 4. Physical and mechanical properties, and weathering susceptibility index of rocks.

Site No	Unit weight γ_d ($\times 10^{-3}$ g/mm³)	Porosity n (%)	Pore size distribution					Pressure P (MPa)	Compressive strength, S_c (MPa)	Tensile strength, S_t (MPa)	WSI P/S_t
			V_{sum} (mm³/g)	V_1 (mm³/g)	V_2 (mm³/g)	V_3 (mm³/g)	V_4 (mm³/g)				
1	2.28	11.9	52.0	7.82	23.2	13.9	7.00	0.710	109	5.79	0.123
2, 3, & 4	1.87	18.8	100.3	17.7	28.7	17.5	36.4	2.588	4.12	0.304	8.51
5	1.85	28.0	151.5	9.11	54.1	69.3	19.0	1.765	59.3	5.98	0.295
6	2.49	8.6	34.6	2.50	3.53	10.3	18.3	1.735	124	16.7	0.104
7	1.75	32.1	183.7	70.2	49.3	25.4	38.7	2.627	4.12	0.441	5.96
8 & 9	1.78	31.8	179.4	75.9	56.6	15.2	31.7	2.170	2.74	0.451	4.81
10, 11 & 12	1.68	28.2	168.2	9.78	40.0	101.1	17.3	1.678	33.1	3.63	0.462
13	2.15	19.9	92.2	3.63	9.58	61.1	17.9	1.870	37.7	6.86	0.263
14	2.68	2.9	10.9	1.10	0.99	1.17	7.60	0.745	176	13.4	0.0556
15	2.23	13.4	60.0	3.02	4.09	20.0	32.9	2.800	109	11.6	0.241
16	2.17	18.2	83.9	8.76	16.9	49.3	8.97	1.098	66.5	6.86	0.160
17	2.18	1.4	6.23	1.32	0.83	1.11	2.97	0.243	270	21.0	0.0116
18	2.56	16.2	63.0	10.6	12.2	23.1	17.1	1.801	(30)*	(0.54)*	3.34
19	2.19	17.6	80.7	26.5	25.9	19.6	8.65	0.857	23.3	1.18	0.726
20 & 21	2.39	10.6	44.4	9.90	15.4	15.0	4.25	0.509	20.9	2.84	0.179
22	2.20	15.6	71.1	5.61	26.5	32.2	6.79	0.810	57.9	4.51	0.180
23	2.35	1.6	6.86	1.33	0.83	2.13	2.57	0.237	113	19.6	0.0121
24	2.25	13.2	58.6	6.34	23.9	14.5	13.9	1.258	70.6	6.08	0.207

* Since the test specimen of Site 18 could not be prepared due to its brittleness, this value was estimated from the R-value.

NER & SNETHLAGE 1982), and that the pore is cylindrical, the crystallisation pressure, p, is expressed (e.g. GAURI et al. 1990) by:

(1) $p = 4\sigma/d$

where d is the diameter of pore and σ is the surface tension between solid and liquid which takes on 9×10^{-3} N/mm for an NaCl solution (GAURI et al. 1990), that is, sea water. Using equation (1) with this σ-value, the crystallisation pressure was calculated for each grade of pores, assuming the median pore diameter to be 10 μm for the large pores (d_1-size), 1 μm for medium pores (d_2-size), 0.1 μm for small pores (d_3-size), and 0.01 μm for very small pores (d_4-size). The results show that $p_1 = 0.036$ MPa for large pores, $p_2 = 0.36$ MPa for medium pores, $p_3 = 3.6$ MPa for small pores, and $p_4 = 36$ MPa for very small pores, indicating that larger pressures are generated in smaller pores.

Multiplying each of V_1, V_2, V_3, and V_4 by the unit weight of rocks, γ_d (g/mm³), gives the pore volume per unit volume of rocks (mm³/mm³) for each pore-size grade. Assuming that crystallisation pressure per unit volume is represented by $p_i \cdot V_i \cdot \gamma_d$ for the i-th pore, the total pressure generated in the unit volume, P, is given by:

(2) $$P = \sum_{i=1}^{4} p_i V_i \gamma_d$$

It should be mentioned that this equation holds on the assumption that there is no time delay in pressure generation as a function of pore size.

Rock fracture is expected to occur when the crystallisation pressure exceeds the "tensile strength" of the rock (e.g. WELLMAN & WILSON 1968). The ratio P/S_t, therefore, gives the susceptibility of rocks to salt weathering. This ratio is called here the weathering susceptibility index, WSI, which can be described by:

(3) $$\text{WSI} = \frac{1}{S_t} \sum_{i=1}^{4} p_i V_i \gamma_d$$

The values for WSI at each site were calculated by substituting the measured values into equation (3). These results are shown in the last row of Table 4.

Fig. 6 shows the plot of WSI (Table 4) versus the mean rate of tafoni growth (Table 3) for each location. Four error bars in this figure show the range between the maximum and minimum rates of tafoni growth in the "four specific locations" described above. The best-fit line through the data points is given by:

(4) $D_d/T = 0.130 \, \text{WSI}^{0.648}$

This relationship indicates that a rock with larger WSI has a higher rate of tafoni growth. For example, tuff and conglomerate, with abundant micro pores and small S_t-values, experience rapid tafoni development at rates of 0.1–1 mm/y. Extrapolation of this rate shows that tafoni 1 m deep will develop on these rocks in 10^3–10^4 y: tafoni of such a size could have developed during the Holocene. On the other hand, granite and basalt, with a small proportion of pore space and large S_t-values, are subject to slow tafoni development of only about 0.01 mm/y. On these rocks, the formation of tafoni 1 m deep requires a period

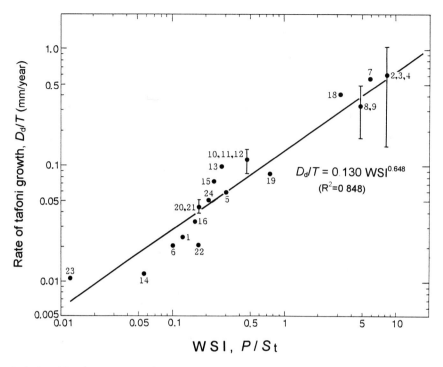

Fig. 6. Relationships between weathering susceptibility index (WSI), P/S_t, and the rate of tafoni growth D_d/T, where P is the total crystallisation pressure, S_t is the tensile strength, D_d is the depth of tafoni, and T is the period for tafoni growth.

of 10^5 y, which is much longer than the Holocene period. This result may be supported by the fact that large tafoni carved in hard rocks with low porosity such as granite are rarely found along the rocky coasts of Japan where rapid surface uplift has raised the rocks far above the spray zone within the past 10^5 y.

Conclusions

No studies have previously been undertaken to elucidate the influence of rock properties on tafoni development. In the present study, relationships between rates of tafoni growth and rock properties were investigated at twenty four locations of rocky coasts in Japan where many tafoni of a range of sizes occur on uplifted shore platforms composed of various lithology. The rate of tafoni growth was defined as the ratio of the mean value of the ten largest depths of thirty tafoni, D_d, for each site to the period of their growth, T, that is, the duration from emergence of the shore platform to the present. The calculated values for rates of tafoni growth, D_d/T, range from about 0.01 mm/y to 1 mm/y. Since the salt crystallisation hypothesis indicates that pore-size distribution and tensile strength affect the rate of tafoni growth, these two rock properties were measured in the laboratory using intact rock samples collected from the studied shore platforms. The ratio of crystallisation pressure which depends on pore sizes to tensile strength was called here the weathering susceptibility index, WSI. The calculated values of WSI were strongly correlated with mean rates of tafoni growth and the relationship between the two was expressed by $D_d/T = 0.130$ $WSI^{0.648}$. This relationship indicates that rock types with a larger WSI have a higher rate of tafoni development, and leads to the conclusion that the rate of tafoni growth in coastal environments is mainly controlled by the combination of tensile strength and the volume of micro pores.

Acknowledgements

We are grateful to Professor Takasuke Suzuki of Chuo University for teaching us the merit of PSD in landform material sciences and for permitting us to use a PSD analyser in his laboratory, to Professor Tsuguo Sunamura of the University of Tsukuba for critical reading the manuscript and providing constructive criticisms, to Mr. Takaaki Tsurukai and students majoring Engineering Geology at Chuo University for help with the PSD tests, and to Dr. Keiji Mizuno, Mr. Kouichi Watase, Mr. Takashi Hirose, Ms. Chiaki T. Oguchi & Mr. Kazuki Ando, staff and students of the University of Tsukuba, for help with the field and laboratory tests. This study was financially supported through the Science Research Fund of the Ministry of Education, Science and Culture (No 05452340) and the fund of University of Tsukuba Project Research.

References

BARTRUM, J.D. (1936): Honeycomb weathering near the shore line. – New Zealand J. Sci. & Tech. 18: 593–600.
CAILLEUX, A. (1953): Taffonis et erosion alveolaire. – Cahier Geologique de Thoiry 16–17: 130–133.
EVERETT, D.H. (1961): The thermodynamics of frost damage to porous solids. – Trans. Faraday Society 57: 1541–1551.
FITZNER, B. & R. SNETHLAGE (1982): Über Zusammenhänge zwischen Salzkristallisationsdruck und Porenradienverteilung. – G.P. Newsletter 3: 13–24.
GAURI, K.L., N.P. CHOWDHURY, N.P. KULSHRESHTHA & A.R. PUNNURU (1990): Geologic features and durability of limestones at the Sphinx. – Environm. Geol. Water Sci. 16: 57–62.
GILL, E.D., R.R. SEGNIT & N.H. MCNEILL (1981): Rate of formation of honeycomb weathering features (small scale tafoni) on the Otway coast, SE Australia. – Proc. Roy. Soc. Victoria 92: 149–154.
Japan Meteorological Agency (1982): Climatic Table of Japan, Part 2, Monthly Normals by Stations (1951–1980). – Japan Meteorological Agency, Tokyo.
MAEMOKU, H. & K. TSUBONO (1990): Holocene crustal movement in the southern part of Kii Peninsula, outer zone of southwest Japan. – J. Geogr. (Tokyo Geogr. Soc.) 99: 349–269. [in Japanese with English abstract]
MATSUKURA, Y. & N. MATSUOKA (1991): Rates of tafoni weathering on uplifted shore platforms in Nojima-zaki, Boso Peninsula, Japan. – Earth Surf. Proc. Landf. 16: 51–56.
MCGREEVY, J.P. (1985): A preliminary scanning electron microscope study of honeycomb weathering of sandstone in a coastal environment. – Earth Surf. Proc. Landf. 10: 509–518.
MOTTERSHEAD, D.N. (1994): Spatial variation in intensity of alveolar weathering of a dated sandstone structure in a coastal environment. Weston-super-Mare, UK. – In: ROBINSON, D.A. & R.B.G. WILLIAMS (eds.): Rock Weathering and Landform Evolution. – 151–174, John Wiley & Sons, Chichester.

MOTTERSHEAD, D.N. & K. PYE (1994): Tafoni on coastal slopes, South Devon, U.K. – Earth Surf. Proc. Landf. **19**: 543–563.

NAKATA, T., M. KOBA, T. IMAIZUMI, W.R. JO, H. MATSUMOTO & T. SUGANUMA (1980): Holocene marine terraces and seismic crustal movements in the southern part of Boso Peninsula, Kanto, Japan. – Geogr. Rev. Japan **53**: 29–44. [in Japanese with English abstract]

OTA, Y., T. MATSUDA & K. NAGANUMA (1976): Tilted marine terraces of the Ogi Peninsula, Sado Island, central Japan, related to the Ogi earthquake of 1802. – Zisin (J. Seism. Soc. Japan) **29**: 55–70. [in Japanese with English abstract]

OTA, Y., Y. MATSUSHIMA, M. MIYOSHI, K. KASHIMA, Y. MAEDA & H. MORIWAKI (1985): Holocene environmental changes in the Choshi Peninsula and its surroundings, easternmost Kanto, central Japan. – The Quart. Res. (J. Japan Assoc. Quat. Res.) **24**: 19–29. [in Japanese with English abstract]

SUNAMURA, T. (1992): Geomorphology of Rocky Coasts. – 302 p., John Wiley & Sons, Chichester.

SUZUKI, T. & Y. MATSUKURA (1992): Pore size distribution of loess from the Loess Plateau, China. – Trans. Japan. Geomorph. Union **13**: 169–183.

TAMURA, A. (1979): Holocene marine terraces and crustal movements of Sado Island, central Japan. – Geogr. Rev. Japan **52**: 339–355. [in Japanese with English abstract]

TRENHAILE, A.S. (1987): The Geomorphology of Rock Coasts. – 384 p., Clarendon Press, Oxford.

WELLMANN, H.W. & A.T. WILSON (1965): Salt weathering, a neglected geological erosive agent in coastal and arid environments. – Nature **205**: 1097–1098.

WELLMAN, H.W. & A.T. WILSON (1968): Salt weathering or fretting. – In: FAIRBRIDGE, R.W. (ed.): The Encyclopedia of Geomorphology. – 968–970, Reinhold Book Co., New York.

Address of the authors: YUKINORI MATSUKURA and NORIKAZU MATSUOKA, Institute of Geoscience, University of Tsukuba, Ibaraki 305, Japan.

Cavernous weathering in the Basin and Range area, southwestern USA and northwestern Mexico

GERD KIRCHNER, Mainz

with 10 figures and 2 tables

Summary. Cavernous weathering forms are investigated in the Basin and Range Province of North America, a large region characterized by predominantly dry climates. There is no principal latitudinal or meridional and little altitudinal influence on the distribution of these forms within the study area but they are restricted by several factors. The majority of tafoni occurs in areas having less than 250 mm of annual precipitation and eight or more arid and semiarid months, in granitic rocks or certain kinds of pyroclastites, in blocks on relatively stable land surfaces, and in short horizontal (< 10 km) and vertical (< 200 m) distances to potential sources for weathering-active salts (usually playas) transported by the wind to the weathering environment. Salt weathering as the main process in creating cavernous weathering forms is concluded from the high salt contents of most samples of weathered material taken from them. The predominant salts are halite and gypsum, in many cases also soda niter; sodium sulfate and sodium carbonate salts are not very abundant. The character of the prevailing salts and the comparatively rare occurrence of high relative humidity values in most parts of the study area suggest that salt crystallization is much more important than salt hydration in disrupting the rocks.

Zusammenfassung. *Kavernöse Verwitterung im Basin-and-Range-Gebiet, südwestliche USA und nordwestliches Mexiko.* – Untersucht werden kavernöse Verwitterungsformen in der nordamerikanischen Basin-and-Range-Provinz, einem großräumigen, überwiegend durch Trockenklimate charakterisierten Gebiet. Es ist kein prinzipieller zonaler oder meridionaler und kaum höhenbedingter Einfluß auf die Verteilung dieser Formen innerhalb des Untersuchungsgebiets festzustellen, jedoch eine Einschränkung durch verschiedene andere Faktoren. Die Mehrzahl der Tafoni kommt in Gebieten mit unter 250 mm Jahresniederschlag und mit mindestens acht ariden und semiariden Monaten vor, in granitischen Gesteinen oder bestimmten Arten von Pyroklastiten, in Blöcken auf relativ stabilen Landoberflächen und in geringen horizontalen (< 10 km) und vertikalen (< 200 m) Distanzen zu playas, die normalerweise als potentielle Liefergebiete für verwitterungsaktive Salze dienen, die durch den Wind zum Ort der Verwitterungstätigkeit transportiert werden. Auf Salzverwitterung als Hauptprozeß bei der Bildung kavernöser Verwitterungsformen wird aus den hohen Salzgehalten in den meisten dort genommenen Proben von Verwitterungsmaterial geschlossen. Die vorherrschenden Salze sind Halit und Gips, in vielen Fällen auch Nitronatrit; Natriumsulfat- und -karbonatsalze sind wenig verbreitet. Der Charakter der überwiegend vorkommenden Salze und das im größten Teil des Untersuchungsgebiets vergleichsweise seltene Auftreten hoher relativer Luftfeuchtigkeitswerte weist darauf hin, daß Salzkristallisation bedeutender für die Gesteinszerstörung ist als Salzhydratation.

Résumé. *Désagrégation caverneuse dans la région Basin and Range, États-Unis sud-ouest et Mexique nord-ouest.* – Les formes de désagrégation caverneuse sont étudiées dans la région Basin and Range nordaméricaine, une large région charactérisée prépondérant par des climats secs. On ne peut pas trouver d'influence principal zonal ou méridional et guère d'influence altitudinal sur la distribution de ces formes au dedans de la région d'étude, mais pourtant une restriction par des autres facteurs divers. Ainsi, la plupart des taffonis éxiste en régions avec moins de 250 mm de précipitation annuelle et avec au moins huit mois arides ou semi-arides; dans des roches granitiques ou dans certains sortes de pyroclastites; dans des blocs sur des surfaces de terrain relativement stabiles; et en petites distances horizontales (< 10 km) et verticales (< 200 m) aux terrains d'origine des sels (normalement de playas) qui sont transportés par le vent au lieu de l'action désagrégative. L'haloclastisme comme processus principal pour le développement des formes de météorisation caverneuse est conclu des contenus élévés de sel dans la plupart des échantillons de materiel désagrégé. Les sels prédominants sont l'halite et le gypse, en beaucoup de cas aussi le nitrate de soude; les sulfates et carbonates de soude sont peu distribués. Le caractère des sels prédominants et l'éxistence comparativement rare de hauts valeurs d'humidité d'air rélative dans la plupart de la région d'étude indique que la cristallisation des sels est plus important pour la désagrégation des roches que l'hydratation des sels.

1 Introduction

Cavernous weathering forms (tafoni and alveoles) exist in almost every climatic region on earth, but they are by far most widespread in warm-dry climates which, for example, are characteristic for the Basin and Range area of southwestern North America.

Regional geomorphological overviews on the one hand are only marginally concerned with weathering forms (e.g. THORNBURY 1965, HUNT 1974, GRAF 1987), detailed studies of cavernous weathering on the other hand usually are limited to relatively small areas with uniform climate and lithology (e.g. DRAGOVICH 1969, BRADLEY, HUTTON & TWIDALE 1978, MARTINI 1978, MUSTOE 1983). The present study attempts to combine the two scales of research in order to provide a general view of the distribution of cavernous weathering forms in the studied region and of the processes causing them. Therefore, the influences exerted by factors such as lithology, altitude, topographic situation, degree of aridity, and type and availability of potential weathering agents are taken into account (KIRCHNER 1991, 1995).

2 The Basin and Range Province

The Basin and Range physiogrpahic province (BRP) (Fig. 1) is a part of the Intermontane Region of North America and is situated between the Pacific mountain ranges in the west and the Rocky Mountains in the east (HUNT 1974: 482, DOHRENWEND 1987: 303). Between the Sierra Madre Occidental and the Sierra Madre Oriental, it extends far south into Mexico (STEWART 1978, HENRY & ARANDA-GOMEZ 1992), but this area will be excluded here.

The large-scale morphology of the BRP is characterized by elongated, more or less parallel mountain ranges and intervening basins. These structures have been produced by block faulting associated with Late Cenozoic extensional tectonics. The characteristic morphology is differently pronounced in the various subregions of the BRP, depending on the age of youngest block faulting: it is most typical in the Great Basin and more altered and subdued in the Mojave and Sonoran desert regions.

As a consequence of the area's varied geological history, the mountains are constituted of a multitude of *rock types*, often changing over small distances. They include a variety of clastic, volcanoclastic and carbonatic sediments laid down in extensive Proterozoic and Paleozoic geosynclines, as well as Mesozoic and Tertiary plutonic rocks, and Tertiary volcanics and pyroclastics.

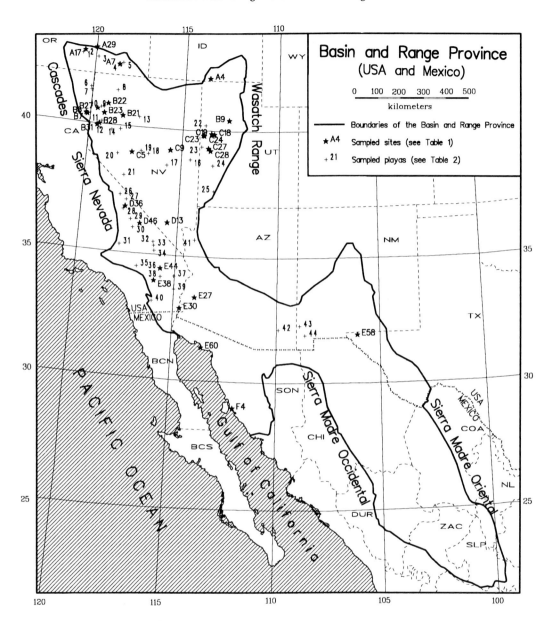

Fig. 1. Location of the study area.

Three main *landform units* can be distinguished in the BRP: ranges, piedmont slopes, and basin centers. The mountains usually have a relatively high relief due not only to their elevation above the neighbouring basins but also to their deeply dissected flanks. Among piedmont slopes, pediments formed by erosion can be differentiated from alluvial fans formed by accumulation. Basin centers in most cases are occupied by playas whose surface structures mainly depend on the hydrologic conditions. Some basins in the marginal parts of the BRP contain more or less saline lakes, a few areas have external runoff.

Present climate is mostly arid to semiarid with mean annual precipitation generally under 250–300 mm (the driest areas are Death Valley and the lower Colorado Valley with less than 50 mm). The seasonal variation of precipitation shows winter maximum in the Great Basin and Mojave Desert and summer maximum in the Sonoran Desert. Mean annual temperatures range from 6–8 °C in the northern Great Basin to 20–24 °C in the southernmost part of the study area, i.e. the Sonoran Desert (and the Death Valley area). Both daily and annual amplitudes average 15–25 °C.

Except for woodland-covered middle and higher mountain slopes, *vegetation* consists of xerophytic shrubs and grasses. Ground-cover is relatively dense in the northern sagebrush-steppe and in the xerophytic woodland of the Sonoran Desert uplands, but quite sparse in the drier rest of the BRP lowlands.

3 Methods

Due to the large areal extent of the BRP, total coverage of the area could by no means have been attempted during fieldwork. Instead, a large number of field study sites have been visited all over the study area. At these sites, existence, character, and abundance of cavernous weathering forms, as well as topographic situation, lithology, and type of underlying landforms were noted and, in most cases, samples of weathered material were taken from within the weathering forms. On playas, samples of surficial sediments were taken. Characterization of weathering forms included aspect and height of the cavern opening above ground surface, aspect and gradient of underlying slopes, relative altitude above basin floor, distance to potential sources of weathering agents, and type of weathered material.

The fine material (< 2 mm) from the samples was analyzed by chemical methods with respect to the amount and types of salts they contained. The salt minerals present in the samples were determined by X-Ray diffractometry.

In order to gain information about the climatic conditions at the tafoni sites, temperature and precipitation data from more than 300 stations all over the BRP were used to create computer generated isolines. The data for each study site were obtained by interpolation between isolines, a method which may not satisfy strict climatological criteria but which was the only means to get at least approximate results given the fact that in a region as topographically diverse as the BRP even the above mentioned number of stations does not create a very dense network over a total area of ca. 10^6 km^2.

For a further description and discussion of methods see KIRCHNER (1991: 133 ff., 1995: 98 ff.).

4 Cavernous weathering forms in the Basin and Range area

4.1 Topographic position

Cavernous weathering forms are found all over the study area, irrespective of latitude (the latitudinal extent of the study area is between 28° and 43 °N). Due to less favourable lithologic, hydrologic, and climatic conditions, they are quite rarely observed in the

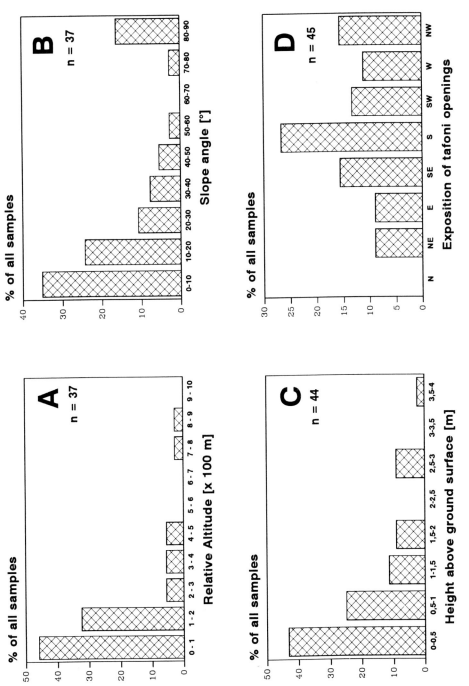

Fig. 2. Topographic position of the studied tafoni. a) Altitude above basin floor, b) Slope angle of landforms underlying tafoni sites, c) Height above ground surface, d) Exposure of tafoni openings.

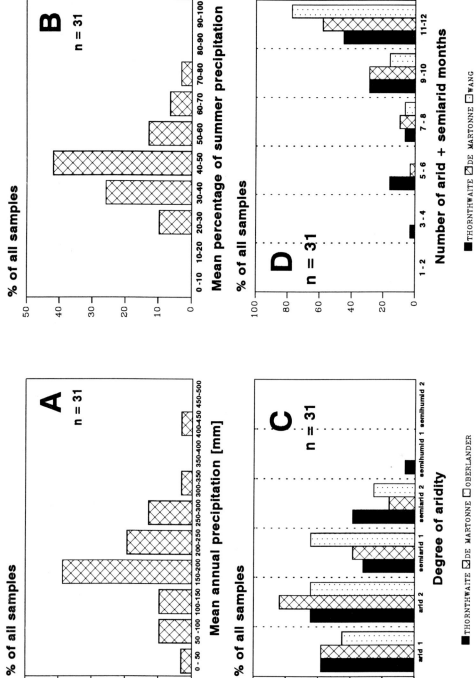

southeastern part (Mexican Highland) of the BRP (KIRCHNER 1995: 184, 234), which is therefore not included here.

Tafoni are found at all altitudes up to about 2500 m, so that a greater influence than by absolute altitude seems to be exerted by the relative height of the study sites above the neighbouring basin floor, which in most cases (78 %) is less than 200 m (Fig. 2a). Basin floor altitudes vary between −86 m and ca. 2100 m. For similar observations in the southwestern USA see BLACKWELDER (1929: 393), in other areas HÖLLERMANN (1975: 389) and MARTINI (1978: 55).

Typically, tafoni are developed in blocks greater than about 0.5 m which are located on gently (< 20°) inclined fan terraces and denudation slopes (Fig. 2b). In some cases they also exist in rock slopes or faces which accounts for the secondary maximum at high angles in the diagram. The lower edge of the opening mostly is in a relatively short distance (< 1 m) above the ground surface (Fig. 2c), a fact also stated for example by DRAGOVICH (1969: 174), KLAER (1970): 76) and GODFREY (1980: 194). The tafoni openings have all possible directions of aspect (except north) but southerly (SE to SW) exposures are dominant (Fig. 2d).

4.2 Climatic influence

As tafoni exist in the cooler northern part of the Basin and Range area as well as in the warmer southern part and over a wide range of altitudes, it is probable that temperatures have only a minor influence on their distribution. It depends much more on annual precipitation, which is at most of the tafoni sites (81 %) less than 250 mm (Fig. 3a). Furthermore, the majority of these sites (67 %) show a relatively low proportion of summer precipitation (May through October) of between 30 and 50 % (Fig. 3b). In areas with a decided summer maximum (> 70 % of annual precipitation), only few cavernous weathering forms were found; it is questionable whether there is a true dependency or the regions of the Basin and Range area with summer rain only accidentally conform with regions that are for other reasons unfavourable for tafoni development.

Using indices to determine the degree of aridity developed by various authors (DE MARTONNE 1926, THORNTHWAITE 1931, OBERLANDER 1979), most tafoni sites (61–87 %) fall into the arid field or in the drier half of the semiarid field (Fig. 3c). Calculation of monthly indices of aridity after DE MARTONNE, THORNTHWAITE, and WANG (the latter as cited in JÄTZOLD 1961) shows that tafoni preferably (74–94 %) develop in areas having at least eight arid or semiarid months (Fig. 3d).

Fig. 3. Climatic conditions at the studied tafoni sites. a) Mean annual precipitation, b) Mean percentage of summer precipitation (May–October), c) Degree of aridity according to indices developed by several authors.

Index of aridity after ...	arid1	arid2	semiarid1	semiarid2	semihumid1	semihumid2
THORNTHWAITE	0–8	8–16	16–24	24–32	32–48	48–64
DE MARTONNE	0–5	5–10	10–15	15–20	20–25	25–30
OBERLANDER	−100 to −83	−83 to −67	−67 to −50	−50 to −33	−33 to −17	−17 to 0

d) Number of arid and semiarid months according to indices of aridity after THORNTHWAITE, DE MARTONNE and WANG.

Fig. 4. Numerous tafoni and alveoles in lapilli tuff, Stansbury Mts., Utah.

A possible objection is that in contrast to the micro-climate inside the weathering hollows the macro-climate is of little relevance for the actual weathering environment. In response, the observation can be stated that there is not much difference in temperature (maximally a few °C) or in relative air humidity (< 10 %) between the inside of tafoni and the free air. this is shown by a few own measurements and confirmed by more detailed data by DRAGOVICH (1967, 1981), HÖLLERMANN (1975), WEINGARTNER (1982).

4.3 Lithologic influence

Cavernous weathering is found in a wide variety of rock types in the Basin and Range area. Apparently, only rocks with relatively closely spaced discontinuities (planes of sedimentation, foliation, jointing) such as shales, slates or suites of strongly interbedded sedimentary rocks principally remain unaffected. Although tafoni were observed in almost all other rock types, a few of them show a much higher susceptibility than the others.

By far the most tafoni have been found in granitoids and in acidic pyroclastics and volcanics (35 % and 30 %, respectively) (Fig. 4–6). Less important, but still relatively frequent are cavernous weathering forms in basic volcanics and volcanic breccias and in carbonate rocks (Fig. 7a).

In addition to massive texture and acidic composition of the rocks, another important factor is the existence of homogenous rock bodies of a certain minimum extent. This is shown by the observation of tafoni exclusively in blocks having diameters of more than approximately 50 cm and preferentially (65 %) in rock faces having widely (d > 1 m)

Fig. 5. Tafoni in a block of granitic rock, Tule Valley, Utah.

Fig. 6. Tafonized block of granitic rock on alluvial fan surface, Death Valley, California.

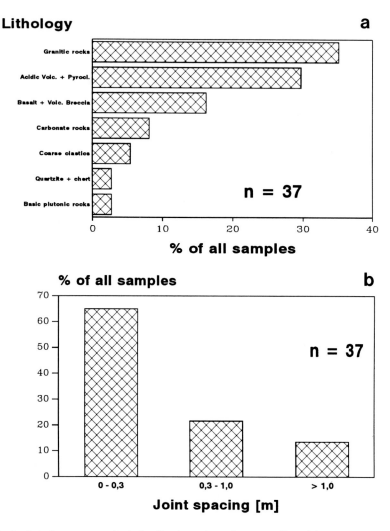

Fig. 7. Lithologic influence on tafoni distribution. a) Rock types affected by cavernous weathering, b) Joint spacing in rocks affected by cavernous weathering.

spaced joints, whereas they are rarely developed in closely jointed rock faces (d < 30 cm) (Fig. 7b).

4.4 *Analysis of weathered material from cavernous weathering forms*

As described in numerous other studies, the inner tafoni walls usually break down into thin, surface-parallel scales, granules, and fine material. The scales are in most cases 1–5 mm thick and consist of fresh or little altered rock, although it may be friable enough to produce the granules and fine material mentioned above. The thickness of the weath-

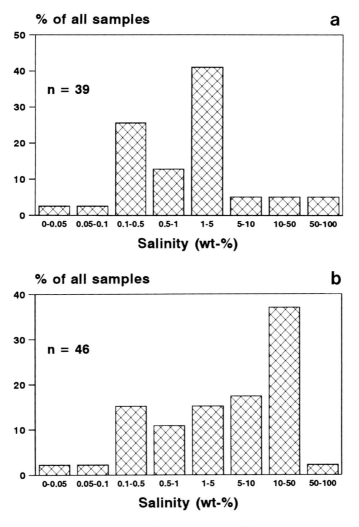

Fig. 8. a) Salt content of weathered material from tafoni, b) Salt content of surface material from playas.

ered layer is between 1 and 10 mm with the greatest thickness and most active scaling at the tafoni ceiling.

These statments mainly are valid for tafoni in the particularly susceptible granitoids and acidic volcanics and pyroclastics, where scaling and granular disintegration produce rounded and relatively smooth interior walls. In other less frequently tafonized rocks such as carbonate rocks, basic volcanics, and quartzites, the tafoni walls are more angular and characterized by protruding edges.

39 samples of weathered material from tafoni, developed in all kinds of lithologies, were analyzed by chemical methods and by XRD. The results show that most of them (69 %) had salt contents of more than 0.5 weight percent (Fig. 8a). The maximal value

was 64.5 weight percent. The preponderance of sodium and calcium among the cations and of chloride and sulfate among the anions (Table 1) was interpreted as sodium chloride and calcium sulfate being the most important salts in the studied tafoni. The predomination was particularly distinct in samples with high total salt content. Sodium nitrate and sodium bicarbonate were also found quite frequently, whereas it is noteworthy that sodium sulfate and sodium carbonate were virtually negligible (with two exceptions).

The chemical analysis was largely confirmed by XRD (only 18 samples had salinities high enough to allow the salt minerals to be identified by XRD). Halite (NaCl) and gypsum ($CaSO_4 \cdot 2H_2O$) were found as the most common salt minerals. Reflections of the other calcium sulfate minerals anhydrite ($CaSO_4$) and bassanite ($CaSO_4 \cdot 1/2 H_2O$) were only found in a few samples.

For comparative purposes (see below), 52 samples of surficial sediments from playas in the relative neighbourhood of the studied tafoni sites were taken. The majority of the samples had salinities over 1.0 weight percent with a maximal value of 76.3 weight percent (Fig. 8b). More data on the salt content of playas and salt lakes in the Basin and Range area have been gathered from other publications in KIRCHNER (1995: Tab. 18).

Sodium was dominant by far among the cations, while chloride usually was the most common anion with bicarbonate and sulfate also being quite abundant (Table 2). Again, XRD mostly confirmed the chemical analysis (28 samples were suitable for inspection) and revealed halite as the predominant salt mineral. Next most common was thenardite (Na_2SO_4) which even prevailed in some samples. A few samples contained gypsum while trona ($NaHCO_3 \cdot Na_2CO_3 \cdot 2H_2O$) was rare or absent in general but predominated in a few samples from playas located in areas of abundant volcanic and pyroclastic rocks.

5   Processes responsible for cavernous weathering in the Basin and Range area

Among the processes that were considered in early studies as responsible for cavernous weathering are marine abrasion, corrasion by wind, and rock disintegration due to short-time temperature variations (for a discussion see WILHELMY 1958: 140–142, 158, FRENZEL 1965: 314–315). KLAER (1956: 54–55) attributed tafoni formation primarily to temperature variations supported in coastal areas by salt efflorescences. WILHELMY (1958) on the other hand explains cavernous weathering by thorough chemical weathering of the rock interior ("core-softening") under a protective crust ("case-hardening") followed by mechanical removal of the softened material to form the typical hollow.

In many recent publications concerned with cavernous weathering forms, however, salts are viewed as the only or at least the most important weathering agent (see for example HÖLLERMANN 1975, BRADLEY, HUTTON & TWIDALE 1978, GODFREY 1980, KLAER & WASCHBISCH 1981, MUSTOE 1982, GOUDIE 1984, BUTLER & MOUNT 1986, FRENZEL 1986, SMITH & MCALISTER 1986, RÖGNER 1989). The results obtained in this study (see Chapter 4) confirm this interpretation.

Admittedly, the destructive effect of salts upon rocks is deduced here from their occurrence in the weathered material. Of course, it is possible to doubt this conclusion because no one has yet observed an expanding salt crystal break off a piece of rock. It may be possible that at least NaCl, by its mere pesence in the weathering environment, accelerates the dissolution rate of silica and in this way promotes the *chemical* weathering

Table 1. Salt content of weathered material from cavernous weathering forms in the Basin and Range Province (concentrations in ppm).

Site	Rock type	Na⁺	K⁺	Ca²⁺	Mg²⁺	Cl⁻	SO₄²⁻	NO₃⁻	CO₃²⁻	HCO₃⁻	Total	Predominant salts as identified by chemical analysis	XRD analysis
A 4 City of Rocks	Granodiorite	172	319	348	21	267	206	0	0	1045	2380	CaCO₃, NaCl	–
A 4 City of Rocks	Granodiorite	1590	730	1580	258	1780	3130	4460	0	0	13500	CaSO₄, NaCl, KNO₃, NaNO₃	–
A 7 Alvord Valley	Rhyolite/Rhyodacite	3090	558	883	658	2250	2630	4060	0	1530	15700	NaNO₃, NaCl, CaSO₄	–
A 17 Summer Valley	Basalt breccia	3300	234	271	17	2400	1660	4380	0	2130	14400	NaNO₃, NaCl	halite(?), soda niter(?)
A 17 Summer Valley	Basalt breccia	28800	1200	91	3	17200	6880	17400	0	153	78600	NaCl, NaNO₃, Na₂CO₃, Na₂SO₄	halite, soda niter(?)
A 29 Alkali Lake Basin	Rhyolite	21200	1310	217	5	732	40100	2380	6900	458	673	Na₂SO₄	thenardite
B 6 Honey Lake Basin	Calcareous tufa	858	688	146	17	19	216	105	900	360	2410	–	–
B 7 Honey Lake Basin	Lahar deposits	1380	375	783	15	87	3740	325	0	4020	10700	NaHCO₃, CaSO₄	gypsum(?), nahcolite(?)
B 7 Honey Lake Basin	Lahar deposits	363	15	15	4	48	264	0	0	522	1230	NaHCO₃, Na₂SO₄	–
B 9 Stansbury Mts.	Lapilli tuff	39000	743	9730	10	72600	16200	760	0	644	140000	NaCl, CaSO₄	halite, gypsum
B 9 Stansbury Mts.	Lapilli tuff	206000	593	9500	118	31600	19800	1080	0	305	553000	NaCl, CaSO₄	halite, gypsum
B 21 Humboldt Range	Granodiorite	495	340	888	37	692	2100	863	0	474	5890	CaSO₄, NaCl, KNO₃	–
B 22 Black Rock Desert	Granodiorite	7280	400	1000	76	9680	4010	450	0	821	23700	NaCl, CaSO₄, Na₂SO₄	halite
B 23 Selenite Range	Granodiorite	48	35	0	2	30	120	15	0	111	361	–	–
B 23 Selenite Range	Granodiorite	164	198	1340	80	51	3300	13	0	152	5300	CaSO₄	–
B 23 Selenite Range	Granodiorite	663	118	59	17	745	384	30	0	0	2020	NaCl, Na₂SO₄	–
B 27 Smoke Creek Desert	Granodiorite	1780	108	1220	10	2770	3050	193	0	258	9390	NaCl, CaSO₄	halite(?), gypsum(?)
B 28 Pyramid Lake Basin	Basalt	308	725	125	24	125	264	1130	0	760	346	KNO₃, NaHCO₃, CaSO₄	–
B 31 Virginia Mts.	Ash-flow tuff	69	85	168	18	65	264	355	0	0	1020	–	–
C 5 Shoshone Mts.	Rhyolite	138	206	2690	300	381	5790	1300	0	0	10800	CaSO₄	–
C 9 White Pine Range	Chert(?)	2440	749	3750	363	1780	7220	4990	0	274	21600	CaSO₄, NaNO₃, NaCl	–
C 18 Black Rock Hills	Pyroclastic rock	4830	210	182	25	6580	744	418	0	152	13100	NaCl	halite
C 19 Fish Springs Range	Carbonate rock	26300	823	27500	700	43800	13900	1450	0	76	115000	NaCl, CaSO₄	halite, gypsum
C 23 Deep Creek Range	Granitic rock	2580	90	533	32	3890	720	358	0	305	8480	NaCl, CaSO₄	anhydrite, gypsum
C 24 Deep Creek Range	Granitic rock	303	113	3020	55	564	7590	195	0	305	12100	CaSO₄	halite, gypsum, bassanite
C 27 Tule Valley	Limestone	223000	4830	17400	5030	322000	56400	15400	150	610	645000	NaCl, CaSO₄	–
C 28 Tule Valley	Granitic rock	88	37	205	16	151	288	3	0	458	1250	CaSO₄, NaCl	–
D 13 Spring Mts.	Fanglomerate	2500	2950	3650	593	4050	1540	13500	0	1590	30400	Ca(NO₃)₂, NaCl	thermonatrite(?), nahcolite(?)
D 36 Saline Valley	Basalt breccia	8200	135	14	1	1380	4730	3090	6320	6430	29900	Na₂CO₃, NaHCO₃, NaCl	gypsum(?)
D 46 Death Valley	Granitic rock	705	215	5180	129	725	7950	348	0	485	15700	CaSO₄, NaCl	–
E 27 Kofa Mts.	Rhyolite	40	60	298	17	28	360	165	0	998	1970	CaCO₃, CaSO₄	–
E 27 Kofa Mts.	Pyroclastic rock	47	69	933	27	117	1850	285	0	2010	5340	CaSO₄, CaCO₃	–
E 30 Cargo Muchacho Mts.	Quarz Monzonite	640	325	11000	363	1380	17100	6810	0	964	3860	CaSO₄, Ca(NO₃)₂	gypsum
E 30 Cargo Muchacho Mts.	Quarz Monzonite	4280	595	7180	308	9080	6460	12500	0	0	4050	Ca(NO₃)₂, NaCl, CaSO₄	halite, anhydrite(?), soda niter(?)
E 38 Joshua Tree Nat. Mon.	Quarz Monzonite	3700	950	1240	300	412	132	12100	0	70	1890	NaNO₃	soda niter(?)
E 44 Bristol Lake Basin	Gabbro	9380	285	1050	40	15700	780	1020	0	1110	29400	NaCl, CaCl₂	halite
E 58 Hueco Tanks	Syenite	51	193	948	43	51	1730	670	0	152	384	CaSO₄, Ca(NO₃)₂	–
E 60 Punta Pelican	Granite	157	25	127	15	268	168	47	0	152	959	NaCl, CaSO₄	–
F 4 Sierra Seri	Silicic tuff	310	116	490	90	659	816	780	0	0	3260	CaSO₄, NaCl, Ca(NO₃)₂	–

Table 2. Salt content of playa surface sediments in the Basin and Range Province (concentrations in ppm).

	Name	Na⁺	K⁺	Ca²⁺	Mg²⁺	Cl⁻	SO₄²⁻	NO₃⁻	CO₃²⁻	HCO₃⁻	Total	Predominant salts as identified by chemical analysis	XRD analysis
Oregon:													
1	Summer Lake	17300	1250	152	738	11300	2500	70	15700	2740	51800	NaCl, Na₂CO₃, NaHCO₃	halite
1	Summer Lake	173000	5780	22	3	40900	23600	45	136000	53500	433000	Na₂CO₃, NaHCO₃, NaCl, Na₂SO₄	trona, halite
2	Alkali Lake	92800	9200	26	4	19200	20700	113	96000	11000	249000	Na₂CO₃, NaHCO₃, NaCl,Na₂SO₄	trona, halite
3	Rabbit Playa	80	49	16	6	34	132	98	0	0	415		
4	Alvord Lake	32800	875	38	4	20800	17400	135	13800	1220	87100	NaCl, Na₂SO₄, Na₂CO₃	halite, thermonatrite(?)
4	Alvord Desert	2980	125	54	66	3500	1260	55	2850	1830	12700	NaCl, Na₂CO₃	
5	Alvord Desert	105000	443	23	3	25400	41900	188	72000	32600	278000	Na₂CO₃, Na₂SO₄, NaHCO₃, NaCl	halite, trona
NE-California:													
6	Middle Alkali Lake	97500	375	35	5	90400	49920	145	25050	4423	267850	NaCl, Na₂SO₄, Na₂CO₃	halite, thenardite
7	Lower Alkali Lake	26600	105	50	12	10100	18500	95	2400	0	57860	Na₂SO₄, NaCl	halite
Nevada:													
8	Black Rock Desert N	723	76	33	3	34	132	75	300	915	2290	NaHCO₃, Na₂CO₃	
9	Black Rock Desert S	56800	1600	1010	15	88900	12600	100	0	1220	162300	NaCl	halite
10	Smoke Creek Desert	2270	115	47	52	498	852	25	1350	458	5670	Na₂CO₃, Na₂SO₄, NaHCO₃, NaCl	
11	Winnemucca Lake	19900	1840	160	12	28500	5290	345	450	610	57000	NaCl, Na₂SO₄	halite
12	Duck Lake	7830	1330	295	1800	9000	2460	328	600	4120	27800	NaCl	
13	Buffalo Valley	58800	10300	26	3	21400	23700	118	39600	12800	167000	Na₂CO₃, NaCl, Na₂SO₄, NaHCO₃	trona, halite, nahcolite(?)
14	Carson Sink	86000	1780	1400	23	131000	9740	413	450	0	231000	NaCl	halite
15	Carson Sink	78500	813	66	12	116000	4370	550	4650	763	206000	NaCl	halite
16	Spring Valley	330	445	343	54	429	780	48	0	763	3190	CaSO₄, NaCl	
17	Railroad Valley	16200	1930	44	388	23200	960	173	3000	305	46200	NaCl	halite
18	Monitor Valley	149000	7080	39	10	45200	241000	35	12000	3660	458000	Na₂SO₄, NaCl	thenardite, halite
19	Big Smoky Playa	86000	4680	29	6	33300	69800	95	43500	10980	248000	Na₂SO₄, Na₂CO₃, NaCl	halite, thenardite
20	Gabbs Valley	18000	433	2140	9	19300	15000	1500	150	610	57200	NaCl, Na₂SO₄, CaSO₄	halite gypsum
21	Columbus Marsh	28400	793	213	41	37700	5260	605	3000	0	76000	NaCl	halite
Utah:													
22	Great Salt Lake Desert	23300	1700	1740	803	39900	936	328	0	0	68800	NaCl	halite
23	Salt Marsh Lake	63500	3730	1090	993	55300	53900	33	0	1680	180300	NaCl, Na₂SO₄	halite, thenardite
24	Wah Wah Valley	1590	283	275	125	1935	276	705	0	1980	7170	–	
25	Parowan Lake	41800	1650	100	31	50600	10100	265	0	3970	108500	NaCl, Na₂SO₄	halite
SE-California:													
26	Deep Springs Lake	11200	10500	36	9	52900	131000	155	16500	4880	328000	Na₂SO₄, NaCl,Na₂CO₃	thenardite, halite
27	Eureka Valley	278	123	210	35	45	216	93	0	1680	2680	NaHCO₃, CaSO₄	
28	Saline Valley	222000	22000	14300	11700	93400	399000	305	0	610	763400	Na₂SO₄, NaCl	halite, thenardite
29	North Panamint Playa	1270	75	53	13	844	492	140	450	460	3800	NaCl, Na₂CO₃, Na₂SO₄, NaHCO₃	
30	South Panamint Playa	47000	2380	8730	1030	75200	27000	105	0	0	162000	NaCl, CaSO₄	halite, gypsum
31	Koehn Lake	114000	1550	4600	543	182000	8200	425	0	305	312000	NaCl, CaSO₄(?)	halite, gypsum(?)
32	Silurian Lake	625	154	116	26	84	168	265	0	1370	2810	NaHCO₃	–
33	Silver Lake	2210	66	125	42	542	300	295	600	1220	5400	Na₂CO₃, NaHCO₃, NaCl	
34	Soda Lake	109000	638	30	3	115000	74300	125	6900	3360	309000	NaCl, Na₂SO₄	halite, thenardite
35	Lucerne Valley	3440	475	89	318	2080	1780	863	1500	152	10700	NaCl, Na₂SO₄, Na₂CO₃	
36	Bristol Dry Lake	35000	303	10500	23	53500	22300	195	0	0	122000	NaCl, CaSO₄	halite, gypsum(?), anhydrite(?)
37	Danby Lake	4830	73	112	18	5340	1920	200	0	915	13400	NaCl, Na₂SO₄	
38	Dale Lake	30100	400	1520	27	46200	7960	3030	0	458	89750	NaCl, CaSO₄	halite
39	Ford Dry Lake	47	109	161	13	46	84	90	0	458	1010	CaCO₃(?)	
40	Clark Lake	12600	140	108	12	16400	1830	185	0	1220	32500	NaCl, Na₂SO₄	halite
Arizona:													
41	Red Lake	29	278	153	32	25	96	45	0	1464	2120	KHCO₃, CaCO₃	–
42	Willcox Playa	2550	1250	270	2270	476	504	450	150	2600	10530	?	–
New Mexico:													
43	Animas Playa	2650	223	80	95	66	180	1260	2100	1070	7720	Na₂CO₃, NaHCO₃, NaNO₃	–
44	Playas Lake	458	700	217	440	275	180	1360	1500	458	5590	?	–

of rocks, as YOUNG (1987) and MOTTERSHEAD & PYE (1994) suppose. Their studies, though, were mostly done in humid coastal environments.

Speaking in favor of the assumption that the salts that occur in cavernous weathering forms are immediately responsible for their development are the results of numerous experiments which testify the principal weathering efficacy of salts (for example GOUDIE, COOKE & EVANS 1970, GOUDIE 1974, 1986, 1993, COOKE 1979, SMITH & McGREEVY 1983, 1988, SPERLING & COOKE 1985, DAVISON 1986, SMITH, McGREEVY & WHALLEY 1987, JERWOOD, ROBINSON & WILLIAMS 1990a, 1990b), theoretical considerations (MORTENSEN 1933, WINKLER & WILHELM 1970, WINKLER & SINGER 1972, SPERLING & COOKE 1980) and SEM photographs of salt crystals in rock cavities (McGREEVY 1985: 513–514, FRENZEL 1986: 147).

A look at the analytical results in this study (see Table 1) shows that sodium sulfate and sodium carbonate rarely appear in weathered material from tafoni in the BRP. These salts are widely used in laboratory experiments on salt weathering and in most cases have been shown to be the most effective in rock destruction, which in turn is probably the reason that they are so often used. Magnesium sulfate, which also had a strong destructive effect in experiments was not detected in mentionable quantities at all.

The efficacy of these salts is primarily attributed to their strong volumetric expansion (up to 314 %) upon hydration from water-free or water-poor phases (thenardite – Na_2SO_4, thermonatrite – $Na_2CO_3 \cdot H_2O$) to water-rich phases (mirabilite – $Na_2SO_4 \cdot 10 H_2O$, natron – $Na_2CO_3 \cdot 10 H_2O$) (SPERLING & COOKE 1985: 544). Theoretical hydration pressures in both above mentioned systems at conditions of low temperature and high relative humidity (WINKLER & WILHELM 1970: 571) are high enough to surpass the tensile strength of any rock. At relative humidities lower than 70–80 % and somewhat higher teperatures (10–20 °C), the respective hydration pressures decrease to distinctly lower values. Consequently, in an experiment by GOUDIE (1993), there was no debris production from samples in temperature/humidity cycles with maximal relative humidities of 53 % or less, little debris production by sodium carbonate in a cycle that reached 85 % relative humidity and only in a cycle where relative humidity attained 100 % were both salts able to cause distinct disintegration.

On the other hand, sodium chloride, calcium sulfate, and sodium nitrate, which were frequently found in weathered material from cavernous weathering forms in the BRP, usually showed little effect in salt weathering experiments. These salts either do not hydrate at all ($NaNO_3$) or with only a small uptake of crystal water ($CaSO_4 \cdot 1/2 H_2O \rightarrow CaSO_4 \cdot 2 H_2O$) or only at temperatures below 0.2°C ($NaCl \rightarrow NaCl \cdot 2 H_2O$).

While the volumetric expansion at the hydration of bassanite to gypsum is not very strong (32 %), the concomitant hydration pressures are extremely high at low temperatures and high relative humidities and still fairly high at high temperature – intermediate humidity conditions (WINKLER & WILHELM 1970: 568).

Crystallization of $NaCl$ and $NaNO_3$ from supersaturated solutions seems to be an effective process for rock destruction in the BRP. Due to their high solubility (265 and 464 g/l, respectively), high amounts of salt can be contained in interstitial waters and subsequently be expelled if conditions of supersaturation ensue. This can happen by evaporation of the water and, in the case of $NaNO_3$, also by cooling of the solution. The crystallization pressures exerted by these two salts surpass the tensile strength of virtually any rock already at a temperature of 0 °C and a supersaturation ratio of 2 (SPERLING & COOKE 1980: Table I); with rising temperature and higher supersaturation the pressures increase distinctly.

The crystallization of $CaSO_4$ causes somewhat lower pressures but primarily, the low solubility of this salt (2 g/l) precludes a sufficient accumulation in relatively short time spans.

Two observations stand against salt hydration as the most important process in cavernous weathering in the BRP. The first is, of course, the low abundance of the strongly hydrating salts Na_2SO_4, Na_2CO_3 and $MgSO_4$ mentioned above. The second is that relative humidity probably does not reach the necessary values very often. At a temperature of 20 °C, thermonatrite ($Na_2CO_3 \cdot H_2O$) only hydrates to natron ($Na_2CO_3 \cdot 10 H_2O$) above a relative humidity of 77 %, the transformation of thenardite (Na_2SO_4) to mirabilite ($Na_2SO_4 \cdot 10 H_2O$) requires a value of 71 %. At a temperature of 0 °C, the hydration of sodium carbonate still needs 54 % relative humidity, for sodium sulfate, no data are available.

There are no data of the average daily maxima of relative humidity in the BRP to my disposition, but estimates based on mean monthly values as well as some own measurements during field work suggest that humidity values above 70 % can be regularly attained only in the winter months. At the temperature-humidity conditions that just allow hydration, the exerted pressures are fairly low, however. In order to reach high hydration pressures, relative humidity has to surpass 80 % or 90 %, depending on temperature. This is confirmed by the above mentioned results of GOUDIE's (1993) salt weathering experiments.

The third possible way for salts to induce mechanical stress in rocks is their large thermal expansivity, especially in the case of NaCl and $NaNO_3$. Unfortunately, there are no clear results concerning the importance of this process. While NOCITA (1987) claims it to be the predominant type of salt weathering in hot deserts, the pure thermal expansion of salts had no effect at all in experiments conducted by GOUDIE (1974).

6 Origin and transport of the salts

Salts can reach the points of their weathering activity principally by three ways:

1) They can already be contained in the affected rock. Most rocks possess a certain primary content of salts or of chemical elements that can form salts after they have been released from the crystal lattice of rock-forming minerals by chemical weathering. Especially the salts in fluid inclusions, the halogens contained in minerals like micas or amphiboles, and the salt minerals mixed into sedimentary rocks during deposition in evaporitic environments or during diagenesis are of importance here.

Literature data give mean values of 130–240 ppm Cl and 270–440 ppm S for granitic rocks, 50–160 ppm Cl and 250–300 ppm S for basalts, 207–354 ppm Cl for metamorphic rocks, 100–180 ppm Cl and 2400–3000 ppm S for claystones and shales, 130–660 ppm Cl and 240 ppm S for carbonate rocks and 10–20 ppm Cl and 240 ppm S for sandstones (PARKER 1967: 13, FIELD 1972, JOHNS 1972, CHRISTIANSEN & LEE 1986). Nitrogen usually has distinctly lower levels in rocks. PARKER (1967: 13) gives mean values of only 20 ppm for basalts and granitic rocks but about 600 ppm for claystones and shales.

2) They can be deposited on the upper surfaces of the rock, either by dry deposition of saline eolian dust or as wet deposition of ions dissolved in precipitation water.

3) They can be transported into the rock from below by capillary rise of soil moisture or, particularly, from the capillary fringe of saline ground water.

To my opinion, the second and third possibilities are far more important than the first. This interpretation is based on the following considerations.

The observation that most cavernous weathering sites are located in relatively short horizontal distance (10 km) (Fig. 9a: 59 %) and small altitudinal difference (200 m) (Fig. 2a: 78 %) to playas or salt lakes suggests that these serve as source areas for the weathering-effective salts. Large amounts of fine-grained saline sediments are available on playa surfaces as can be seen from the data given in Table 2 and, more extensively, in KIRCHNER (1995: Tab. 18). The salts present in the studied weathering forms and the salts occurring in the nearest potential source area mostly show an approximate correspondence (Fig. 10) when plotted together into triangular diagrams for comparison.

The only way to bring these sediments uphill to the weathering sites is, of course, by eolian transport. The material most readily entrained and carried by the wind is the size class commonly addressed as dust.

If the wind is the major agent for transporting the salts to the cavernous weathering forms, they should preferably develop on the lee sides of playas with respect to the main wind direction. However, I found no clearly preferred direction from the studied tafoni to the nearest playa (Fig. 9b). This is not surprising bearing in mind that the data come from all over the BRP and that the local wind directions result from a variety of modifications of the regional air flow which, at least in the northern part of the BRP, in general is from the west. The most important factors here are deviations of the regional winds caused by topography and systems of valley and mountain winds and of up-slope and down-slope winds (HOUGHTON, SAKAMOTO & GIFFORD 1975: 36 ff.).

A much better correspondence is found between the exposition of tafoni openings and the direction to the nearest salt source area: for 40 % of the sites the difference between these two is less than 30° (Fig. 9c). This point will later be addressed in more detail, but first, a look should be taken at the erodibility of the sediments that provide the saline dust.

When the playa surface is protected by an undisturbed salt or clay crust over a certain thickness, the threshold friction velocity necessary for the wind to entrain resting particles is higher than 2.5 m/sec, a value that would correspond to an atmospheric wind velocity of approximately 50–75 m/sec, depending on surface roughness. Winds of this magnitude only very rarely occur in nature (GILLETTE et al. 1982: 9008). If there are no crusts or if the crusts are thinner or disturbed, the threshold friction velocities are greatly reduced to values corresponding to atmospheric wind velocities of 5–15 m/sec. These are attained regularly even in regional winds in the BRP (BRAZEL 1989: 89, HOUGHTON, SAKAMOTO & GIFFORD 1975: 37, 39, DEHARPPORTE 1984). Stronger winds occur in phenomena like dust storms and dust devils.

Dust storms which are observed during the summer months in the southern parts of the BRP at an average rate of up to 5 per year (PÉWÉ et al. 1981: 172, BRAZEL 1989: 71) have mean wind velocities of 9–15 m/sec and mean gust velocities of 17–22 m/sec (BRAZEL 1989: 89). Dust devils in contrast to dust storms only affect very small areas

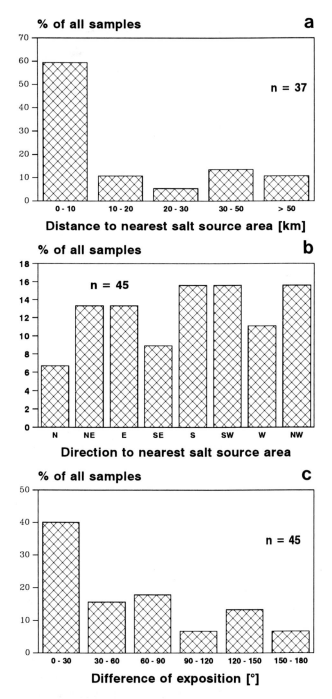

Fig. 9 a) Distance and b) Direction of studied tafoni sites to the nearest salt source area, c) Difference between exposition of tafoni openings and direction to nearest salt source area.

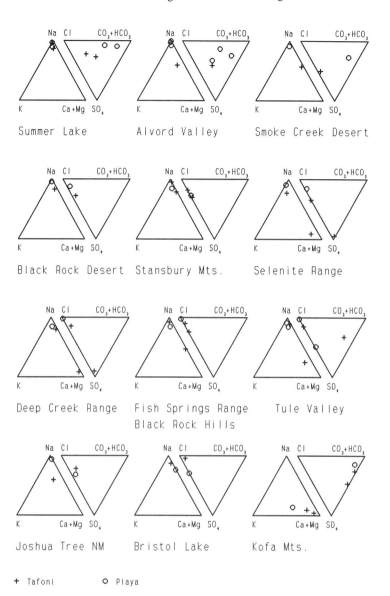

Fig. 10. Some examples for comparison of water-soluble ions in weathered material from tafoni and in surface material from the nearest playas.

but occur quite often in the BRP: HALL (1981: 1932) reports frequencies of 1/km²/d over undisturbed substrates and far more over disturbed substrates. The turbulent wind velocities attained in dust devils are up to 22 m/sec (PYE 1987: 92 ff.).

The above mentioned disturbances are, for example, curled-up edges of contraction polygons as a result of a high content of expandable clay minerals or "salt blisters" that are caused by capillary rise and expose uncrusted material in their centers when they are broken up (BREUNINGER, GILLETTE & KIHL 1989: 50 ff.). Finally, if loose particles are present on the surface of the crust, they can serve as "projectiles" (e.g. by saltation) to raise material from crusts for whose erosion the wind velocity actually is not sufficient.

The saline dust entrained from the playa surfaces is differentially transported by the wind. Coarser grain sizes are redeposited after relatively short distances: the major part of the dust is sedimented within a few hundreds or thousands of meters from the edge of the playa. Measured values of this dry deposition vary strongly: REHEIS & KIHL (1995) report dust deposition rates in the southern Great Basin of 4–16 g/m²/a; the salt content of the dust was between 4 and 25 %. Main dust sources were playas and alluvial surfaces which produced about the same amount of dust per area. Contribution of alluvial sources to total dust flux was greater due to their much larger extension compared to playas. Playa dust, however, usually had higher salt concentrations.

The dust deposition rate of 8.6 ± 0.9 g/m²/a (salt content 20 %) measured by AMUNDSEN et al. (1989: 368) on the eastern piedmont of the Spring Mountains, Nevada, in a distance of ca. 30 km to the nearest playa, corresponds well to REHEIS & KIHL's regional rates. YOUNG & EVANS (1989: 106) recorded dust deposition rates of 22 g/m²/a in a distance of 5 km to the playa while maximum deposition occurred at the eastern playa edge resulting in mean values of more than 2000 g/m²/a.

Only a small part of the dust is carried out of the basin. The finest particles remain in the air as aerosol until they are washed out by rain and brought down as wet deposition. Measurements of the salt content of precipitation water in the BRP gave values between 3 and 15 ppm (JUNGE 1958, JUNGE & WERBY 1958); multiplied by the amount of annual precipitation, this results in an average salt influx by wet deposition of 0.7–1.9 g/m²/a.

The salts contained in the dust deposited on the rock surfaces are dissolved by rain (or by occasional dewfall) and carried into the rock via cracks, along with the salts already present in the precipitation water. Another possibility is the capillary rise of saline soil moisture or ground water into the rock.

The moisture which has entered the rock by either of these two ways is transported to the rock surface by evaporative flux, where it evaporates and the salts contained in it crystallize, accumulating with time as this process repeats. After a model by CONCA & ASTOR (1987) the internal moisture flow is preferentially directed towards the side faces of a rock body because of differential hydraulic conductivity of the different parts of its surface. As the side faces thus attain a higher concentration of salts, salt weathering begins there first, leading to incipient concavities which act as starting points for cavernous weathering. Because rock surfaces that have been attacked by salt weathering have an even higher hydraulic conductivity (CONCA & ASTOR 1987), more saline moisture is transported to these locations. The process thus can be regarded as self-enforcing, eventually resulting in the typical tafoni hollows with salt-enriched inner walls that break down into small scales or granules.

Returning to the observation stated above that most tafoni openings are exposed in the general direction to the nearest salt source area (Fig. 9c), two explanations can be given which are not mutually exclusive.

1) On most land surfaces where cavernous weathering forms have been found, especially on alluvial fan terraces, the general direction of slope is towards the basin

center. Tafoni openings, again, preferably are exposed in roughly the same direction as the slope of the underlying landform: for 60 % of the studied sites, the difference between the two directions was 0–30°, but only for 9 % it was more than 90°, i.e. openings looking up-slope. DRAGOVICH (1969: 177) made similar observations and attributed the fact to the greater stability and to the higher moisture supply of the down-slope side of blocks, a view that should agree with the moisture flow model of CONCA & ASTOR (1987).

2) In most cases, winds that transport saline dust arrive at blocks and rock slopes from the direction of the playa and deposit part of their load on the rock surface. After the results of GOOSSENS & OFFER (1990: 18 f.), eolian dust deposition preferably takes place on the windward side of hills, moisture and vegetation conditions being equal. There were no informations whether this holds for objects of any size or only for larger forms like hillslopes (field measurements were done in an area with a relief of ca. 50 m), but their field results agreed well with a laboratory experiment on a 1:2500-scaled model of the original relief (with a vertical exaggeration of about a factor 2). When considering further that the majority of particles larger than 20 µm is transported in the lowest few meters of the atmosphere (PYE 1987: 53), the conclusion is near that objects of the size of maybe 0.5–10 meters (blocks, tors, small rock faces) can act as topographic obstacles in this sense and that saline dust is therefore preferably deposited on their windward side.

In contrast to the processes outlined above, I rate the importance of salts, respectively salt-forming elements, primarily contained in the substrate rock as relatively low, because their amount should not normally be sufficient to explain the sometimes very high salt contents measured in material from the weathered zone in tafoni. Furthermore, thorough chemical weathering, observed nowhere in the BRP, would be necessary for their complete release.

7   Conclusions

Cavernous weathering forms are found all over the studied region of the BRP and in all altitudes up to ca. 2500 m. However, they are not evenly distributed but usually are restricted to areas that more or less meet the following prerequisites:
- fully arid or dry-semiarid climate, characterized by less than 250 mm of annual precipitation and eight or more arid/semiarid months;
- coarse-grained, acidic, massive rocks (granitoids and certain pyroclastites), in lower abundance in acidic and basic volcanic rocks, carbonate rocks, quartzites;
- distance to a salt-source area (in most cases a playa) of usually less than 10 km in horizontal and 200 m in vertical direction;
- relatively gently sloping land surfaces that can be considered stable.

If at least one of these factors is especially pronounced, the others lose in importance, for example, in rocks that are particularly susceptible, cavernous weathering forms develop in greater distances to salt source areas than usual or, in the contrary, under a high salt influx like right at the edge of a salt-rich playa, they can also form in rocks that usually are very resistant to this type of weathering.

The high salt contents that have in most cases been measured in weathered material from tafoni from all parts of the study area suggests that salt weathering is the primary reason for their development. Among these salts, sodium chloride and calcium sulfate, in many cases also sodium nitrate, distinctly predominated, while sodium sulfate and sodium carbonate showed little abundance.

The majority of salts active in weathering must originate from wind-transported saline playa dust. Evidence for this is given by the proximity of most weathering forms to a playa and by the frequently relatively close correspondence of the salts present in weathered material from tafoni to those in samples from the respective potential source area.

If the playa-surface sediments are not protected by a salt or clay crust, the saline dust can be entrained at wind speeds that occur quite often in the BRP. Dust storms and dust devils, both characterized by high wind speeds, are, of course, even more effective in this respect. Loose particles on the surface can serve as projectiles to break open the otherwise deflation-resistant crusts.

Most of the saline dust is redeposited on rock surfaces in relatively short distances to the playa it was eroded from; only a small percentage leaves the basin. Water from light rains or occasional dewfall dissolves the salts from the dust on the rock surfaces and takes it into the rock via small cracks. The moisture is then transported to the surface again by evaporative flux (which is preferentially directed towards the sides of a block or rock face) where the salts crystallize and accumulate with time. Depending on the respective conditions, saline soil moisture or ground water which enters the rock by capillary rise can be an additional or the more important source for the salts.

The salts enriched near the rock surface mechanically disrupt the rock and thus cause the cavernous weathering forms regarded here. Usually, incipient concavities of the surface are enlarged by flaking or granular disintegration as typical effects of salt weathering which in the BRP chiefly seems to work by crystallization of NaCl or $NaNO_3$. Hydration shattering, on the other hand, should have only minor importance in this area due to the usual lack of strongly expanding salts like Na_2SO_4 and Na_2CO_3 and to unfavorable climatic conditions (insufficient air humidity for the most part of the year).

References

BLACKWELDER, E. (1929): Cavernous rock surfaces of the desert. – Am. J. Sci. 17: 393–399.
BRADLEY, W.C., J.T. HUTTON & C.R. TWIDALE (1978): Role of salts in development of granitic tafoni, South Australia. – J. Geol. 86: 647–654.
BRAZEL, A.J. (1989): Dust and climate in the American Southwest. – In: LEINEN, M. & M. SARNTHEIN (eds.): Paleoclimatology and Paleometeorology: Modern and Past Patterns of Global Atmospheric Transport. NATO Advanced Science Institutes Series C (Mathematical and Physical Sciences) 282: 65–96.
BREUNINGER, R.H., D. GILLETTE & R. KIHL (1989): Formation of wind-erodible aggregates for salty soils and soils with less than 50 % sand composition in natural terrestrial environments. – In: LEINEN, M. & M. SARNTHEIN (eds.): Paleoclimatology and Paleometeorology: Modern and Past Patterns of Global Atmospheric Transport. – NATO Advanced Science Institutes Series C (Mathematical and Physical Sciences) 282: 31–63.
BUTLER, P.R. & J.F. MOUNT (1986): Corroded cobbles in southern Death Valley; their relationship to honeycomb weathering and lake shorelines. – Earth Surf. Proc. Landf. 11: 377–387.
CHRISTIANSEN, E.H. & D.E. LEE (1986): Fluorine and Chlorine in Granitoids from the Basin and Range Province, Western United States. – Econ. Geol. 81: 1484–1494.

CONCA, J.L. & A.M. ASTOR (1987): Capillary moisture flow and the origin of cavernous weathering in dolerites of Bull Pass, Antarctica. – Geology **15**: 151–154.
COOKE, R.U. (1989): Laboratory simulation of salt weathering processes in arid environments. – Earth Surf. Proc. **4**: 347–359.
DAVISON, A.P. (1986): An investigation into the relationship between salt weathering debris production and temperature. – Earth Surf. Proc. Landf. **11**: 335–341.
DEHARPPORTE, D. (1984): West and Southwest wind atlas. – 120 p., Van Nostrand Reinhold, New York.
DE MARTONNE, E. (1926): Aréisme et Indice d'aridité. – C.R. Acad. Sci. **182**: 1395–1398.
DOHRENWEND, J.C. (1987): Basin and Range. – In: GRAF, W.L. (ed.): Geomorphic Systems of North America. – Geol. Soc. Am. Cent. Spec. Vol. **2**: 303–342.
DRAGOVICH, D.J. (1967): Flaking, a weathering process operating on cavernous rock surfaces. – Geol. Soc. Am. Bull. **78**: 801–804.
– (1969): The origin of cavernous surfaces (tafoni) in granite rocks of southern Australia. – Z. Geomorph. N.F. **13**: 163–181.
– (1981): Cavern microclimates in relation to preservation of rock art. – Studies in Conservation **26**: 143–149.
FIELD, C.W. (1972): Sulfur: Element and Geochemistry. – In: FAIRBRIDGE, R.W. (ed.): The Encyclopedia of Geochemistry and Environmental Sciences, p. 1142–1148. – Encyclopedia of Earth Sciences Series, Volume IV A, Van Nostrand Reinhold, New York.
FRENZEL, G. (1965): Studien an mediterranen Tafoni. – N. Jb. Geol. Paläont. Abh. **122**: 313–323.
– (1986): Die granitische Tafoni-Verwitterung. – Heidelberger Geowiss. Abh. **6**: 139–154.
GILLETTE, D., J. ADAMS, D. MUHS & R. KIHL (1982): Threshold friction velocities and rupture moduli for crusted desert soils for the input of soil particles into the air. – J. Geophys. Res. **87**: 9003–9015.
GODFREY, A.E. (1980): Porphyry weathering in a desert climate. – In: PICARD, M.D. (ed.): Henry Mountains symposium. – Utah Geol. Assoc. Publ. **8**: 189–196.
GOOSSENS, D. & Z.I. OFFER (1990): A wind tunnel simulation and field verification of desert dust deposition (Avdat Experimental Station, Negev Desert). – Sedimentology **37**: 7–22.
GOUDIE, A.S. (1974): Further experimental investigation of rock weathering by salt and other mechanical processes. – Z. Geomorph. N.F. Suppl. **21**: 1–12.
– (1984): Salt efflorescences and salt weathering in the Hunza Valley, Karakoram mountains, Pakistan. – In: MILLER, K.J. (ed.): The International Karakoram Project, Volume 2: Proceedings of the International Conference in London, p. 607–615. – Univ. Press, Cambridge.
– (1986): Laboratory simulation of "the wick effect" in salt weathering of rock. – Earth Surf. Proc. Landf. **11**: 275–285.
– (1993): Salt weathering simulation using a single-immersion technique. – Earth Surf. Proc. Landf. **18**: 369–376.
GOUDIE, A.S., R.U. COOKE & I.S. EVANS (1970): Experimental investigation of rock weathering by salts. – Area **4**: 42–48.
GRAF, W.L. (1987): Regional geomorphology of North America. – In: GRAF, W.L. (ed.): Geomorphic Systems of North America. – Geol. Soc. Am. Cent. Spec. Vol. **2**: 1–4.
HALL, F.F. (1981): Visibility reductions from soil dust in the western U.S. – Atmospheric Environment **15**: 1929–1933.
HENRY, C.D. & J.J. ARANDA-GOMEZ (1992): The real southern Basin and Range: Mid- to late Cenozoic extension in Mexico. – Geology **20**: 701–704.
HÖLLERMANN, P. (1975): Formen kavernöser Verwitterung ('Tafoni') auf Teneriffa. – Catena **2**: 385–410.
HOUGHTON, J.G., C.M. SAKAMOTO & R.O. GIFFORD (1975): Nevada's Weather and Climate. – Nevada Bureau of Mines and Geology Spec. Publ. **2**, 78 p.
HUNT, C.B. (1974): Natural Regions of the United States and Canada. – 725 p., W.H. Freeman & Co., San Francisco.

JÄTZOLD, R. (1961): Aride und humide Jahreszeiten in Nordamerika. – Stuttgarter Geogr. Stud. 71, 130 p.
JERWOOD, L.C., D.A. ROBINSON & R.B.G. WILLIAMS (1990a): Experimental frost and salt weathering of chalk I. – Earth Surf. Proc. Landf. 15: 611–624.
– – (1990b): Experimental frost and salt weathering of chalk II. – Earth Surf. Proc. Landf. 15: 699–708.
JOHNS, W.D. (1972): Chlorine: Element and Geochemistry. – In: FAIRBRIDGE, R.W. (ed.): The Encyclopedia of Geochemistry and Environmental Sciences, p. 155–156. – Encyclopedia of Earth Sciences Series, Volume IV A, Van Nostrand Reinhold, New York.
JUNGE, C.E. (1958): The distribution of ammonia and nitrate in rainwater over the United States. – Trans. Am. Geophys. Union 39: 241–248.
JUNGE, C.E. & R.T. WERBY (1958): The concentration of chloride, sodium, potassium, calcium, and sulfate in rain water over the United States. – J. Meteorology 15: 417–425.
KIRCHNER, G. (1991): Physikalische Verwitterung in Trockengebieten am Beispiel des Basin-and-Range-Gebiets (südwestliche USA und nördliches Mexiko). – Unpubl. doctorate thesis, 396 p., Johannes-Gutenberg-Universität Mainz.
– (1995): Physikalische Verwitterung in Trockengebieten unter Betonung der Salzverwitterung am Beispiel des Basin-and-Range-Gebiets (südwestliche USA und nördliches Mexiko). – Mainzer Geogr. Stud. 41: 267 p.
KLAER, W. (1956): Verwitterungsformen im Granit auf Korsika. – Petermanns Geogr. Mitt., Erg.-H. 261, 146 p.
– (1970): Formen der Granitverwitterung im ganzjährig ariden Gebiet der östlichen Sahara (Tibesti). – Tübinger Geogr. Stud. 34: 71–78.
KLAER, W. & R. WASCHBISCH (1981): Neuere Erkenntnisse über den Prozeß der Tafoniverwitterung. – Aachener Geogr. Abh. 14: 67–79.
MARTINI, I.P. (1978): Tafoni weathering with examples from Tuscany, Italy. – Z. Geomorph. N.F. 22: 44–67.
McGREEVY, J.P. (1985): A preliminary scanning electron microscope study of honeycomb weathering of sandstone in a coastal environment. – Earth Surf. Proc. Landf. 10: 509–518.
MORTENSEN, H. (1933): Die Salzsprengung und ihre Bedeutung für die regionale klimatische Gliederung der Wüsten. – Petermanns Geogr. Mitt. 79: 130–135.
MOTTERSHEAD, D.N. & K. PYE (1994): Tafoni on coastal slopes, South Devon, U.K. – Earth Surf. Proc. Landf. 19: 543–563.
MUSTOE, G.E. (1982): The origin of honeycomb weathering. – Geol. Soc. Am. Bull. 93: 108–115.
– (1983): Cavernous weathering in the Capitol Reef Desert, Utah. – Earth Surf. Proc. Landf. 8: 517–526.
NOCITA, B.W. (1987): Thermal expansion of salts as an effective weathering process in hot desert environments. – Geol. Soc. Am. Abstr. with Progr. 19 (7): 790.
OBERLANDER, T.M. (1989): Characterization of arid climates according to combined water balance parameters. – J. Arid Environm. 2: 219–241.
PARKER, R.L. (1967): Composition of the earth's crust. – U.S. Geol. Surv. Prof. Paper 440-D, 19 p.
PÉWÉ, T.L., E. PÉWÉ, R.J. PÉWÉ, A. JOURNAUX & R.M. SLATT (1981): Desert dust: characteristics and rates of deposition in central Arizona. – Geol. Soc. Am. Special Paper 186: 169–190.
PYE, K. (1987): Aeolian dust and dust deposits. – Academic Press, London, 334 p.
REHEIS, M.C. & R. KIHL (1995): Dust deposition in southern Nevada and California, 1984–1989: relations to climate, source area, and source lithology. – J. Geophys. Res. 100 (D5): 8893–8918.
RÖGNER, K. (1989): Geomorphologische Untersuchungen in Negev und Sinai. – Paderborner Geogr. Stud. 1, 258 p.
SMITH, B.J. & J.J. McALISTER (1986): Observations on the occurrence and origins of salt weathering phenomena near Lake Magadi, southern Kenya. – Z. Geomorph. N.F. 30: 445–460.
SMITH, B.J. & J.P. McGREEVY (1983): A simulation study of salt weathering in hot deserts. – Geogr. Ann. 65-A: 127–133.

SMITH, B.J. & J.P. MCGREEVY (1988): Contour scaling of a sandstone by salt weathering under simulated hot desert conditions. – Earth Surf. Proc. Landf. **13**: 697–705.
SMITH, B.J., J.P. MCGREEVY & W.B. WHALLEY (1987): Silt production by weathering of a sandstone under hot arid conditions: an experimental study. – J. Arid Environm. **12**: 199–214.
SPERLING, C.H.B. & R.U. COOKE (1980): Salt weathering in arid environments: experimental investigations of the relative importance of hydration and crystallization processes. I: Theoretical considerations. – Bedford College London, Papers in Geography **8**: 45 p.
– – (1985): Laboratory simulation of rock weathering by salt crystallization and hydration processes in hot, arid environments. – Earth Surf. Proc. Lancf. **10**: 541–555.
STEWART, J.H. (1978): Basin and Range structure in western North America. – In: SMITH, R.B. & G.P. EATON (eds.): Cenozoic tectonics and regional geophysics of the western Cordillera. – Geol. Soc. Am. Memoir **152**: 1–31.
THORNBURY, W.D. (1965): Regional Geomorphology of the United States. – 609 p., John Wiley & Sons, New York.
THORNTHWAITE, C.W. (1931): The climates of North America according to a new classification. – Geogr. Rev. **21**: 633–655.
WEINGARTNER, H. (1982): Tafoniverwitterung in Naxos. – Salzburger Exkursionsber. **8**: 90–107.
WILHELMY, H. (1958): Klimamorphologie der Massengesteine. – 238 p., Georg Westermann Verlag, Braunschweig.
WINKLER, E.M. & P.C. SINGER (1972): Crystallization pressure of salts in stone and concrete. – Geol. Soc. Am. Bull. **83**: 3509–3513.
WINKLER, E.M. & E.J. WILHELM (1970): Salt burst by hydration pressures in architectural stone in urban atmosphere. – Geol. Soc. Am. Bull. **81**: 567–572.
YOUNG, A.R.M. (1987): Salt as an agent in the development of cavernous weathering. – Geology **15**: 962–966.
YOUNG, J. & R. EVANS (1986): Erosion and deposition of fine sediments from playas. – J. Arid Environm. **10**: 103–116.

Address of the author: Dr. GERD KIRCHNER, Geographisches Institut, Johannes Gutenberg-Universität, Saarstr. 21, D-55099 Mainz. Present address: Rubensallee 95, D-55127 Mainz.

Palaeoflood analysis of channel-fill deposits, central Tapti river basin, India

J.N. MALIK and A.S. KHADKIKAR, Baroda

with 5 figures and 2 tables

Summary. The Quaternary sediments exposed on the banks of the river Aner which is a tributary of the river Tapti, near Piloda (district Dhule) reveal three channel-fill structures, spatially at the same level. The left bank of the channel truncates the right bank of the predecessor channel resulting in the partial preservation of the channel-fills. Three orders of bounding surfaces are observed in each of the channel fill deposit. Second order bounding surfaces enclose packets of first order bounding surfaces which have a concave upward geometry. The sediment packets enclosed by the first order bounding surfaces are on an average 5 to 6 cm thick in the central portions of the channel fills and taper out at the margins. The third order bounding surfaces which enclose the second and first order surfaces are the channel scours. The lower order bounding surfaces demonstrate a decrease in their concavity progressively upward through the section and in instances become planar at the apex. The channel-fill deposits are planed off by a higher order (4th order) bounding surface separating the underlying channel deposits from a calcrete horizon. This calcrete horizon lies 1 meter below the present day topography.

The first order bounding surfaces enclose sediments deposited after a monsoon rainfall. The second order bounding surfaces which represent events of scouring document events of higher discharges (?minor floods) which led to the excavation and remobilization of the sediments. The third order bounding surfaces represent events of peak floods or anomalous discharge. Palaeohydraulic calculations for the bankfull discharge give figures around 150 m^3/s. This figure however expresses the minimum discharge. It is difficult to calculate the exact discharge but the channel forming discharge could have been up to a magnitude higher as the predecessor channel was instantaneously eroded. The lateral shifting of channels (avulsion) in the absence of any obvious tectonic influence can best be explained through episodic floods.

1 Introduction

The recognition of sedimentary structures in fluvial sediments identifiable with events of anomalous discharge is vital towards the study of palaeofloods. Rivers cutting through resistant rocks manifest flood events as slack water deposits (BAKER 1988). In terrains of less competent lithologies the manifestations include channel formation accompanied by terrace generation, dumping of coarse detritus in a dominantly finer sediment association and lateral shifting of river channels (BAKER 1977, BAKER & PICKUP 1987, HUCKLEBERRY 1994, KALE et al. 1994, MARTINI 1977, WELLS & DORR 1987). The lateral shifting of river channels, a process termed as river avulsion occurs at both large as well as smaller scales. Large scale avulsions owe their genesis to base level changes (e.g. TORNQUIST 1994, VAN GELDER et al. 1994). The isolation of tectonic influence from climatic signatures/base level influences is quite difficult. WELLS & DORR (1987) have illustrated this difficulty in their study of the phenomenal episodic shifting of Kosi river, India, by an amount of 113 km through a time span of 228 years. Here we present an example of

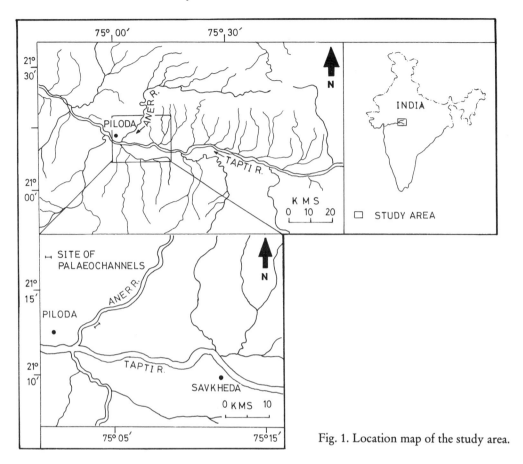

Fig. 1. Location map of the study area.

small scale avulsion from the Quaternary alluvial deposits of the Tapti river basin, western India, and discuss the palaeohydrologic significance.

2 General

The present day Tapti river (also termed Tapi) originates in the Trappean highlands of Betul district in Madhya Pradesh (Fig. 1), and flows westwards into the Arabian Sea, cutting across the Sahyadri mountain ranges and the alluvial plains of south-Gujarat. The Pleistocene sediments of the region are exposed along the banks of the river Tapti and its tributaries where they form cliffs up to 30 m in height. The river Aner meets the Tapti

Table 1. Flood discharges recorded on the Tapti River at Savkheda (after KALE et al. 1994).

Date	Discharge m³/s
11 August 1979	20,845
30 August 1987	24,200
31 July 1991	11,181

Palaeoflood analysis of channel-fill deposits

Fig. 2. Photograph of channel-fill deposits.

Table 2. Palaeochannel characteristics.

Channel number	Width (m)	Height (m)	Width/ Height	Slope S	Area (m²)	Q_b (m³/s)
1	32.78	6.16	5.3	0.00031	132.13	153.5
2	31.46	6.38	4.9	0.00030	140.36	163.8

river east of Piloda (Fig. 1) on its northern bank. The average rainfall in the Tapti river basin is 830 mm. Though not many flood records exist, the discharges for major flood events is given in Table 1.

3 Description

The channel fill sediments are bounded by surfaces which record the time interval of non-deposition. These bounding surfaces depict discrete identifiable geometrical inter-relationships which enables the construction of hierarchies. The ordering of bounding surfaces followed here is similar to the methodology adopted in ALLEN (1983). The lowermost order of bounding surfaces is termed the 1st order bounding surface and higher hierarchies are attributed increasing numerical denominations. The present study revolves around a set of three channel fills (Figs. 2, 3) overlain by a horizontal pedogenised horizon. The exposure is observed on the southern bank of the channel bed of the tributary Aner of the river Tapti, near Piloda, Dhule district, Maharashtra. The extrapolated dimensions and calculated ratios of the channel fills assuming a symmetrical cross section is listed in Table 2.

1st order bounding surfaces

In each of the channel scour these surfaces separate packets of laminar sediments of 5 to 7 cm thickness. The bounding surfaces are concave upwardly arcuate and merge with

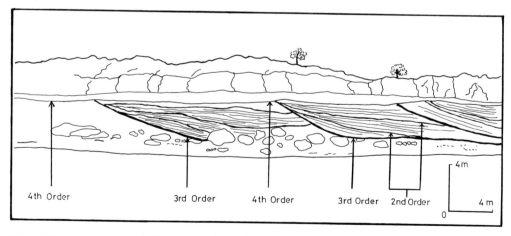

Fig. 3. Tracing overlay of field photograph of channel-fill deposits.

each other on the margin of the channel scour resulting in the crescent shaped morphology of the packets. The bounding surfaces become progressively planar upwards through the section.

2nd order bounding surfaces

These bounding surfaces sandwich packets of sediments having thickness of 20 to 30 cm. Similar to the lower order bounding surface these too are concave upward and become planar progressively upwards. In channel number 2 these surfaces are more prolifically developed the total number aggregating three, while channel number 1 exhibits five such bounding surfaces. Channel number 3 hosts four bounding surfaces, however the surfaces do not significantly plane off at the top and retain their concavity.

3rd order bounding surfaces

The third order bounding surfaces are representatives of the geometry of the channel scours. These surfaces are preserved partially due to the truncation of the channel scour surface of the predecessor channel by the subsequent (temporally speaking) channel. The third order bounding surface show varying degrees of concavity in each channel fill deposit, the most acute being the channel number 2.

4th order bounding surfaces

This bounding surface which truncates the third order and lower order surfaces has a sub planar geometry and separates the array of channel deposits from the overlying pedogenised aeolian silts. The pedogenised aeolian silts consist of a thin Holocene A horizon and at the base a laterally extensive petrocalcic calcrete horizon is observed, the base of which helps in the delineation of the bounding surfaces.

4 *Discussion*

The channel fill deposits formed in a tributary which was part of the drainage network of the Tapti river. The palaeocurrent direction of the channels is towards south. The three orders of bounding surfaces represent events of non deposition between successive episodes of aggradation primarily through suspension settling. The concavity of the bounding surface is interpreted as a manifestation of flow velocities of the surface run-off. The ratio of the width of the channel to the height of the channel is taken as the representative parameter of channel concavity. The channel width is extrapolated assuming a symmetric cross-sectional channel geometry. This assumption we justify owing to the absence of lateral accretion deposits characteristic of shifting streams. Moreover if the width of the channel remains constant, then the channel-width/channel-height ratio directly reflects change in the hydraulic regime. Additionally grain size analysis of the channel fill deposits reveals a mean size of 3 Ø (Fig. 4), which makes it unnecessary to account for grain size variations since the sediment package is homogeneous in context of the same. Thus under conditions of constant channel width, homogeneous grain size distribution, the changes in the dimensional attributes of the channel are solely controlled by the flow velocity which in turn is dependent on precipitation rates. Hence the hierarchy defined in the bounding surfaces based on geometrical inter-relationships document hierarchies in flow

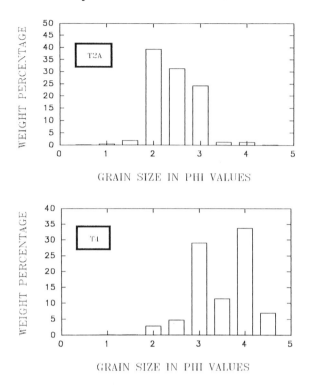

Fig. 4. Grain size distribution of channel-fill sediments showing dominance of finer sediments (3 Ø to 4 Ø).

velocities, with the more concave bounding surfaces representing increased flow velocities. Within this framework the first order bounding surfaces document time of non deposition between successive episodes of post monsoonal suspension settling of the sediments. The second order bounding surfaces represent a higher hierarchy in the flow velocity of the stream which indicates that these might be the end result of minor floods. Peak flood events would have probably caused a reorganization in the tributary channel-flow pattern. The avulsion phenomenon observed in the present study is a effect of such peak flood events. We have calculated the bankfull discharges of the two channels 1 and 2 based on the equation:

$Q_b = 4.0\ A_b^{1.21} S^{0.28}$ (WILLIAMS 1978)

where Q_b is the bankfull instantaneous discharge in m³/s, A_b is the bankfull area in m² and S is the slope which was calculated from the equation by WILLIAMS (1988):

$S = 0.0020\ W_b^{-0.006} D_{max}^{-0.91}$

where S is slope in metres per metre, W_b is the bankfull width in metres and D_{max} is the maximum particle size in mm. The results of the analysis is tabulated in Table 2. However, the estimate does not take into account that the formation of the new channel (e.g. channel 2) requires erosion of part of an existing channel deposit. Thus the actual discharge may

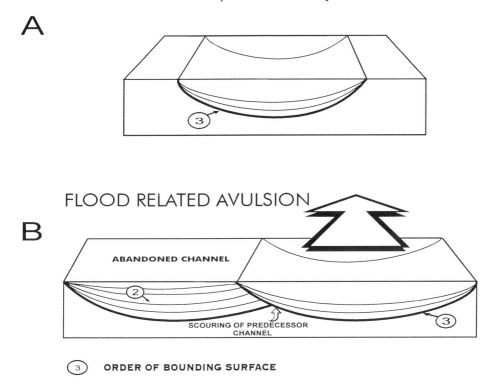

Fig. 5. Sequence of events which led to the formation of the channel fills. a) Formation of channel; b) Flood related avulsion with concomitant scouring of predecessor channel.

have been higher than the derived ~150 m³/s (probably a magnitude more). It is difficult to arrive at this value since the amount of sediment instantaneously excavated is not known but could be of the order of 800 m³/s which is comparable with flood discharges of the Tapti river of 20,200 m³/s at Savkheda (KALE et al. 1994).

Fig. 5 shows the sequence of events which have led to the formation of the channel-fill deposits. Consequently the third order bounding surface documents episodes of anomalous discharge due to extreme precipitation events. The progressive decline in the concavity of the bounding surfaces in each of the channel fill deposit is an indication of phases of progressive decrease in the flow velocity of the tributary channel.

In the absence of an absolute age control on these palaeoflood events, the analysis of the recurrence interval becomes largely speculative. The Tapti river flood records are few, and existing data shows a recurrence of every 10–20 years (KALE et al. 1994).

4 Conclusion

Manifestations of flooding are diverse and recognition of possible representative structures is important in palaeoflood analysis. A suite of three channel fill deposits exhibits small scale avulsion. The instantaneous bankfull discharge calculated on the basis of channel dimensions falls around values of 150 m³/s. However, the avulsions observed can be attributed to peak floods probably having discharges up to 800 m^3/s owing to the fact that the predecessor channel had to be partly eroded. Each channel fill reveals evidences of progressive flow velocity loss in the declining concavity of the first order and second order bounding surfaces. The second order bounding surfaces record a lower hierarchy in flow velocity as compared to the peak flood generated third order bounding surface. Due to the absence of absolute age control not much is commented on the inter-avulsion period, however in the present contet the inter-avulsion period might have been similar to that documented in case of the Tapti River of 10–20 years (KALE et al. 1994).

Acknowledgements

The authors would like to thank Prof. S.S. Merh (Baroda) for providing constant support throughout the study. We would also like to thank Prof. Vishwas S. Kale (Pune) for helping us in making the palaeohydraulic calculations by providing relevant literature. Financial assistance provided by DST (Project no. ESS/044/012/90) to JNM and ASK and a CSIR Senior Research Fellowship to ASK (9/114/(81)/95/EMR-I) is acknowledged.

References

ALLEN, J.R.L. (1983): Studies in fluviatile sedimentation: bars, bar complexes and sandstone sheets (low sinuosity braided streams) in the Brownstone (L. Devonian), Welsh Borders. – Sediment. Geol. **33**: 237–293.

BAKER, V.R. (1977): Stream-channel responses to floods with examples from central Texas. – Geol. Soc. Amer. Bull. **88**: 1057–1071.

– (1988): Quaternary Palaeoflood hydrology in Tropical regions. – Nat. Sem. Rec. Quat. Studies in India: 26–36.

BAKER, V.R. & G. PICKUP (1987): Flood Geomorphology of the Katherine Gorge Northern Tertiary, Australia. – Geol. Soc. Amer. Bull. **98**: 635–646.

HUCKLEBERRY, G. (1994): Contrasting channel responses to flood on the middle Gila river, Arizona. – Geology **22**: 1083–1086.

KALE, V.S., L.L. ELY, Y. ENZEL & V.R. BAKER (1994): Geomorphic and hydrologic aspects of monsoon floods on the Narmada and Tapti rivers in central India. – Geomorphology **10**: 157–168.

MARTINI, I.P. (1977): Gravelly flood deposits of Irnine Creek Ontario, Canada. – Sedimentology **24**: 603–622.

SMITH, N.D., T.A. CROSS, J.P. DUFFICY & S.R. CLOUGH (1989): Anatomy of an avulsion. – Sedimentology **36**: 1–23.

TORNQUIST, T.E. (1994): Middle and late Holocene avulsion history of the river Rhine (Rhine Meuse delta, Netherlands). – Geology **22**: 711–714.

VAN GELDER, A., J.H. VAN DER BERG, G. CHENG & C. XUE (1994): Overbank and channel fill deposits of the modern Yellow River delta. – Sediment. Geol. **90**: 293–305.

WELLS, N.A. & J.A. DORR Jr. (1987): Shifting of the Kosi river, northern India. – Geology **15**: 204–207.

WILLIAMS, G.P. (1978): Bankfull discharge of rivers. – Water Resour. Res. **14**: 1141–1154.

– (1988): Paleofluvial estimates from dimensions of former channels and meanders. – In: BAKER, V.R., R.C. KOCHEL & P.C. PATTON (eds.) (1988): Flood Geomorphology: 321–334; Wiley.

Address of the authors: J.N. MALIK and A.S. KHADKIKAR, Department of Geology, M.S. University of Baroda, Baroda 390 002, India.

Scale and morphometry interaction in the drainage network of badlands areas

CARMELO CONESA-GARCÍA, FRANCISCO LÓPEZ-BERMÚDEZ, and
MARIA ASUNCION ROMERO-DÍAZ, Murcia

with 5 figures and 9 tables

Summary. The effect of the scale variations applied to the morphometry of the drainage networks is examined in a badland basin, located in the semi-arid Mediterranean area of south-eastern Spain. After a primary statistical analysis, based on the correlation and adjustment of the composition and texture variables of the drainage network with the scale denominators that are used, the degree of suitability of each scale factor is established. Moreover, application of multidimensional scaling, taking into account the normalization of Euclidean distances among the values adopted by the morphometrical variables in different scales and spatial unities, allows us to define variations depending on the specific nature and structure of the data.

1 Introduction

The problem of the spatial scale in geographical investigation has already been widely treated (MCCARTY, HOOK & KNOS 1956, HAGGETT 1965, BARRET 1972, CHURCH & MARK 1980, POSTOLENKO 1986). In many cases, the importance of the spatial magnitude understood as a dynamic property according to the object of study is underlined. HAGGET, CHORLEY & STODDART (1965) put special emphasis upon indicating that the spatial magnitude is not a mere passive quality, since from different scales unlike parameters are made dominant, being able to give several generalization levels and even to identify specific problems related to each one of then.

In geomorphology, there have been various attempts to classify landform unities in function of the spatial scale used (TRICART 1965) and at the same time the relationship between scale and survey information is emphasized. PITTY (1982) indicates textually: "Scale influences methods used to collect date and indicates which specialism, from geophysics to sedimentology and soil science, will illuminate the significance of the results."

The effect of the spatial scale in the morphometry of drainage networks is a clear example of this relationship, such as JARVIS (1977) has suggested. According to him, the regularities and systematical relationships that characterize the networks of a given scale in a region generally do not have to be representative of the drainage network on all the scales in this space.

In the semi-arid environment basins, these aspects and the diversity of drainage networks reflect a complex interaction among rainfall, geomorphological characteristics and the uses of the soil of the basins. The study of the morphometry of the drainage network in badland areas has special relevance, on the one hand, because the relationship between the erosion and processes of production of sediments are influenced by the climate and

108 Carmelo Conesa-García et al.

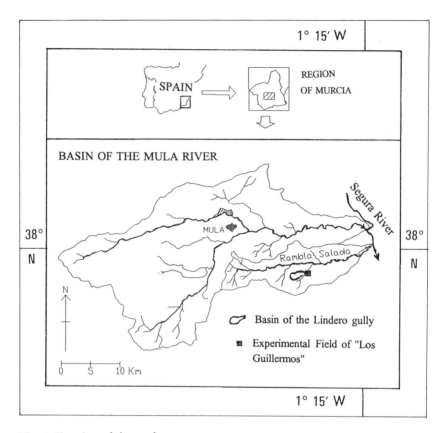

Fig. 1. Situation of the study area.

the characteristics of the basin drained; on the other hand, for the design of digital terrain models in hydrology whose resolution scale is a basic aspect (TARBOTON et al. 1991, MOORE et al. 1991, JENSON 1991). Because of this, important deviations in the desired results and undesirable margins of error, in the morphometry of the network, can be produced by the effect of variations in the spatial analysis scale, or in the size of the pixel.

The cartographic compilation is based on air photographic analysis of different scales (1/18000, 1/12000, 1/3500 and 1/700) corresponding to a flight in 1993. The measurement of the lengths of the courses, in such maps, has been carefully carried out by a digital curvimetre. Before this, the restitution of the drainage network for each one of the scales considered was carried out. The definition of these scales was undertaken by the Geographical Service of the Spanish Army, taking into acount the variations introduced by the irregularity of the terrain. The differential alterations observed in each scale vary with the order to the scale and with the relief ratio, so that, considering the second factor more or less constant, it has been possible to trace directly, based on the photogrametric information, network maps differentiated in function of the scale variation.

2 The natural environment of the study area

The nature of this article and the need to work with large scales, led to the choice of a badland basin of reduced extension (9.5 hectares): the basin of the "*Barranco del Lindero*", partly monitoring through the Experimental Field of "*Los Guillermos*" (LÓPEZ-BER-MÚDEZ et al. 1992). This area is located in the Region of Murcia (South-East Spain), approximately in the centre of the hydrological basin of the "*Rambla Salada*" (Fig. 1). In the geological context, it forms part of a Neogene-Quaternary sedimentary basin, predominantly integrated by marls, clays, sands and conglomerates. Over the whole area a great number of fractures can be observed that show recent tectonic activity, which control the drainage network and accelerate the erosion processes.

The climate is characterized by an annual mean tempereature of 18 °C, an accused annual thermal oscillation (16 °C), scarce and very irregular rainfall (about 300 mm/year), and a high potential evapotranspiration in the summer months (130 mm/month).

Under these climatic and lithological conditions, the natural vegetation is composed of a scrub of romero ("*Rosmarinus officinalis*"), albaida ("*Anthyllis citysoides*"), esparto ("*Stipa tenacissima*") and various cistaciae, with isolated examples of "*Ephedra fragilis*", algarrobo ("*Ceratonia siliqua*"), lentisco ("*Pistacea lentiscus*") and acebuche ("Surges European var. *silvestris*"). The exposition of the hillsides is, also, an essential factor that influences the degree of plant colonization and the erosionability of the materials. On the sunny slopes, bare of vegetation, badland easily develops, while in the shade, generally occupied by grassy scrub and shrubs, there is a less pronounced slope with greater geomorphological stability.

3 Methods and results

3.1 Variables of composition and texture of the drainage network

STRAHLER's ranking system (1964), which modifies HORTON's (1945), has had a huge repercussion on the study of river network quantification, and even today, in spite of the success of other hierarchic classification proposals (SCHEIDEGGER 1965, SHREVE 1967), continues to be the most widely used. This system has been particularly criticized for its application to heterogeneous areas, which often determines that basins of similar drainage order and network of the same magnitude may have a wide range of sizes. In fact, the variability of the size of the basin in function of the current order usually turns out to be smaller than the morphometrical differences existing among the orders. Also, the application, in given cases, of HORTON's relationships calculated with STRAHLER's topological model was been questioned. Thus, the relationship of the mean length of the watercourses of a given order with the said order category, found through this model, are not well-adjusted to the geometric ratio progression Rl or law of stream length established by HORTON. These observations do not mean, however, any sort of restriction upon adopting STRAHLER's ordering scheme in the interaction analysis scale – morphometry of networks, even in areas with different environmental conditions. For their topological structure, it is appropriate to establish differences of relative scale (EBISEMIJU 1985), and, in such a sense, it has been used in the present work.

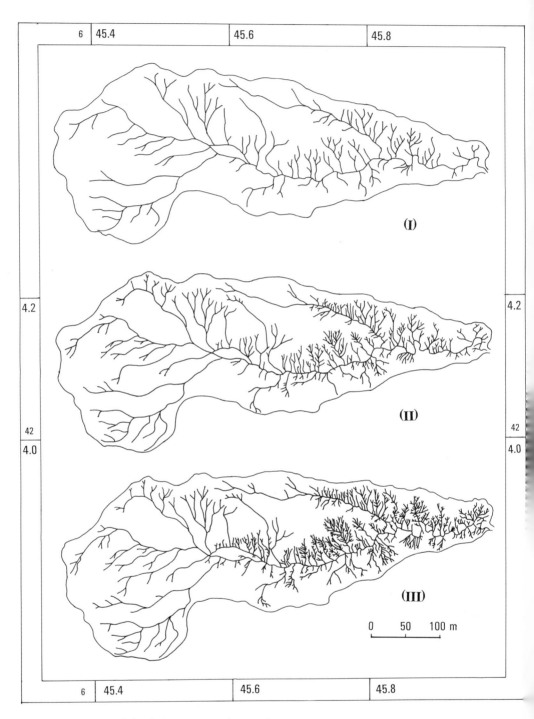

Fig. 2. Restitution of the drainage network to scale 1/12000 (I), 1/3500 (II) and 1/700 (III).

The analyzed variables, as of restitutions to different scales (Fig. 2), refer specifically to the linear elements, composition and texture of the drainage network: number and lengths of the reaches, bifurcation, mean lengths by stream orders, length ratios, drainage density and course frequency. The variables that are independent of the scale variations (shape, area, length of the basin, slope index ...) have been excluded. The exclusion of these parameters, some of them relative indices, has been necessary to avoid unsuitable or redundant information. The seven variables chosen are important properties of the network geometry and have been discussed in numerous geomorphological and hydrological investigations (JARVIS 1977, GARDINER & PARK 1978, CASTILLO-SÁNCHEZ 1986, LÓPEZ-BERMÚDEZ et al. 1988, ROMERO-DÍAZ 1989, CONESA-GARCÍA 1990).

The *number and length of the stream segments* bear a direct relationship to the course frequency and the drainage density respectively. The comparison of the data obtained for these variables to different scales, in erodible lithologies, submitted to different environmental controls (structural, tectonic, topographic, land uses), permits to know the degree of suitability of the scale used in this sort of area.

With the scale change 1/18000 to 1/700, applied to the study basin, the total number of recorded sections is increased by 587 per cent (Table 1). This high relative value can be attributed to the important effect that all increases have in the resolution of the scale in the topological characteristics of a drainage network, such as that of "*Barranco del Lindero*", organized on intensively eroded lands. The detail level that is obtained upon going from 1/18000 to 1/12000 is not very meaningful and insufficient in relationship to the optimum necessary to quantify fine and ultrafine texture networks. The greater increase rates are reached here with the use of the largest scales, adopting a geometric progression trend.

The *mean length of the courses* ($\bar{L}_u = \bar{L}_1 Rl^{u-1}$) is in direct relationship with their number, the basin area, lithology, soil and degree of vegetation cover, the latter strongly conditioned by the orientation. The headwaters sectors, having greater stability, have a mean length of 21.7 m, as compared to the badland areas composed of numerous rills and gullies, whose value is reduced to 5 m on the south-facing slope and to 6 m on the shady slope (Table 2). The effects caused by the use of a particular scale are different according to the stream orders: (1) the reaches of first and second order register mean lengths which progressively diminish with the increase in scale; (2) the values of L_3 do not undergo considerable changes with the scale change, since they keep the same progression as N_3.

Table 1. Number of segments (N_u) and total lengths (L_n) of each stream order obtained through the restitution of the drainage network from different scales.

Scale	Nu						Ln (Km)					
	1	2	3	4	5	Σ	1	2	3	4	5	Σ
1/18000	110	33	6	1	–	150	2.11	1.02	0.32	0.46	–	3.9
1/12000	147	43	9	2	1	201	2.51	1.07	0.38	0.35	0.20	4.51
1/3500	365	100	17	4	1	487	2.74	1.94	0.57	0.24	0.50	5.99
1/700	780	206	39	5	1	1031	3.90	2.57	1.05	0.23	0.52	8.27

Table 2. Mean length of the reaches of each order (Lu) calculated for the scales in study (expr. in m).

Order number	Headwaters				badland hillsides (sunny slope)				hillsides ± established (shady slope)			
	1/18000	1/12000	1/3500	1/700	1/18000	1/12000	1/3500	1/700	1/18000	1/12000	1/3500	1/700
1	32.7	28.5	13.4	12.3	14.8	13.6	5.7	3.8	11.2	11.5	7.1	4.7
2	79.4	43.3	34.6	34.2	16.3	15.2	13.4	7.1	20.6	18.3	11.4	9.9
3	80.0	66.0	60.2	80.8	17.6	15.0	10.9	18.6	–	–	20.0	12.3
4	–	–	78.0	74.5	–	–	29.0	14.5	–	–	–	–
mean	44.5	34.6	23.1	21.7	15.3	14.0	7.4	5.0	12.9	12.6	8.1	6.0

The *bifurcation ratio* ($Rb = N_u/N_{u+1}$) is a morphometrical parameter widely adopted in studies of fluvial geomorphology (SCHUMM 1956, GREGORY & WALLING 1973, FERGUSON 1975, MORISAWA 1985, LÓPEZ-BERMÚDEZ et al. 1988). In order to use it as a prediction index in the hydrological behaviour of a basin, a high degree of reliability is required. On occasions, the scales used do not guarantee an accurate forecast of the hydrological events by not providing an acceptable morphometrical restitution. This fact has particular importance in arid and semi-arid environments which, not having gauging stations, require detailed checking of the physical conditions of their basins. In the basin of the *"Barranco del Lindero"*, only from the scale 1/3500 relatively high values of Rb are reached (between 6 and 7.5), in accordance with the geology of the area, characterized by clay lands, marls and sands, and the presence of fractures from the post-miocene age.

The *length ratio*, defined as $\overline{L}_u/\overline{L}_{u-1}$ is a relationship parameter, dependent on the average length of the courses of each order, therefore its behaviour with the scale variations can be deduced from this variable. In the badland hillsides, the length ratio between the reaches of the first and second orders increases with the scale when this is lower than 1/3500; however, at greater scales this ratio is kept as the mean length of both types of reach diminishes in the same proportion. In the headwater areas, the mean length of different order stream varies much less from one scale to another, but also a increase of the length ratio can be appreciated on going from 1/12000 ($\overline{L}_2/\overline{L}_1 = 1.52$) to 1/3500 ($\overline{L}_2/\overline{L}_1 = 2.58$).

The *drainage density*, considered as a measure of dissection and texture, is the most meaningful morphometrical variable of a basin and, as such, the one which best reflects its hydromorphological reality. The problem is in determining which are the most appropriate working scales to obtain real texture data and, above all, what scale must be chosen under one or other environmental conditions in the network restitution.

The annual mean rainfall being more or less the same all over the basin (≈300 mm), the textural differences of the existing drainage network over its different sectors respond to geological, topographic, biogeographic conditions and, possibly, to historical land use. For the headwater area, generally not very abrupt, covered by a Mediterranean scrub, the drainage density measured varies little from one scale to an other. On the other hand, the area formed by non-cohesive rocks and steep slopes, southfacing, presents a surface bare of vegetation, greatly influenced by linear erosion (*badland*). The drainage density on this sort of formation, easily erodible, is very significant (Table 3). However, due to the high slope of the first order segments, their cartographic representation obliges us to

Table 3. Relationship between the analysis scale and the texture of the drainage network of "*Barranco de Los Guillermos*".

Slope	1/18000		1/12000		1/3500		1/700	
	Dd	If	Dd	If	Dd	If	Dd	If
Headwaters	32.9	7.4	36.3	10.5	47.1	20.4	54.9	25.3
Badland (sunlight)	49.6	32.5	58.8	42.0	84.7	114.8	134.2	267.5
Shady hillsides	15.7	12.2	21.6	17.2	36.6	45.2	67.3	111.7
Total basin	41.0	15.8	47.5	21.1	63.0	51.3	87.0	108.5

Dd = Km/Km²
If = num. of segments/Ha

measure geometric distances which are very inferior to the real ones, which make it necessary to use scales sufficiently large to detect their size and even, sometimes, their presence.

In this study, only scale 1/700 has allowed a restitution adapted for the badland areas, giving as a result a drainage density of 134 Km/Km², that represents a fine and nearly ultrafine texture (MORISAWA 1985). Finally, on the shady badland hillsides, of the same lithology as in the sunny slope, and vegetation covers somewhat denser than in the headwater sectors, a fine texture is registered by scale 1/3500 and a mean texture using scale 1/12000 or other lower scales.

3.2 Statistical analysis

(a) Correlation and regression

To determine the degree of adjustment of the morphometrical variables as a linear function of the scale denominator (D_e), *Pearson's correlation coefficient* (r) has been adopted, as a rule obtaining a high negative correlation, with values frequently higher than –0.80 (Table 4). The total number of courses of the headwaters areas and badlands maintain an excellent negative correlation with D_e (–0.98 and –0.97 respectively), while the worst adjustment is given in the total length values of the first order reaches (–0.70). This reflects the influence of the average lengths of these segments, which become progressively smaller when the scale is increased and the number of segments in the order sections are also increased.

Table 4. Correlation of N_u and L_n with the scale denominator.

Slope	r (N_u-scale denominator)				r (L_n-scale denominator)			
	1	2	3	Σ	1	2	3	Σ
Headwaters	–0.97	–0.98	–0.91	–0.98	–0.70	–0.91	–0.99	–0.97
Badland – sunlight (S)	–0.87	–0.84	–0.82	–0.97	–0.89	–0.96	–0.73	+0.89
Shady hillsides (N)	–0.85	–0.86	–0.75	–0.86	–0.92	–0.85	–0.74	–0.89
Total basin	–0.88	–0.88	–0.86	–0.88	–0.87	–0.94	–0.86	–0.47

Table 5. Correlation and covariance of the values of Dd and If with the scale denominators.

Slope	(Dd-scale denominator)		(If-scale denominator)	
	r	cov	r	cov
Headwaters	−0.971	− 5.8·10⁴	−0.983	− 4.9·10⁴
Badland – sunlight (S)	−0.898	−20.2·10⁴	−0.863	−55.7·10⁴
Shady hillsides (N)	−0.894	−12.3·10⁴	−0.855	−23.3·10⁴
Total basin	−0.925	−11.2·10⁴	−0.879	−22.2·10⁴

Logically, the values of drainage density and frequency indices are also highly interrelated with D_e. The correlation, in all the cases, is negative, the headwaters zone being again the one which shows values closest to −1 and a smaller covariance (Table 5). The badland sectors have r coefficients slightly lower (−0.85 to −0.90) and very high covariances.

The determination coefficients (r^2) of the linear regressions for this sort of variable are, however, somewhat lower than the estimates through non-linear adjustments. The regression equations that have a better adjustment among the total sections measured for each scale (N_u) and the scale denominators (D_e) are non-linear expressions, of potential and logarithmic type, whose terms can vary considerably if stable and homogeneous basins are analyzed.

$N_u = 54962 \, D_e{-0{,}596}$ ($r^2 = 0.988$)
$N_u = -274 \ln D_e + 2894$ ($r_2 = 0.984$)
$N_u = 872 \, e^{-0.0001x}$ ($r^2 = 0.929$)
$N_u = -0.045 \, x + 850.5$ ($r^2 = 0.773$)

In fact, the number of segments of first, second and third order maintain a strong negative alometric relationship with the values of D_e. The determination coefficient is located, in these cases, between 0.98 and 0.99, which indicates an excellent adjustment of this morphometrical variable as an alometric function of the scale.

However, in the interaction analysis between the total length of the segments and the scale, the logarithmic adjustment is more suitable, since its determination coefficient is greater than those calculated for the rest of the regression equations.

$L_n = 37.1 \, D_e{-0.227}$ ($r^2 = 0.993$)
$L_n = 1.33 \ln D_e + 16.9$ ($r_2 = 0.998$)
$L_n = 7.66 \, e^{-0.00004x}$ ($r^2 = 0.919$)
$L_n = -0.0002 \, x + 7.38$ ($r^2 = 0.856$)

The potential relationship $y = ax^b$ also represents a good adjustment between both parameters, being even better than the logarithmic when the scale with the measurements of L_1 and of L_3 is related (Fig. 3).

The drainage density of the whole basin shows, according to the trend of L_n, a geometric progression expressed through the equation $Dd = -6.83 \ln D_e x + 100.7$, being $r^2 = 0.979$. By sectors, the regression curve is logarithmical in the most stable hillsides and indistinctly logarithmical or potential for those of greater erodibility.

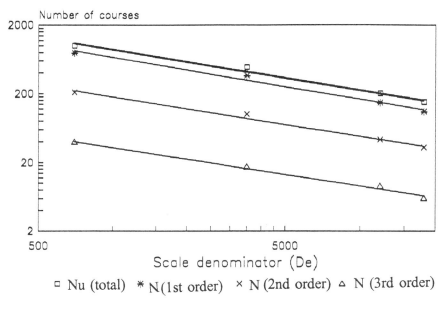

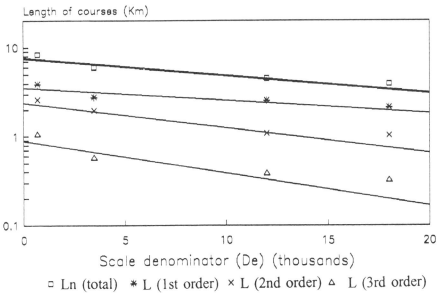

Fig. 3. Adjustment of the variable N_u and L_u with the scale denominators.

(b) Multidimensional scaling

Multidimensional scaling, or combination of multivariant techniques orientated to the analysis of the underlying structure of a system of elements (BORG 1981, COXON 1982,

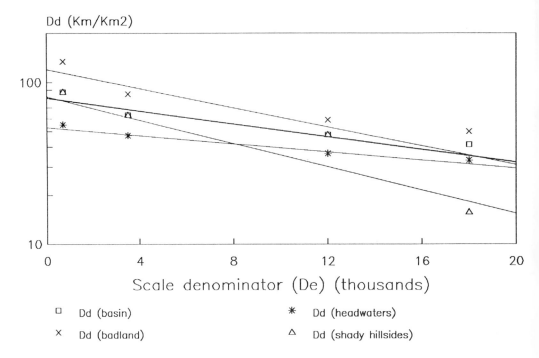

Fig. 4. Regression lines of the drainage density obtained at different scales on their denominators.

EVERITT & DUNN 1983, YOUNG & HAMER 1987, ATO GARCÍA et al. 1990), permits us to represent in a geometric space the proximities (δ) between variable couples through distances (d) in function of a given number of dimensions (k). The application of this technique to the study of the effects of the scale variation of air photographs makes it possible to estimate the existing linear distances between them and to incorporate a second dimension to define intrinsic variations to the relative data of each variable. In this way, it has been possible to measure distances influenced by the different effect that the scale change introduces into the results and to record different order reaches, sorts of exposure and even areas of different lithology.

As variables the morphometric parameters analyzed for each one of the scales of the study: N_u, L_u (for all the stream orders), Rb and Rl for the different slopes) have been adopted.

As a measurement of proximity (δ_{ij}), according to the algorithms used, the normalized Euclidean distance function, from Minkowski's metrics, has been selected:

$$d_{ij} = \left[\sum_{i=1}^{m} (x_{ik} - x_{jk})^2 \right]^{1/2}$$

where x_{ij} and x_{jk} represent the distances of the objects i and j, that is to say, the dissimilarities between the data each pair of scales. This normalization makes the distance func-

Table 6. Euclidean distance matrices normalized for the variations of N_u in relationship to the scale changes.

	N_u (1st order)					N_u (2nd order)			
	(A)	(B)	(C)	(D)		(A)	(B)	(C)	(D)
(A)	0.00				(A)	0.00			
(B)	18.99	0.00			(B)	5.21	0.00		
(C)	137.70	119.00	0.00		(C)	34.50	29.38	0.00	
(D)	367.35	348.72	229.90	0.00	(D)	92.68	87.63	58.63	0.00
	N_u (3rd)					N_u (total basin)			
	(A)	(B)	(C)	(D)		(A)	(B)	(C)	(D)
(A)	0.00				(A)	0.00			
(B)	1.55	0.00			(B)	25.98	0.00		
(C)	5.76	4.29	0.00		(C)	179.38	153.71	0.00	
(D)	18.31	16.96	12.76	0.00	(D)	92.68	455.08	301.82	0.00

(A) Esc. 1/18000; (B) Esc. 1/12000; (C) Esc. 1/3500; (D) Esc. 1/700

tion invariable over the scales and permits the comparison of conglomerates over different sample sizes.

In Table 6 the normalized Euclidean distances are shown for the totals of different order sections, obtained through the use of the scales selected.

The maximum divergences are produced, as is logical, in the first order courses, whose number varies substantially on passing from one scale to an other. The matrix of distances calculated for this order is very different, so that we can find especially high value (367.3) between the smaller scale (1/18000) and the larger (1/700). Considering the disparity existing between pairs of "neighbouring" scales (A-B, B-C and C-D), it is proven that the lowest Euclidean distance (19) is registered upon going from the scale 1/18000 (A) to 1/12000 (B), while the highest (230) is found among the scales 1/3500 (C) and 1/700 (D). In fact, from these latter scales alone it is possible to identify the character of the badland area, typical of the basin of study. The matrices calculated for the number of second and third order reaches show similar relationships to those of the first order and a positive correlation with the scale.

In the case of the distances of L_u obtained by stream orders and scale variations (Table 7), the results are very different, since only in the second order courses a clear negative correlation between the degree of similarity and the work scale is to be seen. The distances, for each pair of correlative scales, vary here from 20.9 in the step 1/18000–1/12000, until 3.75 in the step 1/3500–1/700. This diminishing trend of proximity among the values of L_u according to the increase in scale registers an anomaly in the first order courses, upon going from a scale 1/12000 to 1/3500. Such an obvious change can be explained by the recognition, from 1/3500 and on the sunlight hillsides, of the multitude of small gullies, undistinguishable on other smaller work scales.

The Euclidean distances that the bifurcation and length ratios show, in function of the scale, are expressed in Tables 8 and 9. From them it can be deduced that, for the set of the basin, the greatest degree of differentiation is logically established between the

Table 7. Euclidean distance matrices among the values of L_u from different scales.

	L_u (1st order)					L_u (2nd order)			
	(A)	(B)	(C)	(D)		(A)	(B)	(C)	(D)
(A)	0.00				(A)	0.00			
(B)	4.22	0.00			(B)	20.89	0.00		
(C)	12.55	8.72	0.00		(C)	26.46	6.49	0.00	
(D)	13.90	10.27	1.88	0.00	(D)	27.34	8.54	3.75	0.00

	L_u (3rd order)					L_u (total basin)			
	(A)	(B)	(C)	(D)		(A)	(B)	(C)	(D)
(A)	0.00				(A)	0.00			
(B)	10.07	0.00			(B)	5.77	0.00		
(C)	14.78	5.02	0.00		(C)	13.46	8.08	0.00	
(D)	0.91	10.77	15.55	0.00	(D)	14.98	9.85	2.01	0.00

extreme scales (1/18000 and 1/700) (d(Rb) = 4.28 and d(Rl) = 0.59) and that the widest disparity due to the correlative variation of the used scales is found between 1/12000 and 1/3500 (2.81 and 0.57 for Rb and Rl respectively), which is in accordancy with the existing distances between the scale denominators.

Analyzed for different sort of hillsides, the measurements of similarity represent matrices of different structure. For example, the increase in scale from 1/3500 to 1/700 does not provide any new data for the shady hillside, and, on the other hand, it means a considerable increase in the values of Rb for the sunny slope, where it is possible to find a deep badland on bare soil.

By sectors, the maximum effect introduced by the correlative change of scales in the values of Rl can be seen in the scale step (B)–(C) for the headwaters zone (d = 0.79),

Table 8. Euclidean distance matrices among the values of Rb obtained for different scales and spatial unities.

	Headwaters sectors					Sunlight hillside (badlands)			
	(A)	(B)	(C)	(D)		(A)	(B)	(C)	(D)
(A)	0.00				(A)	0.00			
(B)	0.67	0.00			(B)	0.21	0.00		
(C)	1.08	1.26	0.00		(C)	1.04	0.96	0.00	
(D)	2.06	2.51	1.35	0.00	(D)	1.37	1.20	0.65	0.00

	Shady hillside					Global basin			
	(A)	(B)	(C)	(D)		(A)	(B)	(C)	(D)
(A)	0.0				(A)	0.00			
(B)	0.9	0.0			(B)	0.39	0.00		
(C)	0.7	1.6	0.0		(C)	2.94	2.81	0.00	
(D)	0.7	1.6	0.0	0.0	(D)	4.28	4.16	1.36	0.00

Table 9. Euclidean distances normalized among the length ratios (Rl) estimate for the different sectors and work scales.

	Headwaters sectors					Sunlight hillside (badlands)			
	(A)	(B)	(C)	(D)		(A)	(B)	(C)	(D)
(A)	0.00				(A)	0.00			
(B)	0.73	0.00			(B)	0.07	0.00		
(C)	0.52	0.79	0.00		(C)	0.88	0.86	0.00	
(D)	1.03	1.12	0.52	0.00	(D)	1.20	1.26	1.30	0.00
	Shady hillside					Global basin			
	(A)	(B)	(C)	(D)		(A)	(B)	(C)	(D)
(A)	0.0				(A)	0.00			
(B)	0.2	0.0			(B)	0.17	0.00		
(C)	0.2	0.0	0.0		(C)	0.48	0.57	0.00	
(D)	0.3	0.5	0.5	0.0	(D)	0.59	0.54	0.10	0.00

and in the step (C)–(D), for the zone of badland (sunlight) (d = 1.3) and vegetated hillside (shady) (d = 0.5) (Table 9).

The number of dimensions to be used has been estimated by an objective function that measures the discrepancy between the proximities and the distances adjusted by Kruskal's stress formula (1964).

$$S = \left[\frac{\sum_{i=1}^{n} \sum_{j=1}^{n} (\delta'_{ij} - d'_{ij})^2}{\sum_{i=j}^{n} \sum_{j=1}^{n} d'^2_{ij}} \right]^{\frac{1}{2}} \quad (i \neq j)$$

where δ'_{ij} and d'_{ij} are the disimilarities and distances respectively, calculated from an optimization algorithm.

After testing different dimensional solutions, among those proposed by KRUSKAL & WISH (1978), it was decided to use two dimensions, since in all the cases this solution offers a lower value of the objective function (stress). For the set of the morphometric applications studied here, the final configuration of the stress in the two-dimensional solution is 0, which means perfect quality in the adjustment.

The scaling method adopted is the model SHEPARD (1962), based on a monotonic relationship between proximities and distances, so that the data of disimilarity take the form of the Euclidean distance function, by

$$\delta_{ij} = f(d_{ij}) = \left[\sum_{k=1}^{k} (x_{ik} - x_{jk})^2 \right]^{\frac{1}{2}}$$

f being a monotonic function such that

$$d_{ij} < d_{i'j'} \rightarrow f(d_{ij}) < f(d_{i'j'})$$

for each of i, i', j and j'.

Generally, dimension 1 reflects the variations of the analyzed parameters, as a direct function of the scale changes. Except for the relationship of length, the value that each variable takes, according to such changes, keeps, within this dimension, the same position order as the scale factors (A-B-C-D). The mean length (L_u) is the parameter, whose variation from the scale translates more exactly the unidimentional distances among the scale denominators; in fact, the positions that they hold in dimension 1 are practically the same as those which define the disimilarities among the D_e

De	18000	12000	3500	700
	A	B	C	D
DIM1	−1.38	−0.49	0.72	1.15

The number of reaches scarcely varies from A to B, and, on the other hand, registers an extraordinary increase in resolution upon going from C (1/3500) to D (1/700).

Dimension 2 itself brings in deviations or residual distances with respect to the variations observed in dimension 1, that is to say, it explains to some extent the different behaviour of the spatial unities considered before the scale change and distinguishes some scales from others in function of the nature and structure of the data (Fig. 5).

In the case of N_u, the residue of C, corresponding to the scale 1/3500, may be explained by the unequal progression that, at this level, the number of courses of first and second order undergo in the headwaters sectors and badland slopes. At the same time, in relationship to the mean length of the reaches, this dimension differenciates the scales that show a greater contrast between their stream orders (A, 1/18000 and D, 1/700) and those that, influenced by the important relative weight of the values obtained in headwaters, give some less differentiated average lengths for the basin (B and C).

The relative spatial distribution of the points that define the bi-dimensional plan of N_u and L_u constitutes an accurate reflection of the scaling of N_1 and L_1, a fact in total agreement with the disproportionate increase in the first order segments, in relation to the rest of the orders, typical of the branching networks, of very ramified headwaters, which develop in badland areas.

The bi-dimension of Rb keeps the same scale succession as N_u, but slightly separates the intermediate scales (B and C) with respect to those which offer the minimum and maximum detail. In this case, the same as in the bi-dimensional of Rb, calculated for the badlands sectors, the greatest resolution, strictly dependent on the scale change (dimension 1), is found between B (1/12000) and C (1/3500). However, dimension 2, in badlands, increases the difference between the more detailed scales (C = −0.49 and D = 0.34), suggesting once again the better quality of the scale 1/700 for morphologies of this type.

Finally, the relationship of length shows a scaling uncommon in comparison with the rest of the morphometric variables. Over the whole of the basin, the maximum and minimum value is reached using the intermediate scales, there not being any progression in function of the scale. On the badland slope the influence is only seen, in a diminishing direction with D_e, of the order of magnitude of the scales, scale 1/700 again being that which gives a greater resolution.

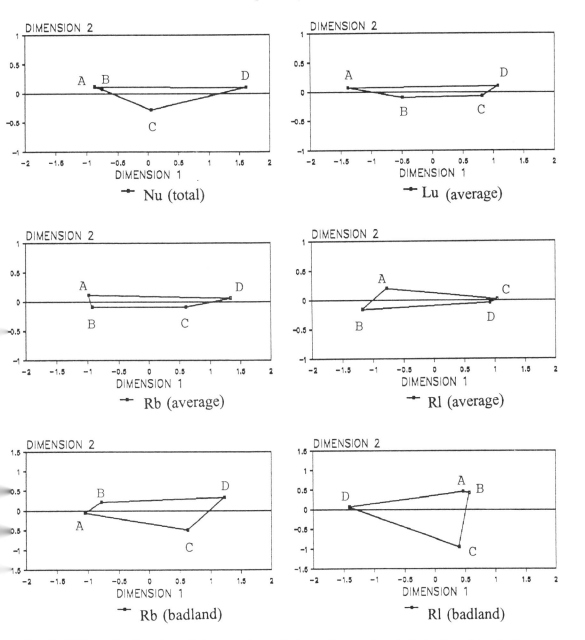

Fig. 5. Bi-dimensional scaling of the morphometrical variables in the whole basin and badland sectors.

4 Conclusions

The *composition and texture parameters* of the drainage network are especially changeable with the variation of scale. At scales smaller than 1/12000, the morphometrical data obtained by restitutions are equivocal and they do not reflect the real morphology of the areas with high erosion. In sectors of relatively stable incision, such as the headwaters of the ravine studied, the scale 1/3500 provides accurate results. On the other hand, in badland areas the use of greater scales is needed. For example in reconstituting the morphometry of the "*Los Guillermos*" badland at scale 1/700, a drainage density of 134 has been obtained, that practically triples the value calculated from the scale 1/18000.

Statistical analysis reveals that there exists a high negative correlation between the morphometrical data extracted from the scales selected and their denominators. The regressions that show a better adjustment in this sense are determined by non-linear expressions. The number of segments, above all those of smaller orders, keep a regression of potential type with the scale denominators, defined by a strong negative alometric relationship and a very high determination coefficient ($r^2 > 0.98$). The variation of the total stream length and of the drainage density shows a good logarithmic adjustment in function of the scale, though in some cases the potential relationship can also be considered acceptable.

With the calculation of the *Euclidean distance matrices* among the values that the variables taken on at different scales, it can be seen that the maximum dissimilarities are produced in the first order courses, upon moving to the scales 1/3500 and 1/700, which shows that, even at smaller scales than these, the morphometrical characterization of the badland areas can be unclear and questionable.

The application of *multidimensional scaling*, developed from these matrices, has allowed the detection of the variation of the effects of the scale in two directions, one longitudinal that reflects the direct influence of the scale denominators, and the other transversal that shows anomalies and differentiated behaviour of the morphometrical variables, caused by the structure of the data used (stream orders, drained areas). The distance that, in dimension 1, separates scale 1/12000 from 1/3500, implies a considerable increase in resolution upon going from one to another. With dimension 2, this increase is more accentuated on the badland hillside, where a disproportionate increase in the reaches of the first order in relationship to the rest of the orders can be seen. On the contrary, considering both dimensions, the smaller scales (1/18000 and 1/12000) are very close to each other and, at the same time, quite removed from the quality that the network restitution at 1/3500 offers.

The knowledge of these interrelationships has a certain theoretical and practical interest, insofar as it constitutes a validation parameter which is very useful in the morphological characterization of particularly fragile and unstable environments, such as badlands areas.

Acknowledgements

This work has been carried out under the auspices of and sponsored by the 'Spanish National Plan for the Environment and Natural Resources, Land Ecosystems', of the CICYT. Projects NAT 89-1072-C06-05 (1990–92) and AMB93-084-4-C06-03 (1993–95).

References

ATO GARCÍA, M., J.A. LOPEZ PINA, A.P. VELANDRINO NICOLAS & J. SANCHEZ MECA (1990): Estadística avanzada con el paquete SYSTAT. – Universidad de Murcia, Murcia, 213–244.
BARRET, E.C. (1972): Geography from Space. – 98 pp., Pergamon Press, Pergamon Topic Geographies, Oxford.
BORG, I. (1981): Multidimensional Data Representations: When and Why. – Ann Arbor, MI: Mathesis Press.
CASTILLO-SANCHEZ, V. (1986): Estudio cuantitativo de paisajes fluviales. E.T.S.-I.M., Universidad Politécnica de Madrid, Tesis Doctoral.
CHURCH, M. & D.M. MARK (1980): On size and scale in geomorphology. – Progr. Phys. Geogr. **4**: 342–390.
CONESA-GARCÍA, C. (1990): El Campo de Cartagena. Clima e hidrología de un medio semiárido. – 450 pp., Universidad de Murcia, Ayuntamiento de Cartagena, Comunidad de Regantes del Campo de Cartagena, Murcia.
COXON, A.P.M. (1982): The User's guide to multidimensional scaling. – Heinemann Educational Books.
EBISEMIJU, F.S. (1985): Spatial scale and drainage basin morphometric interaction. – Catena **12**: 261–270.
EVERITT, B.S. & G. DUNN (1983): Advanced methods of data exploration and modelling. Heinemann, London.
FERGUSON, R. (1975): Network elongation and the bifurcation ratio. – Area **7**: 121–124.
GARDINER, V. & C.C. PARK (1978): Drainage basin morphometry: review and assessment. Progr. Phys. Geogr. **2**: 1–36.
GREGORY, K.J. & D.E. WALLING (1973): Drainage Basin Form and Process: A Geomorphological Approach. – Arnold, London.
HAGGETT, P. (1965): Scale components in geographical problems. – In: CHORLEY, R.J. & P. HAGGETT (eds.): Frontiers in Geographical Teaching. – 164–185, Methuen, London.
HAGGETT, P., R.J. CHORLEY & D.R. STODDART (1965): Scale standards in geographical research: a new measure of areal magnitude. – Nature **205**: 844–907.
HORTON, R.E. (1945): Erosional development of streams and their drainage basins; hidrophysical approach to quantitative morphology. Bull. Geol. Soc. Amer. **56**: 275–370.
JARVIS, R.S. (1977): Drainage network analysis. Progr. Phys. Geogr. **1**: 271–295.
JENSON, S.K. (1991): Applications of hydrologic information automatically extracted from digital elevation models. – Hydrol. Proc. **6**, 1: 31–44.
KRUSKAL, J.B. (1964): Nonmetric multidimensional scaling: a numerical method. – Psychometrika **29**: 115–129.
KRUSKAL, J.B. & M. WISH (1978): Multidimensional scaling. – Sage University Paper series on Quantitative Applications in the Social Sciences, 07-11, Beverly Hills and London.
LÓPEZ-BERMÚDEZ, F., F. NAVARRO-HERVAS, M.A. ROMERO-DÍAZ, C. CONESA-GARCÍA, V. CASTILLO-SANCHEZ, J. MARTÍNEZ-FERNANDEZ & C. GARCÍA-ALARCON (1988): Geometría de cuencas fluviales. Las redes de drenaje del Alto Guadalentín. – Proyecto LUCDEME IV. ICONA, Monografía 50, 239 pp., Madrid.
LÓPEZ-BERMÚDEZ, F., F. ALONSO-SARRIA, M.A. ROMERO-DÍAZ, C. CONESA-GARCÍA, J. MARTÍNEZ-FERNANDEZ, J. MARTÍNEZ-FERNANDEZ (1992): Caracterización y diseño del Campo Experimental de Los Guillermos (Murcia) para el estudio de los procesos de erosión y desertificación en litologías blandas. – In: LÓPEZ-BERMÚDEZ, F., C. CONESA GARCÍA & M.A. ROMERO DÍAZ (eds.): Estudios de Geomorfología en España. – Sociedad Española de Geomorfología. Universidad de Murcia, 151–160.
MCCARTY, H.H., J.C. HOOK & D.S. KNOS (1956): The measurements of association in industrial geography. – Iowa State University, Department of Geography. Report Num. 1.
MOORE, I.D., R.B. GRAYSON & A.R. LADSON (1991): A review of hydrological, geomorphological and biological applications. – Hydrol. Proc. **5**, 1: 3–30.

MORISAWA, M.E. (1985): Rivers. Form and process. – Geomorphology Texts 7, Longman, London, 222.
PITTY, A.F. (1982): The nature of Geomorphology. – 161 pp., Methuen, London.
POSTOLENKO, G.A. (1986): Large scale geomorphological mapping. – Vestnik-Moskovskogo Universiteta, Seriya Geografiya, **4**, 58–64.
ROMERO-DÍAZ, A. (1989): Las cuencas de los ríos Castril y Guardal (Cabecera del Guadalquivir). Estudio hidrogeomorfológico. – 285 pp., Ayuntamiento de Huescar, Universidad de Murcia.
SCHEIDEGGER, A.G. (1965): The algebra of stream-order number. – United States Geological Survey Professional Paper, 525-B.
SCHUMM, S.A. (1956): The evolution of drainage systems and slopes in badlands at Pearth Amboy, new Jersey. – Bull. Geol. Soc. of Amer. **67**: 597–646.
SHEPARD, R.N. (1962): The analysis of proximities: Multidimensional scaling with an unknown distance function, I and II. – Psychometrika **27**: 125–140, 219–246.
SHREVE, R.L. (1967): Infinite topologically random channel networks. – J. Geol. **75**: 17–73.
STRAHLER, A.N. (1964): Quantitative geomorphology of drainage basins and channel networks. – In: CHOW, V.T. (ed.): Handbook of applied hydrology, 39-4, 76.
TARBOTON, D.G., R.L. BRAS & I. RODRIGUEZ-ITURBE (1991): On the extraction of channel networks from digital elevation data. – Hidrol. proc. **6**, 1: 81–100.
TRICART, J. (1965): Principes et méthodes de la géomorphologie. – Masson, Paris.
YOUNG, F.W. & R.M. HAMER (1987): Multidimensional Scaling: History, Theory and Applications. – N.J.: LEA, Hillsdale.

Address of the authors: Department of Physical Geography, University of Murcia, Campus de La Merced, E-30001 Murcia, Spain.

A possible origin for a mega-fluting complex on the southern Alberta prairies, Canada

DAVID J.A. EVANS, Glasgow

with 14 figures

Summary. The geomorphology and stratigraphy of a mega-fluting complex in the lower Red Deer River drainage basin of southern Alberta, Canada, are used in conjunction with adjacent stratigraphic sequences to reconstruct a regional depositional history and genetic model relating to ice advance during the last glacial maximum. A basal grey till (LFA 3) is loaded by coarse gravels (possibly originating as esker sediments) and subglacial cavity fill deposits (LFA 4/5) which may have provided the stable core beneath a deforming till layer (lower LFA 5 and/or lower LFA 6). North of the Red Deer River the fluting complex contains several small sub-flutings which may relate to streamlined esker beads or outer bends of esker meanders deposited during an earlier glaciation or a phase of increased subglacial meltwater discharge. The fluting was therefore initiated by linear development rather than a point source. Future interpretations of the most recent subglacial depositional forms on the Earth's surface must account for possible pre-existing landforms and sediments and their influence on glacier dynamics and substrate rheology.

Zusammenfassung. *Zur möglichen Entstehung eines großen stromlinienförmig überformten Glazialkomplexes in der Prärie von Südalberta.* – Die Geomorphologie und Stratigraphie eines sehr ausgedehnten stromlinienförmig überprägten Glazialkomplexes im Tal des unteren Red River in Südalberta, Canada, werden in Verbindung mit benachbarten Profilen benutzt, um die regionale Sedimentationsgeschichte während des letzten Hochglazials und ein genetisches Modell des Eisvorrückens zu rekonstruieren. Eine graue Basalmoräne (LFA 3) wird von groben Schottern (wahrscheinlich als Os-Sedimente abgelagert) und subglazialen Hohlraumfüllungen (LFA 4/5) überlagert, die möglicherweise als stabiler Kern unter einer deformiert werdenden Grundmoräne gebildet haben (untere LFA 5 und/oder untere LFA 6). Nördlich des Red River enthält der stromlinienförmig überprägte Komplex mehrere kleine, untergeordnete Strömungsformen, die mit stromlinienförmig veränderten Osrückenabschnitten oder den äußeren Bogenteilen von Osmäandern in Verbindung gebracht werden könnten, die aus einer früheren Phase der Vergletscherung mit höherem subglazialen Schmelzwasserabfluß abgelagert wurden. Das "Fluting" wurde also durch lineare Entwicklung und nicht durch eine punktförmige Quelle verursacht. Zukünftige Interpretationen der jüngsten subglazialen Ablagerungsformen müssen die Auswirkung älterer Reliefformen und Sedimente auf die Gletscherdynamik und die Rheologie des Substrats mit einbeziehen.

Résumé. – *Une origine possible pour un complexe macro-cannelure sur les prairies d'Alberta du sud, Canada.* – La géomorphologie et la sratigraphie d'un complexe macro-cannelure glaciaire dans le bassin inferieur du Red Deer River dans le sud d'Alberta, Anada, sont exploités en combinaison avec les séquences stratigraphiques sous-jacentes pour reconstituer une histoire de déposition régionale et un modèle génétique relatif a l'avancée de la glace pendant le dernier maximum glaciaire. Un till basal gris (LFA 3) est chargé de graviers grossiers (peut-être d'origine de sédiments d'esker) et des dépôts sous-glaciaires de remplissage de cavité (LFA 4/5 qui ont peut-être fourni le coeur stable en desous d'une couche déformante de till (LFA 5 inférieur et/ou LFA 6 inférieur). Au nord de la Red Deer River le complexe de cannelure contient plusieurs petites sous-cannelures qui peu-

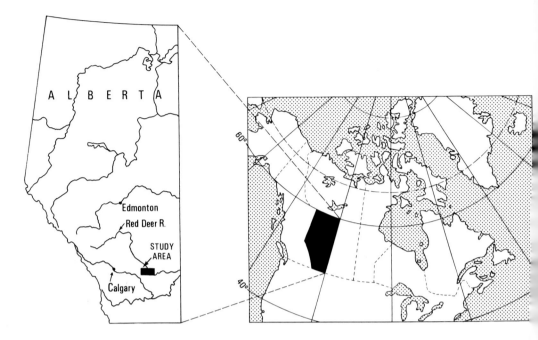

Fig. 1. Map of Alberta and the location of the Red Deer River study area.

vent être en rapport avec les eskers en chapelet d'écoulement laminaire ou les curvatures extérieures des méandres d'eskers déposés pendant une glaciation précédente ou pendant une phase de débit accru sous-glaciaire. La cannelure a donc été initiée par un développement linéaire plutôt qu'une origine ponctuelle. Les interprétations futures des formes de déposition les plus récentes sur la surface de la terre doivent tenir compte des topographies et des sédiments et de leur influence sur les dynamiques des glaciers et la rhéologie substrate.

Introduction

Some of the Earth's most spectacular drumlins and flutings occur in western Canada, especially in the province of Alberta (Figs. 1 and 2). Depite this, few maps of drumlins and flutings exist for the province (cf. GRAVENOR & MENELEY 1958, SHAW 1980, EVANS 1985, Fig. 2). Some detailed local maps (e.g. STALKER 1960, 1973, ROED 1975 and JONES 1982) are used in reconstructions of the last (Late Wisconsinan) advance of the Laurentide Ice Sheet. In addition, the contrary hypotheses of drumlin formation presented by SHAW (1983), SHAW & KVILL (1984) and SHAW et al. (1989; subglacial cavity fills by meltwater) versus BOULTON (1987; differential deformation of materials with varying rheological properties) include the drumlin fields of northern Saskatchewan as examples. SHAW (1975) and JONES (1982) explained the orientations of many fluting fields in Alberta by subglacial topographic constraints, whereas ROED (1975), BOYDELL (1978) and KVILL (1984) interpreted them as the product of deflection of Cordilleran glacier lobes by the Laurentide Ice Sheet. Due to the fact that many flutings appeared to be orientated parallel to buried "preglacial" valeys of the precursor Red Deer River drainage network, STALKER & CRAIG (1956) suggested that glacier flow was strongly controlled by local "buried" topography.

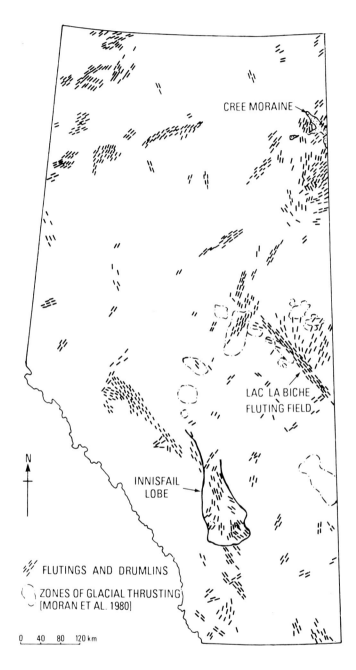

Fig. 2. Map of Flutings and drumlins in Alberta from EVANS (1985) and based on information from GRAVENOR & MENELEY (1958) and SHAW (1980).

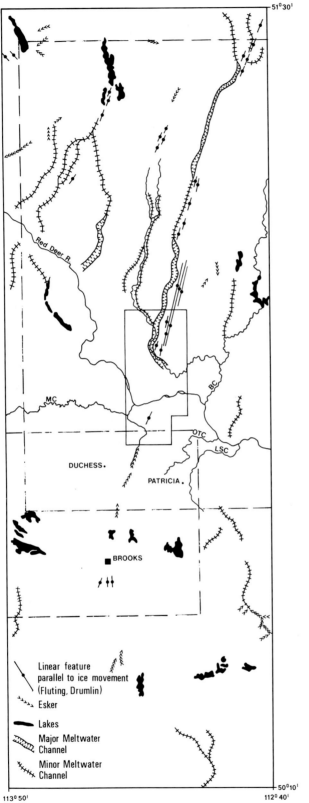

Fig. 3. Extract of geomorphological information from SHETSEN (1987). Solid line boxed area represents coverage of Fig. 8. Broken line boxed areas represent coverage of Fig. 4 and 5. North to south Esker drainage directions, especially south of Brooks, are contrary to the interpretations of this paper. BC = Berry Creek; LSC = Little Sandhill Creek; OTC = One Tree Creek; MC = Matzhiwin Creek.

This paper discusses a complex of mega-scale lineations or mega-flutings in the lower Red Deer River drainage basin and provides a working hypothesis on the formation of such features based upon the regional geomorphology and local glacial stratigraphy. The mega-fluting complex cross-cuts the "preglacial" topography and grades down-ice into streamlined bedrock. An exposure through the basal stratigraphy of the central part of the fluting complex suggests that the sediments may have played a critical, but indirect, role in the formation of the fluting. Sediments and landforms indicate that the formation of the fluting complex involved sand and gravel deposition, possibly as an esker, followed by differential subglacial deformation caused by the different rheological properties of the substrate. The product is a resistant gravel core and a surface carapace of subglacially deformed diamicton.

Geomorphology of the mega-fluting complex

Although early geomorphic mapping in Alberta recognized the most prominent glacially streamlined landforms, SKOYE (1990) and SKOYE & EYTON (1992) have only recently identified more subtle flutings by satellite imagery. They identified numerous low-relief flutings associated with what were originally considered to be isolated mega-flutings. One such mega-fluting complex, which possesses multiple crests, is 80 km long. Located on the lower Red Deer River, north of the city of Brooks, it previously was included on published maps as an esker (BERG & MCPHERSON 1972) and a fluting (SHETSEN 1987) (Fig. 3). SHETSEN maps the fluting complex only discontinuously south of the Red Deer River, where it grades into what is interpreted by her as an esker between Duchess and Brooks. South of Brooks, Shetsen maps three small flutings which form a fan with its apex in the city limits.

Landsat 4 satellite images display detailed geomorphic information in areas of low relief, especially when manipulated for low sun angles on a thin snow cover. Using Landsat 4 images, the mega-fluting complex can be traced from a series of transverse moraines with superimposed flutings, 65 km north of the Red Deer River, to the city of Brooks, 35 km south of the Red Deer River (Figs. 4 and 5). South of the river the fluting complex is dissected by postglacial flood tracks (EVANS 1991) (Fig. 6). Flood waters, draining west to east through a wide and braided spillway track, stripped off surface sediment, but a lineation parallel to the fluting complex continues as small bedrock flutings which fan out to the south of Brooks largely above the 762 m contour (e.g. SHETSEN's fan of flutings; Fig. 5). These features resemble erosional marks associated with turbidites, albeit at a much larger scale (cf. DZULINSKI & WALTON 1965), and therefore prompt a direct comparison with forms identified by SHAW (1983).

The fan crosses through a gap in a wide arc of transverse moraines (Fig. 5b), which coincides with a bedrock ridge forming the watershed between the Red Deer River and Bow River drainages. At the highest point on the fan of flutings directly south of Brooks (762 m asl), flood waters did not remove postglacial lake sediments deposited directly after ice retreat (Fig. 7). These lake sediments lie directly on moulded bedrock. Whether or not a larger fluting composed of deformable sediments existed below 762 m asl prior to flood-water scouring can not be ascertained. SHETSEN (1987) has interpreted an area of sinuous landforms to the south of Lake Newell as eskers associated with subglacial

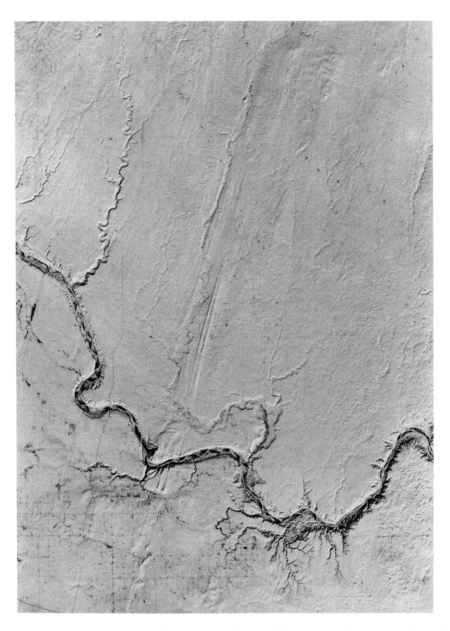

Fig. 4a. Extract from Landsat 4 satellite imagery of part of the lower Red Deer River provided by Dr. I.A. CAMPBELL, University of Alberta. Note different scale to Fig. 5.

Fig. 4b. Map of geomorphology interpreted from Fig. 4a. →

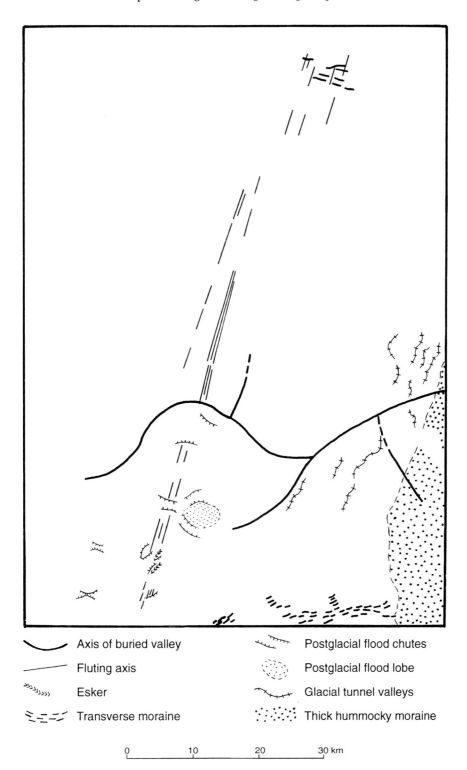

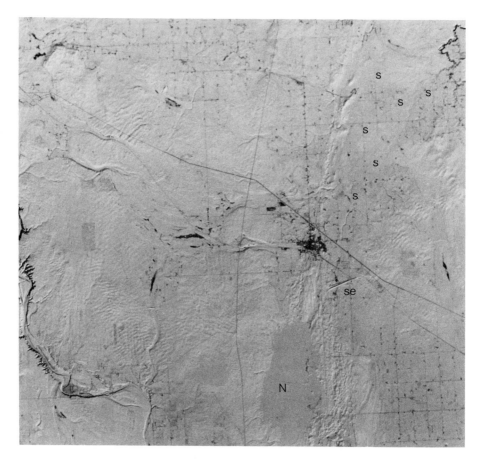

Fig. 5a. Extract from Landsat 4 satellite imagery of Brooks and surrounding prairie provided by Dr. I.A. CAMPBELL, University of Alberta. Brooks is at right of centre and the Brooks Aqueduct is to the southeast of the city. Highway 36 is aligned north-south at the centre of the image. N = Lake Newell; s = postglacial scour area of EVANS (1991); se = southeast postglacial flood channel.

discharge northwards. The northern part of this area is characterized by scoured bedrock with a veneer of boulders rather than esker deposits.

Stratigraphy and sedimentology

A pipeline excavation, immediately south of the Red Deer River, exposed sediment underlying the fluting complex (A–D, Figs. 8 and 9). Although the sections are 30–45 m below the summit of the landform, the sediments and structures provide important information on depositional history and later deformation. The only information available on the upper sediments in the fluting complex comes in the form of ploughed surface material. This consists of a brown diamicton known locally to have been deposited during the last (late Wisconsinan) glaciation (EVANS & CAMPBELL 1992, EVANS 1994).

A possible origin for a mega-fluting complex

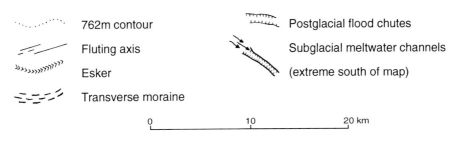

Fig. 5b. Map of geomorphology interpreted from Fig. 5a.

The local Quaternary stratigraphy based upon extensive river cliffs in the One Tree Creek and Little Sandhill Creek drainage basins and along the Red Deer River (Figs. 3 and 6) consists of even main lithofacies associations (LFA's) which occur as tabular stratigraphic bodies (EVANS & CAMPBELL 1992, EVANS 1994). These have been subjected to

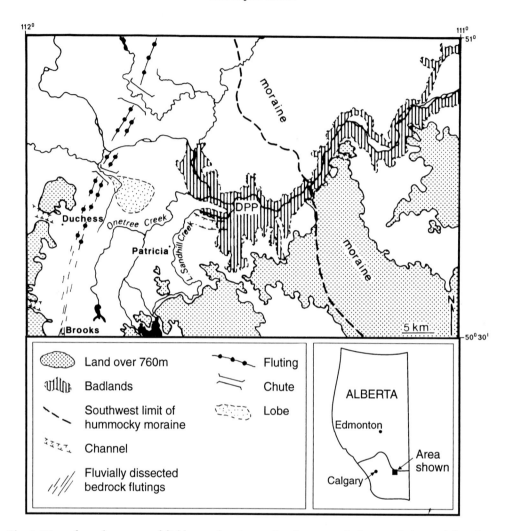

Fig. 6. Map of northern part of field area showing regional geomorphology and chutes, lobe, channels and fluvially dissected bedrock flutings relating to postglacial flooding. DPP = Dinosaur Provincial Park.

glacitectonic disturbance during two glacial overriding phases as indicated by overfolds, thrust slices, shear zones, attenuated lenses and re-oriented clast fabrics of pre-existing beds:

LFA 7   Postglacial lacustrine and fluvial sediments.
LFA 6 (upper)   Brown till with distorted sand and gravel lenses (supraglacial).
LFA 6 (middle)   Brown till (melt-out followed by deformation).
Glacitectonic disturbance phase of LFA's 1–6.
LFA 6 (lower)   Brown (deformation) till.

Fig. 7. Postglacial lacustrine sediments infilling hollows between fluvially scoured bedrock drumlins above 762 m asl, south of Brooks. Sediments fine upwards from laminated fines with dropstones and lag gravels, through silt/clay rhythmites, to massive pond muds. Scale bar is 1 m.

LFA 5 (upper)   Heterogeneous diamicton (subglacial cavity/canal fills).
LFA 5 (lower)   Massive grey-brown (lodgement or deformation) till.
Glacitectoic disturbance phase of LFA's 1–4.
LFA 4   Weathered or stained supraglacial and pro-glacial fluviatile sands and gravels.
LFA 3   Grey till.
LFA 2   Preglacial and pro-glacial lacustrine rhythmites.
LFA 1   Preglacial fluviatile sands and gravels.

The lowest glacigenic sediment exposed in the fluting sections is a grey massive, matrix supported diamicton (Dmm; lithofacies codes after EYLES et al. 1983) with a moderately strong A-axis fabric indicating ice movement from the north and northwest (section A in Fig. 10). This local diamicton has been interpreted by EVANS & CAMPBELL (1992) and EVANS (1994) as a till (LFA 3) which has undergone severe glacitectonic deformation as indicated by attenuated smudges or ingestions of underlying LFA's 1 and 2, a thin basal shear zone intercalating the till and underlying sediments (deformation till) and the formation of attenuated and isoclinally folded slices of LFA 3 within glacitectonically stacked sequences of LFA's 2–5. A small outcrop of massive sandy-silt to the

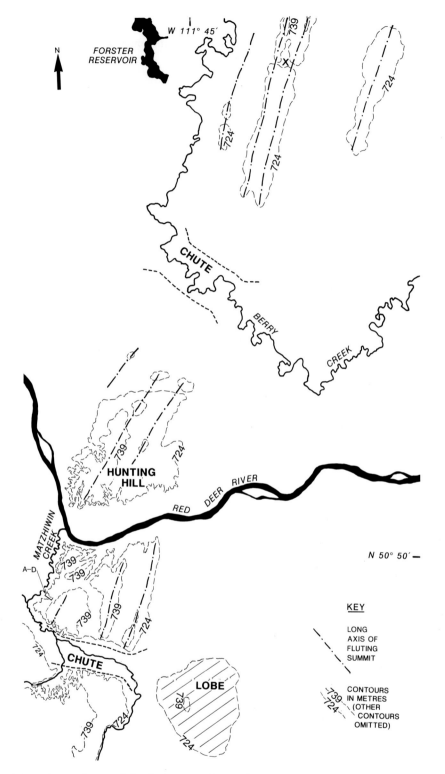

Fig. 8. Map of central part of the fluting complex in the vicinity of Matzhiwin Creek. Chutes and lobe are postglacial flood features from EVANS (1991). A–D indicates location of sections in Fig. 9 and 10. X = road cutting section.

Fig. 9. Overview of sections A–D in Fig. 8 and 10. The mega-fluting surface is visible in the distance at the centre of the photograph. The flat surface which truncates the sections is a postglacial flood channel. Postglacial dissection was responsible for the gullies between the sections.

right of section A is probably part of the underlying proglacial lake sediments (LFA 2), which are extensively exposed in the field area and which widely display shear layers or deformation tills at the junction with overlying LFA 3. This outcrop of LFA 2 could either be a glacitectonized raft or in situ preglacial lacustrine sediments. Glacitectonic deformation of LFA's 1–4 was initiated during the last (late Wisconsinan) ice advance over the area but the exact age of these lithofacies remains to be elucidated (EVANS 1994).

Overlying the grey diamicton (LFA 3) in the fluting sections A and B is a moderately well-sorted cobble gravel which fines upwards into interbedded sands and gravels with diamicton, sand and silt/clay lenses. These deposits have been disturbed from below by intruding diamicton (LFA 3) diapirs and are folded and faulted in association with structures in the underlying diamicton (Figs. 10 and 11). A sub-horizontal, faulted sand lens truncates one till diapir and associated deformed gravel beds but is cross-cut by another, indicating intraformational gravel deposition and substrate disturbance. The fining-upwards succession in sections C and D includes interbeds of laminated fines with dropstones, graded diamicts and ripple, trough and planar cross-bedded sands with minor gravel lags and diamicton balls (Figs. 12 and 13). Although considerable vertical displacement appears to have taken place to the right of section Ci, the degree of disturbance diminishes towards the top of the sections. Extensive but small scale, intraformational normal and thrust faulting has taken place in the finer sediments at the centre of section Ci (Fig. 13). Brown till (LFA 6) crops out on the fluting surface but is not observed in contact with the top of the sediments in sections A–D. One further section through the

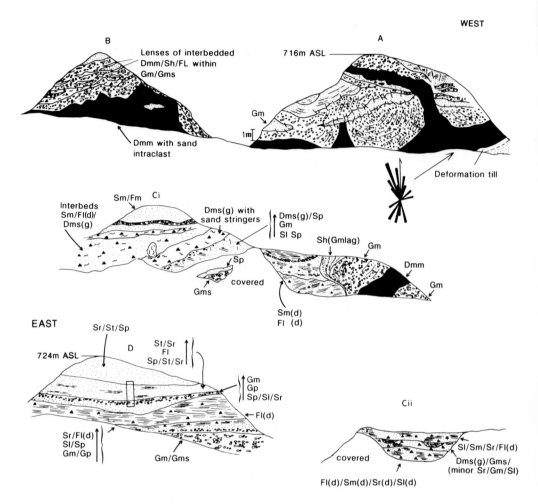

Fig. 10. Stratigraphy of sections A–D in Fig. 8 and 9. Boxed area in section D shows location of Fig. 12.

fluting complex is provided by a road cutting. 14 km north of the Red Deer River (X on Fig. 8), where approximately 3 m of interbedded sands and gravels are capped by 1 m of brown till.

The fining-upwards sequence of gravels, sands, fines and diamictons which overlie grey till in the fluting sections have the characteristics of both LFA 4 and upper LFA 5 of EVANS & CAMPBELL (1992) and EVANS (1994). LFA 4 sands and gravels were originally interpreted as supraglacial in origin, because of their graded contact with LFA 3 (grey till) and their similarity to braided outwash sediments (EVANS & CAMPBELL 1992). Because LFA 4 was discoloured and thus interpreted as a subaerial deposit, it was considered to be of possible interglacial or interstadial age although groundwater staining was suggested as an alternative interpretation. On the other hand, the occurrence of interbedded

Fig. 11. Section A with some of the main lithofacies outlined.

diamicts, sands with bedforms and laminated fines all containing dropstones and gravel lags prompts a comparison with the regionally extensive upper LFA 5 of EVANS & CAMPBELL (1992). Based upon its sedimentology and stratigraphic context, the heterogeneous diamicton of upper LFA 5 has been interpreted by EVANS & CAMPBELL (1992) and EVANS (1994) as a subglacial cavity fill deposit although it could also be the product of "canal" fills associated with deforming layer tills (CLARK & WALDER 1994, EVANS et al. 1995).

The considerable disturbance in sections A–D can be explained by sediment loading in either a pro-glacial environment or in subglacial cavities. The intraformational nature of the till diapirs and associated disturbances in the gravels and sands strongly suggests that meltwater was depositing coarse material directly onto a deformable substrate.

The few small exposures of LFA 4 do not provide sufficient information to discount a subglacial origin similar to LFA 5. Based upon the extensive nature of upper LFA 5, EVANS (1994) suggests that it was deposited by meltwater in subglacial cavities probably during a period of stagnation at the ice sheet margin. This eventually may have triggered the instability and localized re-activation of the ice sheet as documented by glacitectonic disturbance associated with LFA's 1–5 and lower LFA 6. Continued instability resulted in the deposition of middle LFA 6 which records either phases of melt-out interrupted by deformation and pebble fabric re-orientation (EVANS & CAMPBELL 1992, EVANS 1994) or the accretion of deformation tills and canal-fill deposits (CLARK & WALDER 1994, EVANS et al. 1995). The exposure in the fluting complex may serve to demonstrate that the subglacial cavity filling phase involved the deposition of thick cobble gravel deposits at some locations. An esker origin for the core material, which would support earlier

Fig. 12. The normal and reverse-graded sequence at the centre of section D. Massive and planar-bedded gravels (Gm, Gp) capped by trough cross-bedded sands (St), cross-laminated rippled sands (Sr), rhythmites (Fl), rippled sands with starved ripple forms (Sr) and then planar and trough cross-bedded sands (Sp, St).

interpretations of the landform in the literature, explains outcrops of sands and gravels at the core of the fluting complex both north and south of the Red Deer River (Fig. 8) as well as the absence of thick sequences of such sediments from the surrounding prairie surface (Fig. 14). The deposition of such a linear gravel and sand assemblage would require a river flowing upslope if it was to be interpreted as proglacial in origin, because ice sheet retreat was towards the north and northeast in the direction of lower terrain. The interpretation of the more sinuous parts of the linear feature as an esker between Duchess and Brooks and between 18 and 28 km south of Brooks (Fig. 3) by SHETSEN (1987) suggests that this former fluviglacial feature has been modified by glacial streamlining north of Duchess.

Although LFA's 4 and 5 can be differentiated stratigraphically by an intervening phase of glacitectonic deformation and till deposition in the One Tree Creek area (lower LFA 5; EVANS 1994), it is uncertain as to which LFA (4 or 5) the subglacial waterlaid sediments in the fluting must be allocated. However, as the diapiric structures in the underlying till require underconsolidated conditions, it is unlikely that a phase of subglacial deformation (lower LFA 5) occurred after the deposition of the grey till. This implies that the waterlaid sediments overlying LFA 3 should be equated with LFA 4. This local thickening of LFA 4 as a linear feature stretching to the north and south of

Fig. 13. Low angle cross-bedded and planar bedded sands (Sl, Sp) with massive gravel lags (Gm), diamicton balls and minor lenses of graded diamictons (Dmg) from the centre of section Ci. The sediments have undergone minor scale but extensive normal and thrust faulting.

the Red Deer River, compared with the very small outcrops elsewhere in the field area, and the oblique trend of the thick linear assemblage across the regional drainage is best explained by esker deposition. The fact that the undisturbed waterlain sediments at the top of the fluting occur at altitudes well above the base of LFA 6, which would have been the approximate pre-depositional surface for them, and also generally above the altitude of the surrounding prairie surface suggests that they required an ice tunnel or cavity for support when being deposited. This is further verified by the fact that the positive relief linear assemblage does not coincide with the underlying bedrock depressions (Fig. 14). Although overlaps in deposition periods of LFA's 4, 5 and 6 across the field area are possible (cf. EVANS 1994), poor exposures in the upper part of the fluting complex make it difficult to resolve the exact sequence of depositional events.

Formative processes and regional geomorphology

Although the depositional sequence of events cannot be entirely resolved, some interpretations of fluting genesis and distribution in the field area can be attempted. Given the stratigraphy and glacitectonic history of sediments in the lower Red Deer River area, the fluting and drumlin forming hypothesis advocated by BOULTON (1987) requires discussion. This hypothesis invokes differential subglacial deformation and moulding for the production of drumlins and flutings containing coarse-grained waterlain sediments.

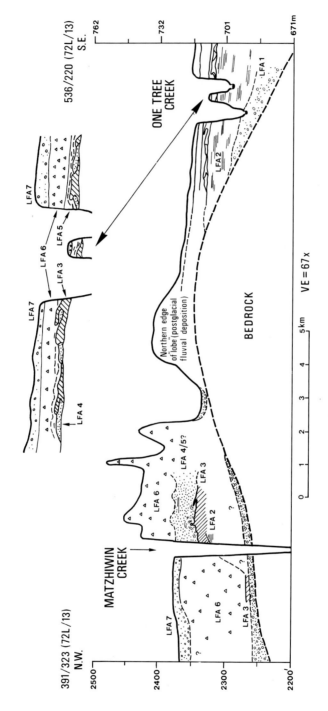

Fig. 14. Topographic cross profile through the mega-fluting complex immediately south of the Red Deer River and adjacent buried valleys with schematic distribution of LFA's 1–6. Bedrock topography (broken line) is interpolated from CARLSON (1970). The buried valleys trend towards the east-northeast and the mega-fluting complex trends north to south. For location of cross profile see Fig. 4b.

A difference in the rheological properties between coarse-grained, permeable material and surrounding till is cited as the best explanation of less-deformed drumlin cores of water-laid gravels and sands. The organization of drumlin fields and individual drumlin forms are thus related to the sedimentary characteristics of pro-glacial outwash. Presumably, drumlins and flutings could also form where pre-existing glacifluvial features such as eskers and kames become moulded by an overriding deforming till layer. ALLEY (1991) suggested that fine-grained tills, such as are common in southern Alberta, may be products of subglacial sediment deformation and EVANS (1994) has presented evidence for glacitectonic disturbance and the production of deformation tills in the One Tree Creek drainage basin.

The basal stratigraphy of the mega-fluting complex in the lower Red Deer River area contains a thick sequence of coarse gravels thought to have been deposited as an esker onto an underconsolidated grey till. The uppermost outcrop of this grey till is at the same altitude (705–710 m asl) at the base of the fluting complex and under the lower terrain of the One Tree Creek and Little Sandhill Creek drainage basins to the east (Figs. 6, 8 and 14). The difference in altitude between the summit of the fluting complex and the prairie surface to the east is 60 m, suggesting that a considerable thickening of Quaternary sediment occurred after the deposition of the grey till. Because the north/south alignment of the linear form is inconsistent with proglacial outwash and unaltered esker forms have been identified by previous researchers to the south of the Red Deer River, its most likely origin is thought to be due to esker sedimentation. Deposition of the stratified sediments (LFA 4 or 5) occurred either during deglaciation prior to the Late Wisconsinan or during evacuation of meltwater reservoirs via subglacial conduits during the Late Wisconsinan. This was followed by subglacial deformation/fluting formation when LFA 6 (brown till) was streamlined into the fluting form around a well-drained core of gravels and sands. Therefore, BOULTON's (1987) hypothesis of subglacial streamlining of well-drained coarser material by a deforming till layer of the kind recognized locally by EVANS & CAMPBELL (1992) and EVANS (1994; lower LFA 5 or lower LFA 6) is regarded as appropriate at this location. LFA 6 appears to have thickened considerably along the flanks of the fluting although extensive exposures are rare to the west of Matzhiwin Creek (Fig. 14). Whether or not the stratified sediment at the core of the fluting dates to a pre-Late Wisconsinan glaciation, an explanation is required for the preservation of the esker form south of Brooks. It appears that the streamlining produced by southward flowing ice was not effective beyond the higher topography south of Brooks.

Postglacial flood waters have extensively eroded the fluting complex to small remnants in the area between the Red Deer River and Brooks (cf EVANS 1991). The remnants may represent thicker esker segments or beads and may be manifest in the non-eroded mega-fluting complex as numerous shorter sub-flutings. To the south of Brooks the fluting complex meets higher topography and takes on the form of a fan of smaller bedrock flutings (Fig. 5). These resemble the erosional marks identified by DZULINSKI & WALTON (1965) and used by SHAW (1983) in his form analogy between subaqueous erosional features and drumlins. These bedrock flutings then grade into a series of meltwater channels and finally back into eskers on the surface of the higher terrain to the south. The large southward indentation in the 762 m contour, hosting Lake Newell immediately south of Brooks, is also thought to be a component of this erosional suite of landforms which document considerable upslope (and therefore subglacial) drainage of water. Although the bedrock flutings north of Brooks were covered by esker sediments prior to pro-glacial flooding, those on the high ground to the south are outside the flood tracks and were therefore never originally covered by coarse gravels. This suggests that although subglacial discharges remained high enough to cut flutings in the weak Cretaceous bedrock on increasingly higher ground, they did not have the capacity to carry coarse debris south of Brooks. This carrying capacity did not increase again until the subglacial stream attained the summit of the higher ground to the south.

The mega-fluting complex can also be put into the context of the regional glacial geomorphology. An arc of moraine segments has been mapped from the satellite imagery revealing an alignment which is transverse to former glacial flow (Figs. 4 and 5). The exact origin of these moraines is unknown but they have been previously interpreted as hummocky moraine or undulating till blanket (cf. SHETSEN 1987). The subtle relief changes obtained by the manipulation of the satellite imagery reveal the transverse pattern. Because these moraines coincide with the higher topography to the south of the fluting complex, they most likely document compressive flow in this part of the Laurentide Ice Sheet. They could relate to either subglacial/proglacial glacitectonic deformation in the frozen toe zone or to supraglacial deposition with ridges marking the positions of debris-rich folia. Deformation could have taken place during ice advance or during a period of sillstand/readvance in response to glacier re-activation. Supraglacial deposition would have taken place during retreat and the positions of the transverse moraines could indicate the areas where compressive flow was bringing material to the glacier surface. Till cover in the area of transverse moraines is generally less than the amplitude of the hummocks and glacitectonically deformed bedrock has been observed in several places around the southern margins of the Dinosaur Provincial Park badlands (EVANS & CAMPBELL 1992), suggesting that subglacial/proglacial thrusting is the most likely origin of the moraines.

Regardless of which hypothesis on transverse moraine genesis is most accurate, the moraines provide outlines of the ice lobes that coalesced in the field area during the last glaciation. To the west of Fig. 5 the moraines mark the former terminus of STALKER's (1973) Innisfail Lobe (Fig. 2) which was defined as a separate ice body using the lithological content of its tills and the distribution of drumlins and flutings. EVANS (1994) has suggested that the Innisfail Lobe may represent an ice stream within the Laurentide ice mass. Because flutings to the south on Fig. 2 clearly show a continuation of the southeasterly flow in the ice stream and, therefore, the transverse moraines drape the flutings, the moraines must mark a retreat position of the ice stream. A similar retreat position of an ice stream is outlined by the arc of transverse moraines to the east and west of the fluting complex (Fig. 5). The eastern margin of this ice stream is demarcated by thick hummocky moraine on Fig. 4. An absence of transverse moraines coincides with the small bedrock flutings to the south of Brooks. This can be explained by postglacial flood erosion only below the 762 m contour. Above this the lack of transverse moraine coincides with the area of subglacial meltwater scouring and thus can be explained by the removal of subglacial material and/or structures by glacifluvial erosion.

The mega-fluting complex is positioned at the centre of a former ice stream, in the area of fastest flow, as demarcated by transverse moraines which record either advance/re-advance or retreat positions of the ice stream. The deposition of the esker probably relates to a phase of meltwater evacuation along the ice stream centre in a southerly direction and against the regional drainage. The water was forced upslope to the south of Brooks where it breached the higher topography marked by the 762 m contour on Fig. 5. The break in the 762 m contour to the south-west of Brooks was produced by postglacial flooding. Before and after esker deposition ice was streaming in a north north-east to south south-west direction. Although the lower grey till and esker sediments could relate to a previous glaciation there is no firm evidence to support such a reconstruction. Therefore, the stratigraphy and geomorphology of the mega-fluting complex documents periods of fast glacier flow with associated substrate deformation and meltwater evacuation along the ice stream axis which produced multiple stacked diamictons and glacifluvial sediments during one glacier advance in the lower Red Deer River area. Although subglacial meltwater drainage systems can remain well developed during periods of substrate deformation producing tunnel valleys (BOULTON & HINDMARSH 1987), stratigraphic and sedimentological data from this area of southern Alberta have been used by EVANS (1994) to suggest that ice streams within the southwestern Laurentide Ice Sheet may have undergone periods of mass instability and fast flow or even surging. Such behaviour could have been in response to large discharges of subglacial meltwater from up-ice reservoirs as envisaged by SHAW (1989), SHAW et al. (1989) and SHOEMAKER (1991). Subglacial tunnel valleys have been identified underlying low relief hummocky moraine to the east of the field area and are discussed by EVANS (1994; Fig. 4).

Conclusion

The mega-fluting complex in the lower Red Deer River drainage area is explained as a glacially moulded esker which was originally deposited over an underconsolidated grey till. After the deposition of the grey till (LFA 3), increased subglacial meltwater discharge was responsible for the deposition of esker gravels and then subglacial cavity fills (LFA 4 or 5). A further phase of fast glacier flow, which may date to a later period of glaciation, was responsible for the streamlining of the well-drained esker deposits beneath a deforming till layer (LFA 6). Smaller sub-flutings on the flanks of the main fluting mass are most likely to be glacially deformed beads or the outer meanders of the esker. While subglacial deformation was continuing over the higher topography formed by the esker, subglacial cavity fills (LFA 5) and tunnel valleys (Fig. 4) may have been developed penecontemporaneously on the lower topography to the east (cf. BOULTON & HINDMARSH 1987).

Thick sequences of deformation, melt-out and lodgement tills (LFA 6) were deposited at a later date in the east when ice began to thin over the mega-fluting complex (cf. EVANS 1994). This is verified by changing till fabrics in the One Tree Creek/Little Sandhill Creek area indicating a shift in ice flow direction from north/south to east north-east/west south-west (EVANS & CAMPBELL 1992).

The mega-fluting complex is situated at the former centre of a large ice stream within the south-west sector of the Laurentide Ice Sheet. Transverse moraines probably relating to sub-/pro-glacial deformation delineate the pattern of advance or re-advance of ice stream snouts in this area. The development of an esker which was later streamlined and the preservation of tunnel valleys under a thin cover of hummocky moraine in the east of the field area indicate that subglacial discharges of meltwater occurred prior to and immediately after till deformation and glacial streamlining. Whether or not subglacial deformation and tunnel valley cutting occurred contemporaneously (cf. BOULTON & HINDMARSH 1987) can only be resolved by the further investigation of the relationships between landform/sediment assemblages in this area of the former southwestern Laurentide Ice Sheet. A highly deformable substrate was probably responsible not only for dynamic and perhaps fast flowing ice streams within the south-western Laurentide Ice Sheet (cf. FISHER et al. 1985), but also for a low profile ice sheet margin (MATHEWS 1974). If discharges varied in magnitude and position over time it would be possible that this margin of the Laurentide Ice Sheet never attained an equilibrium profile. This would help to explain the difficulties DYKE et al. (1982) experienced with ice sheet asymmetry when they attempted to reconstruct dispersal centres for the western Laurentide Ice Sheet.

Finally, it was suggested by EVANS (1994) that the basal grey till and overlying sands and gravels (LFA's 3 and 4) in the One Tree Creek drainage basin were deposited during a pre-Late Wisconsinan glaciation and that their glacitectonic disturbance occurred in the frozen snout zone during the Late Wisconsinan advance phase of the Laurentide Ice sheet. This tectonic disturbance was then followed by deposition of grey-brown lodgement till and subglacial cavity fills (LFA 5) and then brown deformation, melt-out and lodgement tills (LFA 6) as temperate ice passed over the site. It is conceivable that the grey till (LFA 3) and sands and gravels (LFA 4) also date to Late Wisconsinan ice advance and that the esker deposits at the core of the mega-fluting complex are coeval with LFA's 4 and 5 but glacitectonic disturbance did not affect them. LFA 4 would also be subglacial in this scenario. This would imply that ice stream thinning occurred after the deposition of LFA's 3 and 4, at which time the ice became temporarily frozen to its bed (cf. Fig. 11i in EVANS 1994). Such a depositional sequence would increase the number of multiple stacked diamicton units identified by EVANS (1994) to be the products of one glacier advance.

Acknowledgements

Research for this paper was supported by a Natural Sciences and Engineering Research Council of Canada grant awarded to Dr. I.A. Campbell (University of Alberta), a University of London Central Research Fund Grant and funds from the Department of Geography and Topogrpahic Science, University of Glasgow. Several landowners in the Brooks area gave kind permission to access their property. R. Skoye (University of Alberta) provided helpful discussion on the satellite imagery. Thanks to I.A. Campbell for his continuing support and discussion and for providing the satellite imagery. The figures and photographs were reproduced by Yvonne Wilson and Les Hill of the Department of Geography and Topographic Science, University of Glasgow. Professor Ian Thompson (University of Glasgow) kindly translated the abstract into French.

References

ALLEY, R.B. (1991): Deforming-bed origin for southern Laurentide till sheets? J. Glaciol. 37: 67–76.
BERG, T.E. & R.A. MCPHERSON (1972): Surficial Geology Medicine Hat. – Alberta Research Council, Map NTS 72L, scale 1:250,000.
BOULTON, G.S. (1987): A theory of drumlin formation by subglacial sediment deformation. – In: MENZIES, J. & J. ROSE (eds.): Drumlin Symposium. – 25–80, A.A. Balkema, Rotterdam.
BOULTON, G.S. & R.C.A. HINDMARSH (1987): Sediment deformation beneath glaciers: rheology and geological consequences. – J. Geophys. Res. 92: 9059–9082.
BOYDELL, A.N. (1978): Multiple Glaciations in the Foothills, Rocky Mountain House Area, Alberta. – Alberta Research Council, Bull. 36.
BRYAN, R.B., I.A. CAMPBELL & A. YAIR (1987): Postglacial geomorphic development of the Dinosaur Provincial Park badlands, Alberta. – Canad. J. Earth Sci. 24: 135–146.
CLARK, P.U. & J.S. WALDER (1994): Subglacial drainage, eskers, and deforming beds beneath the Laurenride and Eurasian ice sheets. – Geol. Soc. Amer. Bull. 106: 304–314.
DYKE, A.S., L.A. DREDGE & J.-S. VINCENT (1982): Configuration and dynamics of the Laurentide Ice Sheet during the late Wisconsin maximum. – Geogr. Phys. Quatern. 36: 5–14.
DZULINSKI, S. & E.K WALTON (1965): Sedimentary Features of Flysch and Greywackes. – Elsevier, Amsterdam.
EVANS, D.J.A. (1985): Wisconsin ice dynamics in Alberta: A review of prevalent hypotheses and the development of recent ideas. – In: DESHAIES, L. & R. PELLETIER (eds.): Proceedings of the Annual Meeting of the Canadian Association of Geographers: 13–33.
– (1991): A gravel/diamicton lag on the south Albertan prairies, Canada: Evidence of bed armoring in early deglacial sheet-flood/spillway courses. – Geol. Soc. Amer. bull. 103: 975–982.
– (1994): The stratigraphy and sedimentary structures associated with complex subglacial thermal regimes at the south-west margin of the Laurentide Ice Sheet, southern Alberta, Canada. – In: WARREN, W.P. & D.G. CROOT (eds.): Formation and Deformation of Glacial Deposits. – 203–220, Balkema, Rotterdam.
EVANS, D.J.A. & I.A. CAMPBELL (1992): Glacial and postglacial stratigraphy of Dinosaur Provincial Park and surrounding plains, southern Alberta, Canada. – Quatern. Sc. Rev. 11: 535–555.
EVANS, D.J.A., L.A. OWEN & D. ROBERTS (1995): Stratigraphy and sedimentology of Devensian (Dimlington Stadial) glacial deposits, east Yorkshire, England. – J. Quatern. Sci. 10: 241–265.
EYLES, N., C.H. EYLES & A.D. MIALL (1983): Lithofacies types and vertical profile models: an alternative approach to the description and environmental interpretation of glacial diamict and diamictite sequences. – Sedimentology 30: 393–410.
FISHER, D.A., N. REEH & K. LANGLEY (1985): Objective reconstructions of the Late Wisconsinan Laurentide Ice Sheet and the significance of deformable beds. – Geogr. Phys. Quatern. 39: 229–238.
GRAVENOR, C.P. & W.A. MENELEY (1958): Glacial flutings in central and northern Alberta. – Amer. J. Sci. 256: 715–728.
JONES, N. (1982): The formation of glacial flutings in east central Alberta. – In: DAVIDSON-ARNOTT, R., W. NICKLING & B.D. FAHEY (eds.): Research in Glacial, Glaciofluvial and Glaciolacustrine Systems. – 49–70, Geobooks, Norwich.
KVILL, D.R. (1984): The Glacial Geomorphology of the Brazeau River Valley, Foothills of Alberta. – Unpubl. PhD Thesis, University of Alberta.
MATHEWS, W.H. (1974): Surface profiles of the Laurentide Ice Sheet in its marginal areas. – J. Glaciol. 13: 37–43.
ROED, M.A. (1975): Cordilleran and Laurentide multiple glaciation, west-central Alberta, Canada. – Canad. J. Earth Sci. 12: 1493–1515.
SHAW, J. (1975): The formation of glacial flutings. – In: SUGGATE, R.P. & M.M. CRESSWELL (eds.): Quaternary Studies. – 253–258, Quat. Roy. Soc. New Zealand, Wellington.
– (1980): Drumlins and large scale flutings related to glacier folds. – Arctic and Alpine Res. 12: 287–298.

SHAW, J. (1983): Drumlin formation related to inverted meltwater erosional marks. – J. Glaciol. 29: 461–479.
– (1989): Drumlins, subglacial meltwater floods and ocean responses. – Geology 17: 853–856.
SHAW, J. & D.R. KVILL (1984): A glaciofluvial origin for drumlins of the Livingstone Lake area, Saskatchewan. – Canad. J. Earth Sci. 21: 1442–1459.
SHAW, J., D.R. KVILL & R.B. RAINS (1989): Drumlins and catastrophic subglacial floods. – Sediment. Geol. 62: 177–202.
SHAW, J. & D.R. SHARPE (1987): Drumlin formation by subglacial meltwater erosion. – Canad. J. Earth Sci. 24: 2316–2322.
SHETSEN, I. (1987): Quaternary Geology, Southern Alberta. – Alberta Research Council Map, scale 1:500,000.
SHOEMAKER, E.M. (1991): On the formation of large subglacial lakes. – Canad. J. Earth Sci. 28: 1975–1981.
SKOYE, K.R. (1990): Digital Analysis of low-Relief Topography in a Landsat Snowcover Scene in South-Central Alberta. – Unpubl. MSc thesis, University of Alberta.
SKOYE, K.R. & J.R. EYTON (1992): Digital analysis of low-relief topography in a landsat snow cover scene in south central Alberta. – Canad. J. Remote Sensing 18: 143–150.
STALKER, A. MACS. (1960): Surficial Geology of the Red Deer – Stettler Map Area, Alberta. – Geol. Surv. Canada, Mem. 306.
– (1973): Surficial Geology of the Drumheller Area, Alberta. – Geol. Surv. Canada, Mem. 370.
STALKER, A. MACS. & B.G. CRAIG (1956): Use of indicators in the determination of ice-movement directions in Alberta, Canada: A discussion. – Geol. Soc. Amer. Bull. 67: 1101–1104.

Address of the author: Dr. DAVID J.A. EVANS, Department of Geography and Topographic Science, University of Glasgow, Glasgow G12 8QQ, Scotland, U.K.

Likelihood of erosive rains in Lesotho

BENGT CALLES and LENA KULANDER, Uppsala

with 14 figures and 3 tables

Summary. Precipitation data for 8 stations in the lowland and foothill regions of Lesotho were analysed to characterise the geomorphological significance of precipitation from a soil erosion perspective. Lacking long series of recording rain gauge data, short series of intensity recordings were used to establish a relationship between daily amounts and maximum storm intensity. Applying this information on longer series of daily rainfall data made it possible to estimate the frequency with which intensive storms occur that are likely to contribute to soil erosion. Frequency analysis based on the partial duration technique made it possible to determine the recurrence intervals of rainstorms of various 30-minute magnitudes.

The results show that for all sites it can be expected that rainstorms with a maximum 30-minute intensity exceeding 25 mm/h are likely to occur at least once each year. The likelihood of high intensities is highest during December–February with considerably longer recurrence intervals for the onset and end of the wet season. recurrence intervals are longer for sites located in the foothill and mountain regions of the country.

Considering the severe status of soil erosion in Lesotho it is unlikely that rainfall intensity conditions should be regarded as the fundamental cause of soil erosion.

Introduction

Lesotho may justly be regarded as one of the most severely eroded countries in the world. The result of the erosion induced by precipitation is clearly visible in the landscape as extensive gully systems and areas that have been exposed to severe surface wash. The fundamental importance of precipitation conditions for soil erosion has long been recognised. It is therefore justified to investigate if the precipitation climate is unusually extreme and likely to be a major factor causing soil erosion. Annual amounts of precipitation in Lesotho have previously (CHAKELA & STOCKING 1988) been used to assess the erosion hazard in the country.

The annual amounts of precipitation varies over Lesotho. It has so far not been established if the variation in amounts are accompanied by a variation in intensity of precipitation. Information on the spatial distribution of intensive rainstorms and the likelihood of intensive rains to occur in various parts of Lesotho will improve the possibilities of assessing erosion hazard in the country.

This study is a further development of an investigation of rainstorm intensities at Roma, Lesotho (CALLES & KULANDER 1994) and should be regarded as a complement to plot studies of soil erosion carried out by the National University of Lesotho. The main aim is to describe the rainstorm erosivity from a temporal and spatial point of view based on available precipitation data.

Background

Soil erosion is a combination of different erosional processes such as splash, inter-rill, rill and gully erosion each with its individual threshold value to be exceeded for the process to start. It is evident that the thresholds are not the same for the different processes and that a good threshold measure for one process is not a suitable parameter for other processes. Splash erosion may be regarded as a function of the kinetic energy of the rain whereas inter-rill and rill erosion are strongly dependent on other factors such as intensity and amount of precipitation and soil moisture conditions. The cover effect from vegetation acts both as a protection against splash and influences water velocities and water path ways during sheet wash. Soil moisture conditions are influenced by antecedent precipitation events, grain size, hydraulic conductivity and position on a slope segment. In all sloping segments of a landscape a lateral flow of water will occur that increases soil moisture in the concave portions of a landscape. This induces a decrease in infiltration capacity in the concave areas. In areas where the soil is close to saturation the erodibility is higher than in unsaturated or dry areas. The degree of soil saturation will be more important for the likelihood of surface wash to commence than rainfall intensity. In areas with shallow soils, the soil water storage potential is limited and saturated conditions in the soil ocurs more frequently during the rainy season. A common denominator extractable from precipitation data that can serve as a threshold value for different geomorphologically active processes is therefore unlikely to exist.

Precipitation thresholds for soil erosion

A variety of parameters to characterise when erosive conditions exist have been presented. The profound influence of rainfall on soil erosion are conventionally related to the erosive power of a rainstorm. One group of criteria is based on maximum rainfall intensity measures since the relation between a soil's infiltration capacity and the intensity of a rain are commonly regarded as closely related to whether or not runoff will occur. Other criteria are based on the amount of precipitation necessary for surface runoff to occur.

The erosivity index used for the Universal Soil Loss Equation (USLE) EI_{30}, is the product of total kinetic energy and maximum 30-minute intensity (WISCHMEIER & SMITH 1958). Energy is calculated from a simple semi-logarithmic energy-intensity equation. The USLE was not developed to estimate soil loss by specific events but serves to estimate long term soil loss amounts (WISCHMEIER 1977). Other algebraic equations based on intensity or kinetic energy of the rain have been proposed by KINELL (1973) and HUDSON (1963).

The basis for choosing a certain numerical threshold value is either more or less arbitrary, based on visual observations or based on data obtained from plot studies. The selection of one or more characteristic values that have to be exceeded is often not a free choice but is limited by the type of background information available. The most simple type of threshold criteria consists only of a maximum intensity value such as $I_{30} \geq 25$ mm/hour which is regarded as an erosive rain (ATHESIAN 1974, STOCKING & ELWELL 1976, STOCKING 1978, MOORE 1979, ULSAKER & ONSTAD 1984). Most threshold values are based the precipitation conditions which have been observed to produce surface runoff. For southern Africa, HUDSON (1981) presented the KE > 1 index based on observations that for intensities lower than 1 inch/hour (25 mm/hour) splash was negligible. For practical applications the maximum 30-minute intensity has gained a widespread acceptance as a characteristic value which, sometimes based on local or site specific conditions, has been given different numerical threshold values.

Storm intensity and storm amount has also been related to kinetic energy (KOWAL & KASSAM 1976). For Nigerian conditions LAL (1976) related both storm amount and maximum 30-minute intensity to kinetic energy. LAL developed the AI_m-index from studies in Nigeria which includes both storm amount and maximum 7.5-minute intensity. Other amount and maximum intensity based criteria have been presented by STOCKING & ELWELL (1976), STOCKING (1978) and ROWNTREE (1988).

It is obvious that the time resolution for which relationships between precipitation and soil erosion are established will influence the erosivity criteria selected. For many areas in the world only information on daily amounts of precipitation are available. This causes a problem of estimating the maximum intensity during a day. It is also just as obvious that for the limited number of sites where recording rain gauge data are available the type of instrumentation and the time resolution are important factors to be considered when a specific parameter is to be selected. Sometimes this has resulted in selecting a parameter for which data is usually not available at other sites than research fields (e.g. maximum intensities during 7.5 or 5 minutes). Standard recording equipment for rainstorm intensity will in most cases not have a time resolution that enables such detailed determinations.

The problem of defining an appropriate erosivity criterion for all different processes contributing to soil erosion makes it necessary to make a choice. Since maximum short-term intensity measures during 15 (I_{15}), 30 (I_{30}) and 60 minutes (I_{60}) have been widely used for many conventional soil erosion equations (e.g. USLE, SLEMSA), this study has accepted this kind of simple measures as criteria for defining erosive rainfall events. These parameters have the obvious advantage of being fairly easy to determine and are often used for standard evaluation of recording rain gauge data by meteorological offices.

If rainfall intensity is accepted as a measure of erosivity it should, however, be understood that this measure is merely a substitute for more complex hydrological parameters that determine when surface runoff ocurs which may cause soil erosion.

Precipitation conditions in Lesotho

Lesotho is situated on the eastern part of the Southern African sub-continent's interior plateau, with its eastern national border along the Great Escarpment. Altitudes range between 1500 and 1800 m a.s.l. in the western lowland portion and between 1800 and 3400 m a.s.l. in the larger eastern mountainous portion.

Over the southern African sub-continent there is a marked trend of increasing annual rainfall from the arid west (less than 200 mm) to more than 800 mm in the east. Average rainfall in Lesotho ranges between 600–800 mm in the lowlands and often exceeds 100 mm in the mountains. Approximately 80 % of the annual total precipitation falls during summer (October–April) with a pronounced dry season during winter.

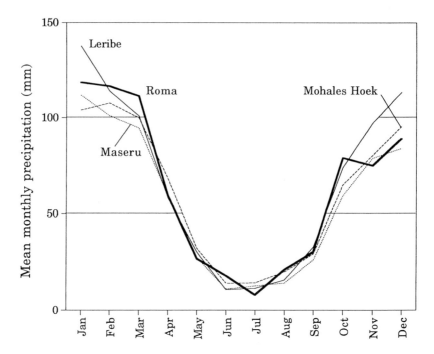

Fig. 1. Monthly distribution of precipitation at Maseru (1886–1989), Leribe (1886–1989), Mohales Hoek (1880–1989) and Roma (1962–1994).

Long series of daily precipitation ar available from some stations in the Lesotho lowlands. The mean monthly amounts from Maseru (1886–1989), Leribe (1886–1989), Mohales Hoek (1880–1989) and Roma (1962–1994) are compiled in Fig. 1.

The inter-annual variation in precipitation amounts is large (Fig. 2), but the difference between mean annual amounts is not very pronounced with long-term means for Maseru 680 mm, Leribe 795 mm, Mohales Hoek 727 mm and Roma 751 mm.

The rains vary from long frontal and orographic drizzle to hard convective downpours (WILKEN 1978). A marked feature is the high temporal variability of rainfall. Spatial variability is high and markedly influenced by local topography (cf. TYSON 1986).

Rainstorm data

A fairly large number of observation points for daily amounts of precipitation exist in Lesotho, but very little information is available on rainfall intensities. The number of stations equipped with recording rain gauges are limited and the length of the existing series of data varies considerably. Some sites have only had recording rain gauges in operation for short periods (one or two years) and at least for some stations recordings have only been made during the rainy season.

Since soil erosion is a problem mainly for agriculture, this study is focused on the conditions prevailing in the Lesotho lowland and foothill regions. The location of sites

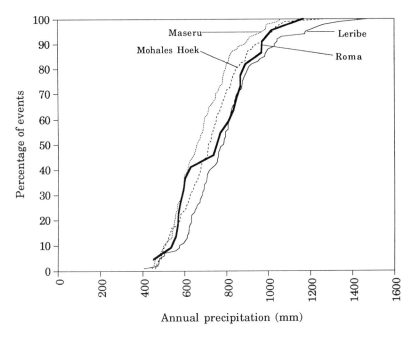

Fig. 2. Distribution of annual precipitation for Maseru 1886–1988 (dotted line), Leribe 1886–1989) (full line), Roma 1962–1994 (thick line) and Mohales Hoek 1880–1989 (dashed line).

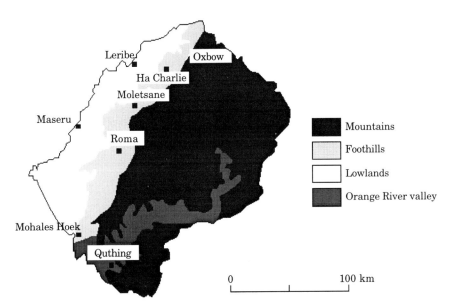

Fig. 3. Location of sites from which records of rainfall intensities were studied. From topographical considerations the country has been divided into lowlands, foothills, Orange river valley and mountain regions.

Table 1. Rainstorm intensity data determined for the study. (x indicates data availability.)

Station	Years 19..	Number of rains	I_{15}	I_{30}	I_{60}	Storm amount	Duration	Daily amount	Mean intensity
Oxbow	58–59 60–79 81–83 91	1263	x	x	x			x	
Leribe/	78 90S93	204		x	x	x	x	x	x
Moletsane	77–81 83–84	286		x	x	x	x	x	x
Maseru	78–81 86 89–91 92–94	272		x	x	x	x	x	x
Ha Charlie	77–85	302	x	x	x	x		x	
Roma	85 88–93	320	x	x	x	x	x	x	x
Mohales Hoek	77–79 80–83 84–86	265		x	x	x	x	x	x
Quthing	90 91–92	52		x	x	x	x	x	x

from which information have been used is shown in Fig. 3. The spatial distribution of the 8 stations covers the major part of the lowland and foothills areas. The available number of data and type of information that has been determined for the different sites are compiled in Table 1.

Daily amounts observed during the period for which intensity data were collected reveal a similar distribution of the frequencies of different amounts for all stations (Fig. 4). A maximum is noted for the daily amount intervals 2–5 and 4–10 mm.

Storm rainfall / daily rainfall relationships

A maximum 30-minute intensity in excess of 25 mm/h may perhaps serve as a crude mean value for erosive rains. Data series from which rainstorm intensities can be determined are, however, often short. Lack of information on a specific intensity parameter may be overcome by using generalised mean relationships between maximum intensity and other precipitation parameters such as rainstorm amount. A previous study of precipitation data from Lesotho related storm amounts to intensity parameters using the data from the National University of Lesotho at Roma (CALLES & KULANDER 1994).

For standard meteorological stations the only information available is daily amounts of rain. To be able to transfer information on storm intensities to sites where only daily

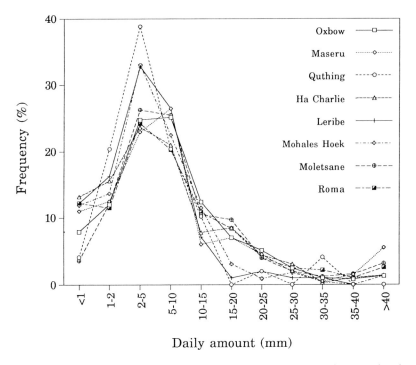

Fig. 4. Frequency distribution of daily amounts of precipitation during the periods when intensity recordings were made at the 8 sites in the study.

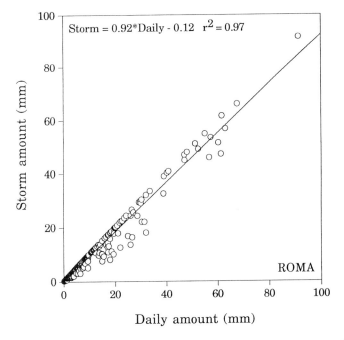

Fig. 5. Linear relationship between daily amount and maximum storm amount at the National University of Lesotho at Roma.

amounts are observed, a relationship between daily amount and amount in individual storms has to be established. For the climatic conditions in Lesotho this study has defined a single rainstorm as one that is separated from the next precipitation event by more than 3 hours of dry spell.

Some of the sites in this study lack information on storm amounts, which has made it necessary to relate intensity parameters to daily amounts. This is not a major drawback since the correlation between daily amount and storm amount was found to be very high (Fig. 5). Very few rainy days have more than one storm. Daily amounts were therefore accepted as a basis for approximating maximum intensities during a day. This facilitates the use of the results of the analysis for stations where only daily amounts are measured.

It should be stressed that all attempts to relate daily amounts to maximum storm intensity will only represent mean conditions. Individual storms may show intensities much higher or lower than the relationship defined by a regression line. The reliability of a regression line will fundamentally depend on the range of values used for the determination. For stations with intensity records only covering low intensities or low daily amounts an extrapolation of the relationship to high amounts will be necessary which will considerably increase the uncertainty.

Simple linear relationships between daily amounts and individual storm amounts were established by a least square fit of data. This simple relationship gives a crude picture of the average proportion of the daily precipitation which can be attributed to the largest storm each day.

Storm amount / intensity relationship

The original charts stored at the Meteorological Office in Maseru were used for data extraction. They were supplemented by data from Roma Campus, stored at the National University of Lesotho. For each storm the total storm amount, maximum amount during 60 and 30 minutes and, where possible 15 minutes were noted. Daily amounts were determined by summing the amount in all storms during 24 hours (normally 8 a.m.– 8 a.m.). A further step in the treatment of data included a comparison of storm amount and maximum intensities observed during 15 (I_{15}), 30 (I_{30}) and 60 minutes (I_{60}).

For the stations Ha Charlie and Oxbow no charts were available for analysis. Instead the values for maximum intensity during 15, 30 and 60 minutes and daily amount (Oxbow) as well as storm amount (Ha Charlie) computed from the original charts by the Meteorological Office were used.

Relationships between daily amounts and maximum intensities can be established in a number of ways. The simple linear regression is naturally closest at hand. However, it is easily shown that if the logarithms of the observed values are used a considerably higher correlation coefficient is obtained (cf. BÄRRING 1988). This is not surprising since the procedure means that a numerically high value is given a lower value when transformed to a logarithm. The influence of the extreme values will thus be subdued.

As a complement to using simple linear regression, the regression equations were also determined for the logarithms of the data. In most cases this gives a better fit of the regression line as shown by a considerably higher correlation coefficients. As an example the result of using the two methods on the information from Maseru is shown in Fig. 6.

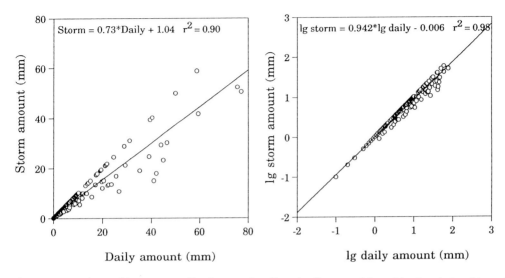

Fig. 6. Comparison of least square fit of regression lines for linear and logarithmic relationships of daily amount and storm amount for Maseru.

Table 2. Regression equations and correlation coefficients for the relationship: intensity = a * daily amount + b based on individual observations at 8 stations in Lesotho.

Station	I_{15}			I_{30}			I_{60}		
	a	b	r^2	a	b	r^2	a	b	r^2
Oxbow	0.56	5.92	0.36	0.49	2.84	0.46	0.33	1.50	0.59
Leribe				0.60	1.72	0.57	0.39	0.80	0.71
Moletsane				0.47	3.22	0.44	0.32	1.93	0.54
Ha Charlie	0.68	3.68	0.42	0.50	2.37	0.46	0.35	1.33	0.57
Maseru				0.56	3.19	0.48	0.31	1.96	0.63
Roma	0.75	4.85	0.51	0.57	3.18	0.50	0.35	1.68	0.62
Mohales Hoek				0.53	2.15	0.53	0.36	1.03	0.53
Quthing				0.99	0.79	0.81	0.77	–0.96	0.83

A comparison shows that a regression line through the logarithms gives a higher numerical value of the slope factor of the regression line than the linear method. The use of logarithmic values will however result in an underestimation of the intensities during days with high amounts. Since high daily amounts are usually combined with high intensities the use of logarithms will result in a misjudgement of the expected intensities in spite of the temptingly high correlation coefficients obtained for logarithmic values. For this reason logarithms of daily amounts and intensities were discarded as a tool for analysing the relationship between daily amounts and maximum intensities.

The factors in the regression equations determined for the different sites using a linear relationship between daily amount and maximum storm intensities are compiled in Table 2.

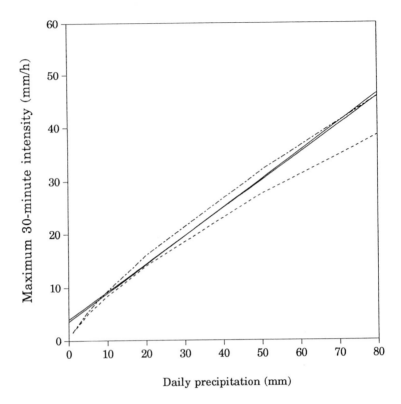

Fig. 7. Comparison of regression lines showing the relationship between daily amount and maximum 30-minute intensity. Full lines show linear regression through individual data (upper line) and group mean data. Stipled lines show regression lines for logarithms of data transformed to linear values. Upper line: grouped data, lower line individual data.

The frequency distribution of daily amounts has a maximu in the low amount range. This cluster of low amount data may influence the position of a regression line through the data points. To decrease this effect the data may be grouped into classes and class mean values determined. Mean values for amounts and the corresponding mean values for observed maximum intensities were computed for all stations using the classes: , 1–2, 2–5, 5–10, 10–15, 15–20, 20–25 and each 5 mm interval up to 40 mm, leaving an high magnitude classe of 40 mm. With a frequency distribution shown in Fig. 4, it is reasonable to use class limits which, at least in the low amount range, resembles a logarithmic series. It should be stressed that the use of group mean values will subdue the effect of high intensities which normally occur during days with high precipitation amounts.

For the grouped mean data regression lines were determined by a least square fit of a regression line. Primarily linear regression was used, but a relationship based on decadic logarithms was also tested. A comparison between the two methods for individual data and class mean data on maximum 30-minute intensities from Roma is shown in Fig. 7.

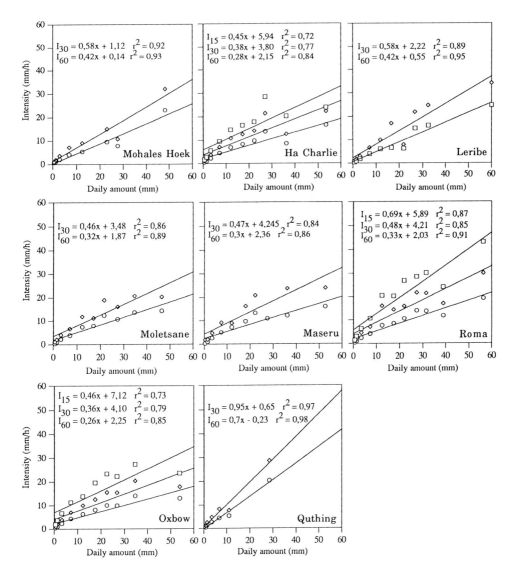

Fig. 8. Maximum storm intensity (during 15, 30 and 60 minutes) as a function of total daily precipitation amount for the stations used in the study. Linear relationships are based on class mean values.

Comparing either storm amount or daily amount with recorded maximum intensities gives the characteristics of a single precipitation event at a site. Regression lines for computed class means of daily amount and intensities at the 8 sites are shown in Fig. 8.

To facilitate a comparison between stations the regression lines for different intensity measures at the eight sites are compiled in Fig. 9. 15-, 30- and 60 minute maximum intensities combine to give a characterisation of the relative proportion of maximum intensities during a storm. The deviating pattern for the station Quthing is likely to be

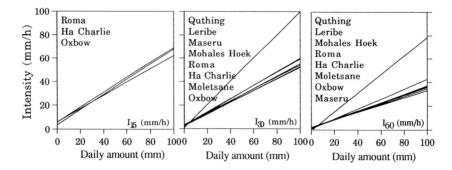

Fig. 9. Mean relationships for various maximum intensities in relation to total daily amount of precipitation based on class mean values. Only 3 stations have information on maximum intensities during 15 minutes.

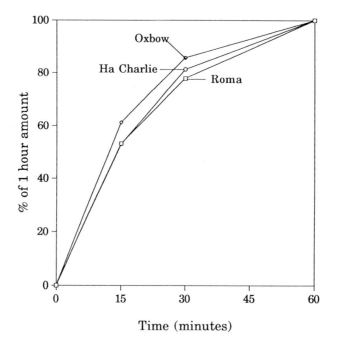

Fig. 10. Approximate maximum intensity distribution of the most intensive part of a storm at 3 sites in Lesotho.

caused by the limited number of observations especially for high daily amounts. All other stations reveal a similar pattern for the relationships between daily amount and maximum storm intensity. The quotient I_{15}/I_{30} can by definition not be smaller than 1, and I_{30}/I_{60} can not exceed 1. The closer to 1 the quotient is the more uniform is the intensity during the 60 minutes during a rain when the intensity is at a maximum. A high value for I_{15}/I_{30} indicates a short very intensive period during a storm.

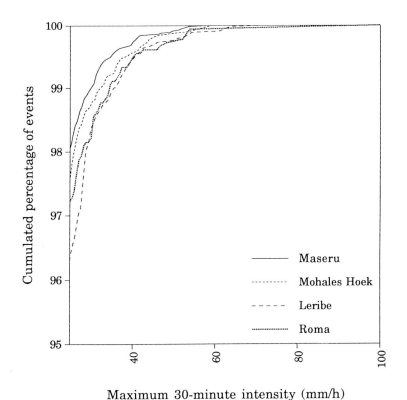

Fig. 11. Frequency of precipitation events with $I_{30} \geq 25$ mm/h. Information is based on the long series of daily precipitation transformed to intensities by the relationships in Table 2.

For three sites the available data made it possible to relate I_{15} and I_{60} values to I_{30} values by using a least square fit of the data and determinign the equation of the regression line through the data set. The relationships does not vary substantially for Oxbow, Roma and Ha Charlie. Assuming that the normal distribution pattern of intensities during a storm involves a high intensity at the onset of a rain and decreasing intensities as the rain progresses an approximation of the mean distribution of rainfall during the most intensive part of a storm can be made (Fig. 10).

The frequency of 'erosive' rains

The information concerning the character of the rain based on intensity determinations is, however, of limited use if the temporal pattern of the precipitation is not analysed. This can be described in terms of the likelihood for certain extreme values to be exceeded.

A comparison of the frequency of various amounts of daily precipitation at Maseru, Mohales Hoek, Leribe and Roma, for which long-term data exist, reveal only limited differences. The good relationship obtained between high daily amounts and high storm intensities justifies a closer look at the frequency distribution of the highest daily amounts. The long data series on daily amounts from four stations were transformed to intensity measures (I_{30}) using the established mean relationships for the four stations (Fig. 11). All daily values corresponding to an I_{30}-intensity equal to or exceeding 25 mm/hour were included. No considerable difference between the four stations studied exists. The rainfall erosivity climate appears to be similar over large areas of lowland Lesotho.

Surface runoff which is an essential component in soil erosion is generally attributed to precipitation events exceeding a threshold value. Lacking a better parameter we accepted $I_{30} \geq 25$ mm/h as a threshold. The frequency with which these events occur is essential for assessing how often events likely to cause soil erosion can be expected. When a long series of data on rainfall is available the probability for an annual extreme value to be exceeded can be determined by intensity frequency analysis (DUNNE & LEOPOLD 1978). The distribution of extreme values can best be described by a Gumbel-distribution or extreme value distribution (GUMBEL 1954). Annual or monthly maximum values may be used to determine the recurrence interval of extreme values for an entire year or individual months. Such an analysis does, however, not take into account the second or third highest values for a month in a year with high maximum values despite the fact that these values may be higher than the maximum value for other years.

If an event exceeding a threshold value is regarded as morphologically active, i.e., the definition of an erosive rain, all such events should be included in a frequency analysis. The data may then be treated as partial duration series. This method involves determination of high magnitude values on an annual or monthly basis. All observations are included i.e. also days with no rain. The total number of data in the new series is thus equal to the number of days in the record and considerably higher than the number of extreme values which only equals the number of years in the series.

The partial duration series does not fit the Gumbel-distribution which is applicable to an annual maximum series. Instead the recurrence interval and magnitude can be determined by a semi-logarithmic relationship, which means that a much simpler method than the Gumbel distribution can be used. Comparisons have shown (AHNERT 1987) that the discrepancy between the two methods is negligible or at least tolerable. Using a semi-logarithmic regression makes it easy to understand the meaning of the factors in the equation. With a relationship $P = a * \lg(RI) + b$, the regression constant (b) is equal to the amount with a one year recurrence interval ($a * \lg(1) = 0$), the sum of the regression constant and the value of the regression gradient (b + a) gives the 10 years return value ($a * \lg(10) = a$) and (b + 2a) the 100 years return value. The numbers obtained from the semi-logarithmic regression line in fact gives a simple characteristic of the probability of rains of different magnitude to occur.

The reliability of recurrence intervals will depend on the number of years on which the statistics is based. Short series of data may include extreme values and will result in recurrence data that deviate from long term calculations. This is likely to be more pronounced in an environment with large inter-annual and intra-annual variations. Fig. 12 shows a comparison between a 20 year period of daily precipitation at Leribe and the same data split into two 10 year periods or four 5 year periods.

The recurrence intervals determined from either of the two 10-year periods closely resemble the 20-year period data, whereas one of the four 5-year periods gives a rela-

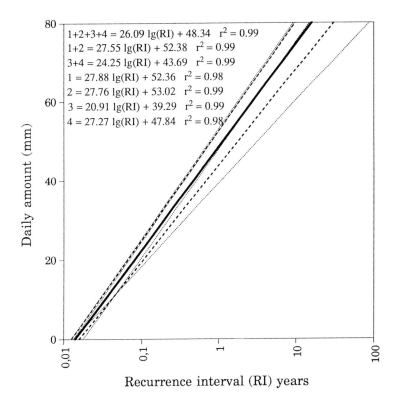

Fig. 12. Recurrence interval of daily precipitation amounts at Leribe during a 20-year period (full thick line). Stippled line shows the same period split into two consecutive 10-year periods and dotted line shows the period split into 4 5-year periods.

tionships clearly deviating from the long-term relationship. This is naturally a sign of caution when using short time series for extrapolation of information. If e.g., an intensity series has by chance been collected during two years with extremely high intensity values compared to long term mean conditions, the recurrence interval for a specific intensity value will be much shorter than the value based on a long series of data. Short series of observations will thus increase the uncertainty of recurrence intervals. It has been suggested (AHNERT 1987) that extrapolation of recurrence intervals should not be considered for periods exceeding 10 times the length of the data record.

It is assumed that the shorter series of daily amount/maximum intensity relationships are representative and applicable to a longer record of daily precipitation amounts. The data series on daily amounts of precipitation at Maseru, Mohales Hoek and Leribe were converted into likely maximum 30-minute intensities using the relationships in Table 2. Data were split into dry season and wet season and the recurrence intervals were determined (Fig. 13). By excluding years with missing values for some months the series comprise approximately 70 years of data.

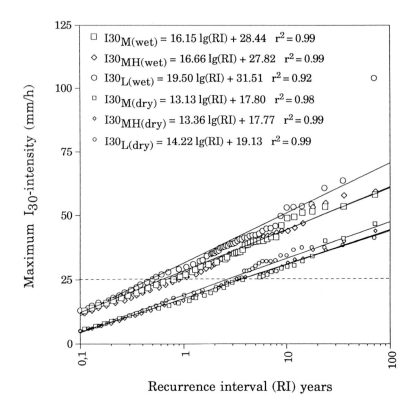

Fig. 13. Recurrence interval of I₃₀. Data split into wet and dry seasons for Maseru, Mohales Hoek and Leribe.

Results show that rainstorms with a maximum 30-minute intensity exceeding 25 mm/h are 5–6 times as frequent during the wet season compared to the dry season. For the dry season (May–October) the threshold value is exceeded every 2.6–3.5 years whereas the wet season is likely to experience such rains 1.5–2.2 times each year. Lower intensities e.g., 10 mm/h will occur on 12.5–14 times each year during the wet season and 3.8–4.3 times in the dry season. Extreme events like 50 mm/h will occur with a recurrence interval of 9–22 years in the wet and 150–280 years in the dry season. For a station like Quthing with only a very short record, the number of data is not sufficient to arrive at an acceptable level of reliability in the determinations.

A direct comparison of recurrence intervals computed for the different stations is difficult to do since background data differ considerably in length of record and type. Maseru, Leribe and Mohales Hoek have long daily records whereas the intensity records are rather short and limited to the wet season. The recurrence interval of maximum 30-minute intensities only during the wet season were computed. The length of record from Oxbow is 20 years, containing both daily amounts and intensities determined by the Meteorological Office in Maseru. For roma a series of daily precipitation amounts of approximately 20 years was available, but the series of observed intensities is much shorter (3–7 years depending on month).

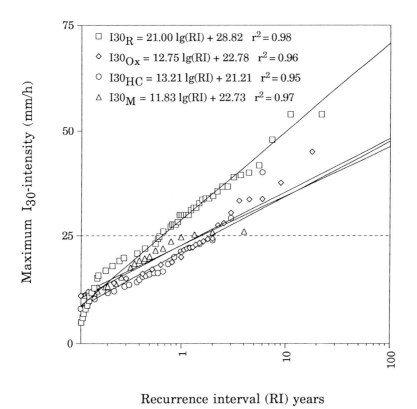

Fig. 14. Recurrence interval of maximum 30-minute intensities at Roma ($I30_R$), Oxbox ($I30_{Ox}$), Ha Charlie ($I30_{HC}$) and Moletsane ($I30_M$). Information based on wet season conditions.

The recurrence interval for a selection of I_{30}-intensities during the rainy and dry seasons at each station are compiled in Table 3. The method of computing recurrence intervals using a combination of daily amounts and amount/intensity relationships obviously result in recurrence intervals that should be regarded as maximum values since the influence of extreme intensities is subdued by the regression method. It is therefore likely that actual recurrence intervals may be shorter.

The results show that it is likely that rains exceeding the threshold value fo 25 mm/h can expected to occur every year at Maseru, Leribe, Mohales Hoek and Roma. The number of expected events each year is obtained by inverting the recurrence interval value. For all sites with a long data record at leat 1–2 events exceeding 25 mm/h can be expected each year.

The recurrence intervals at Oxbow, Ha Charlie and Moletsane are approximately twice the value of the sites with long data series. It is not possible to state whether this is just an effect of the method used or if the geographical setting in the foothills and mountain region is reflected.

The number of erosive rains each year is not strikingly high and there is reason to assume that the erosive power of the rains are not likely to be the major factor responsible

Table 3. Recurrence intervals (years) for maximum I_{30}-intensities exceeding 10, 25 and 50 mm/h in Lesotho. Statistics based on long series of daily precipitation combined with amount/intensity relationships for shorter time intervals. For short recurrence intervals information is transferred to number of expected events/year.

Station	Interval years 10 mm/h	Events/ year 10 mm/h	Interval years 25 mm/h	Events/ year 25 mm/h	Interval years 50 mm/h
Maseru					
wet season	0.07	13.8	0.62	1.6	21.7
dry season	0.25	3.9	3.5	0.3	+284
Leribe					
wet season	0.08	12.7	0.46	2.2	8.9
dry season	0.23	4.4	2.6	0.4	148
Mohales Hoek					
wet season	0.09	11.7	0.68	1.5	21.5
dry season	0.26	3.8	3.5	0.3	259
Roma					
wet season	0.13	7.9	0.66	1.5	10.2
Oxbow					
wet season	0.10	10.1	1.5	0.7	137
Ha charlie					
wet season	0.14	7.1	1.9	0.6	151
Moletsane					
wet season	0.08	11.9	1.6	0.6	202

for the extremely degraded landscape It should, however, be kept in mind that due to the method of computation the recurrence interval may be shorter than the information given in Table 3. An improvement of the reliability of the recurrence intervals will demand longer series of observed values on maximum intensities. This would make it possible to use observed data as the sole basis for partial duration analysis and make estimates of the relationship between daily amounts and maximum storm intensity unnecessary.

As previously stated $I_{30} \geq 25$ mm/h and its frequency of occurrence may not be the best measure for defining erosive conditions. In fact, many observations show that surface wash may occur at much lower intensities. If the prerequisites for surface runoff are considered this is not surprising. Surface runoff can not start unless the soil-water storage capacity at the soil surface is filled. Whether this occurs as total saturation of the soil or only saturation of the uppermost portion of the soil is not important.

Surface runoff occurring in connection with low intensity or low amount storms wil only take place if antecedent events have brought the surface soil close to saturation. To be able to apply this manner of identifying erosive events demands another kind of data than the fairly easily determined $I_{30} \geq 25$ mm/h threshold. This will require information on soil moisture fluxes over time as well as information on the hydraulic conductivity and infiltration capacity of different soils exposed to soil erosion. Since it is unlikely that this kind of information will be available in the foreseeable future it is likely that until alternative methods of treating antecedent storm conditions and their influence on soil moisture the conventional definition of erosive rains will prevail.

Conclusions

The available data on daily precipitation amounts were transformed into maximum 30-minute intensities. These series were treated as partial duration series to estimate the recurrence interval of rains having an intensity high enough to cause soil erosion by surface runoff. The expected recurrence interval of rains exceeding this value was determined for the rainy season for all sites and for the dry season where available information made it possible.

From a temporal point of view the analysis of data for individual months shows long recurrence intervals for months at the onset and end of the wet season. The most likely period for erosive rains is December–February when high temperatures enhances the conditions for convective precipitation. For the entire wet season the data indicate that for stations with a long data record, 1.4 to 2.2 days can be expected to have rains with an intensity high enough to start surface runoff which will contribute to soil erosion.

The spatial differences between lowland stations are small, but stations situated in the northern mountain region and the foothills show recurrence intervals twice as long as for the lowland stations.

Even if intensive rains can be expected to occur each year the number of events is not so high that the natural precipitation conditions can be regarded as the major factor for the status of soil erosion in the country.

References

AHNERT, F. (1982): Untersuchungen über das Morphoklima und die Morphologie des Inselberggebietes von Machakos, Kenia. – In: AHNERT, ROHDENBURG & SEMMEL: Beiträge zur Geomorphologie der Tropen (Ostafrika, Brasilien, zentral- und Westafrika. – Catena Suppl. 2.
– (1987): An approach to the identification of morphoclimates. – In: GARDINER, V. (ed.): International Geomorphology 1986, Part II.
ATHESIAN, J.K.H. (1974): Estimation of Rainfall Erosion Index. – Journ. Irrigation and Drainage Div., ASCE, 100, IR 3: 293–307.
BÄRRING, L. (1988): Aspects of Daily Rainfall Climate Relevant to Soil Erosion in Kenya. – Doctoral thesis Dept. of Physical Geography, Univ. of Lund, Sweden.
CALLES, B. & L. KULANDER (1994): Rainfall erosivity at Roma, Lesotho. – Geogr. Ann. 76A (1–2): 121–129.
CHAKELA, Q.K. & M. STOCKING (1988): An improved methodology for erosion hazard mapping. Part II: application to Lesotho. – Geogr. Ann. 70A.
DE PLOEY, J., M.J. KIRKBY & F. AHNERT (1991): Hillslope erosion by rainstorms – A magnitude-frequency analysis. – Earth Surf. Proc. and Landforms 16: 399–409.
DUNNE, T. & L.B. LEOPOLD (1978): Water in environmental planning. – W.H. Freeman & Co.
GUMBEL, E.J. (1954): Statistical theory of extreme values and some practical applications. – U.S. Bureau Standards Appl. Math., Ser. 33.
HUDSON, N.W. (1963): Raindrop size distribution in high energy storms. – Rhodesian J. Agric. Res. 1: 6–11.
– (1981): Soil conservation. – BT Batsford Academic and Educational Ltd., London, England.

KINELL, P.I.A. (1973): The problem of assessing the erosive power of rainfall from meteorological observations. – Soil Sci. Soc. Amer. Proc. **38**: 657–660.
KOWAL, J.M. & A.H. KASSAM (1976): Energy and instantaneous intensity of rainstorms at Samaru, northern Nigeria. – Tropical Agriculture **53**: 185–198.
LAL, R. (1976): Soil erosion problems on an Alfisol in western Nigeria and their control. – Monograph 1. Int. Inst. of Tropical Agric., Ibadan, Nigeria, 208 pp.
MOORE, T.R. (1979): Rainfall erosivity in East Africa. – Geogr. Ann. **61A**: 147–156.
ROWNTREE, K.N. (1982): Storm rainfall on the Njemp flats, Baringo District, Kenya. – J. Climatology **8**: 297–309.
STOCKING, M. (1978): Interpretation of stone-lines. – South African Geogr. Journ. **60**: 121–134.
STOCKING, M. & H.A. ELWELL (1976): Rainfall erosivity over Rhodesia. – Inst. Brit. Geogr., Trans. New Series **1**: 231–245.
TYSON, P.D. (1986): Climatic change and variability in southern Africa. – Oxford Univ. Press, Cape Town.
ULSAKER, L.G. & C.A. ONSTAD (1984): Relating rainfall erosivity factors to soil loss in Kenya. – Soil Sci. Soc. Amer. J. **48**: 891–896.
WILKEN, G.C. (1978): Agroclimatology of Lesotho. – LASA Paper I.
WISCHMEIER, W.W. (1977): Use and misuse of the universal soil loss equation. – In: Soil erosion: Prediction and control. – Soil Cons. Soc. Amer., Spec. Publ. **21**: 371–378.
WISCHMEIER, W.H. & D.D. SMITH (1958): Rainfall energy and its relationship to soil erosion. – Trans. Am. Geophys. Union **39**: 285–291.

Address of the authors: BENGT CALLES and LENA KULANDER, Institute of Geoscience, Physical Geography, Uppsala University, Nordbyvägen 18 B, S-753 26 Uppsala, Sweden.

Late Quaternary hillslope erosion rates in Japanese mountains estimated from landform classification and morphometry

TAKASHI OGUCHI, Tokyo

with 6 figures and 1 table

Summary. This paper presents a method for estimating long-term hillslope erosion rates in rugged mountains, based on landform classification and morphometry. Hillslopes in mountain river basins in Japan can be divided into smooth slopes and incised slopes. The smooth slopes were formed by freeze-thaw actions around the Last Glacial Maximum that were followed by stabilization after the Late Glacial. The incised slopes, which cut into the smooth slopes, have been formed by landslides and gullying since the Late Glacial due to the onset of a climate with frequent heavy rains. Thus, hillslope erosion rates since the Late Glacial can be estimated from the contour restoration of the incised slopes. The estimated rates for fifteen basins vary from 0.20 to 0.73 mm/yr or 540 to 1,970 t/km²/yr. Although these rates are considerably higher than the average contemporary erosion rates in the world, they are almost equivalent to the contemporary rates in Japanese mountains. This finding indicates that there was no marked acceleration of erosion in Japan during the latest Holocene despite widespread human interference. In other words, natural conditions such as large hillslope inclination and frequent heavy rains have played a dominant role in determining the high erosion rates in Japan.

Introduction

The relation between hillslope erosion rates and landform development is a major subject in geomorphology. The methods for estimating hillslope erosion rates fall into three groups: 1) monitoring sediment discharges in rivers and converting them into slope erosion rates of upstream areas, 2) calculating sedimentation rates in depositional zones and also converting them, and 3) direct estimation from hillslope forms, materials, and processes. Contemporary erosion rates have generally been deduced using the former two methods, i.e., measuring the sediment discharge in rivers or surveying sedimentation rates in reservoirs (YOSHIKAWA 1974, LI 1976, ADAMS 1980). Past, long-term erosion rates have often been determined by the second method, i.e., from the accumulation rates of fan sediments (AKOJIMA 1982), talus deposits (YOSHINAGA et al. 1989), lake deposits (ADAMS 1980), delta and fjord sediments (ADAMS 1980, MILLIMAN et al. 1987), colluvium wedges (RENEAU et al. 1990), fills in small closed depressions (POETERAY et al. 1984), and units of geological formations (LI 1976). These estimations are generally based on the assumption that all the sediments supplied have been trapped in the deposits. This appears valid for the sedimentation in static environments such as lakes and artificial reservoirs. As for river sedimentation, however, only the coarser fractions of the supplied sediments generally play a role, while the finer fractions are transported further downstream. In addition, the erosion of materials from depositional landforms can occur along with deposition. Thus, to improve the estimation of long-term hillslope erosion rates, the use of hillslope forms, materials, and processes is more suitable than the use of accumulated sediments. Such an approach, however, has been confronted with difficulties. Contemporary erosion rates have been derived from the field measurements of slope lowering during a few to several years (SCHUMM 1956), as well as a set of aerial photos taken in different periods (ANIYA 1985). These methods, however, cannot be applied to assess long-term erosion rates extending over 10^2 years. Although isotope measurements have provided a tool for assessing the long-term slope erosion rates (DETHIER et al. 1988, ALBRECHT et al. 1993), their applicability to wide areas such as a whole river basin is limited, because they give point-to-point erosion rates. Long-term erosion rates of partially dissected volcanoes have been estimated based on the contour restoration of the dissected parts (RUXTON & MCDOUGALL 1967, SUZUKI 1969). Such morphometric methods, however, have rarely been applied to rugged, non-volcanic mountains, because it is difficult to separate eroded hillslopes from non-eroded ones, unless detailed surveys on hillslope forms and materials are carried out.

In previous studies I developed a morphological and chronological classification of hillslopes and fluvial surfaces in mountain river basins of Japan and mapped the results (OGUCHI 1988a, 1988b, 1992, 1994a, 1994b). In this paper, firstly, the average slope lowering rates in the basins since the Late Glacial are estimated based on morphometric analyses of the hillslope classification maps, and, secondly, the estimated long-term erosion rates are compared with the present rates.

Late Quaternary geomorphic development of mountainous river basins in Japan

Fifteen river basins in Japan were selected for this study. They belong to three mountainous regions called the Matsumoto, Yamagata, and Kitakami regions from south to north (Fig. 1). The basins have alluvial fans in their lower reaches. The basins above the fan apex consist mostly of steep valley-side slopes, with a minor fluvial terracing along the trunk stream and major tributaries. The basin area above the fan apex ranges from 12 to 78 km^2, and the relief ratio from 0.061 to 0.29 (Table 1), the latter being defined by SCHUMM (1956) as basin relief divided by basin length.

OGUCHI (1988a, 1988b, 1994a, 1994b) discussed the late Quaternary geomorphic development of the fifteen basins. The fluvial surfaces and hillslopes were classified based on the interpretation of aerial photos, the analysis of topo-sheets, and the stratigraphy of more than 400 outcrops (Figs. 2 and 3). The results revealed a landform development common to all the basins:

1) Around the Last Glacial Maximum, dominant hillslope processes consisted of widespread production and transportation of periglacial debris, forming smooth slopes with low declivity covered with angular gravels. With the onset of the Late Glacial and Holocene warm climates, dominant hillslope processes changed to intensive erosion by running water, leading to shallow failure and bedrock gullying mainly in the slope depressions. The shallow failure and gullying left behind two slope units: a hollow-shaped depression and a deep gully surrounded by the depression. These two units, which are designated as upper and lower incised slopes, can be identified by tracing the convex breaks of hillslopes (Fig. 4).

2) The Late Glacial and Holocene slope incision resulted in abundant sediment supply to the trunk stream. Because the stream above the fan apex carried large flows,

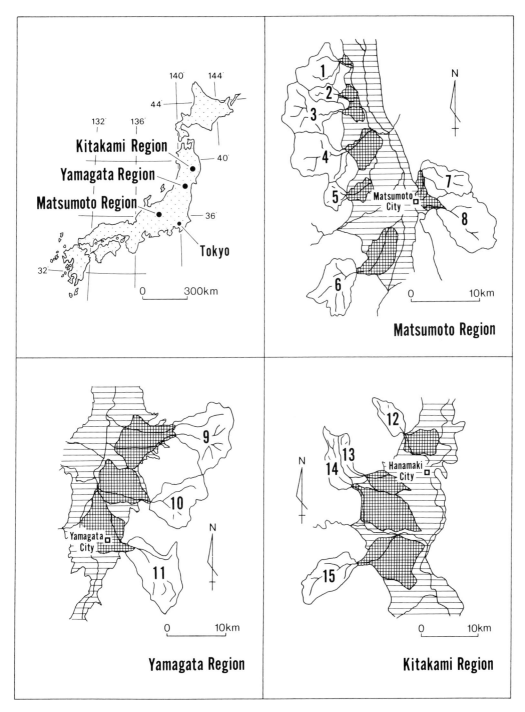

Fig. 1. Location of river basins studied. Light hatch: Floor of intermontane basin, Heavy hatch: Alluvial fans of river basins studied.

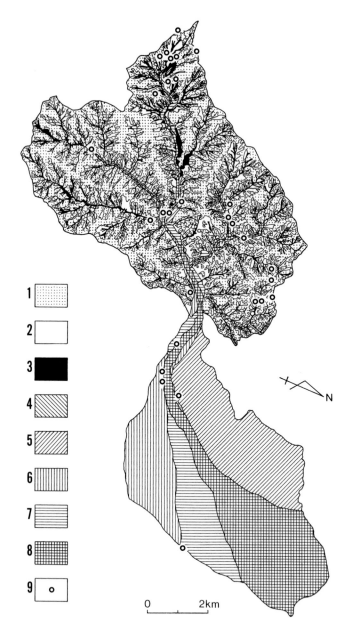

Fig. 2. Landform classification map of No. 6 river basin (Kusari river basin, Matsumoto region) (Simplified after OGUCHI 1988a).
1: Smooth slopes, 2: Upper incised slopes, 3: Lower incised slopes, 4: Q1 (highest) terraces, 5: Q2 terraces (fill terraces formed in the Last Glacial age), 6: Q3 terraces (erosional terraces formed in the Last Glacial age), 7: Q4 terraces (erosional terraces formed in the Late Glacial and Holocene), 8: Active surface (forming at present), 9: Locations of outcrops surveyed (Stratigraphy is described by OGUCHI 1988a).
The areal ratio of the incised slopes to all mountain slopes is 43 percent. The average hillslope erosion rate since the Late Glacial is estimated to be 0.25 mm/yr.

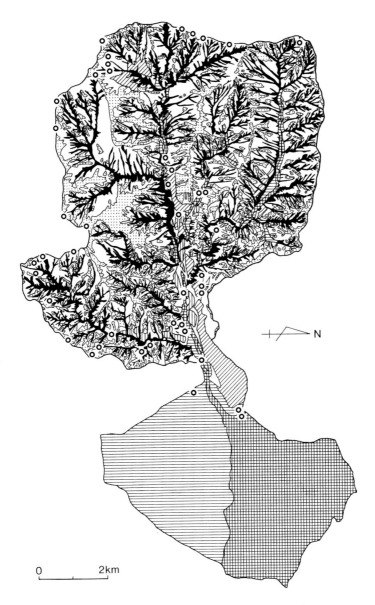

Fig. 3. Landform classification map of No. 4 river basin (Karasu river basin, Matsumoto region) (Simplified after OGUCHI 1988a). The legends are the same as those used in Fig. 2. The characteristics of terraces, such as fill and erosional, are also the same as those in No. 6 basin of Fig. 2, except that Q4 terraces in the alluvial fan are aggradational surfaces formed in the Late Glacial and Holocene. The areal ratio of incised slopes to all mountain slopes is 79 percent. The average hillslope erosion rate since the Late Glacial is estimated to be 0.58 mm/yr.

almost all the sediments supplied were transported downstream and deposited on the alluvial fan. The large flows also led to flood plain incision above the fan apex, leaving erosional terraces behind.

Fig. 4. Hillslope units in No. 6 river basin, Matsumoto region.
S: Smooth slope, U: Upper incised slope, L: Lower incised slope. Note that the area experienced recent logging. The incised slopes developed previously, under the naturally forested condition.

The above landform development shows that both mountain slopes and fluvial surfaces above the fan apex have been incised since the Late Glacial. The incision can be attributed to a marked increase in erosive force during the Pleniglacial–Late Glacial transition. At present, the Japanese islands are subject to heavy rains with the world's highest intensity (MATSUMOTO 1993), because they lie beneath the Polar frontal zone in early summer and typhoon tracks in late summer and autumn. Before the Late Glacial, however, the southward shift of frontal zones resulted in a considerably reduced frequency of heavy rains (SUGAI 1993).

Estimation of hillslope erosion rates since the Late Glacial

The method for estimating long-term hillslope erosion rates was developed, based on the landform development and the result of a preliminary study for five basins (OGUCHI 1991). As noted, the incised slopes have cut into the smooth slopes since the Late Glacial, while the undissected smooth slopes have remained stable. Thus, contour restoration of the incised slopes enables the reconstruction of the slope topography soon before the Late Glacial. In this context, the magnitude of hillslope erosion since the Late Glacial was evaluated.

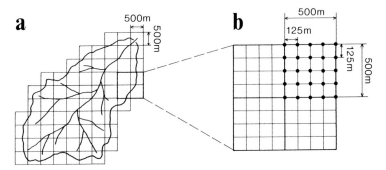

Fig. 5. Grid system used for morphometric analyses.

First, the basins above the fan apex were subdivided into 500 × 500 m square cells (Fig. 5a). Next, the cells suitable for the contour restoration of the incised slopes were selected using the following criteria: 1) the distribution of the preserved smooth slopes was sufficiently continuous to allow contour restoration, and 2) the areal ratio of fluvial surfaces to hillslopes was less than 0.25, since the analysis here focused on hillslope erosion. The number of selected cells was 298 for the Matsumoto region, 127 for the Yamagata region, and 94 for the Kitakami region, which appears to be sufficient for estimating the averaged erosion rates. Using the enlargement of 1:25,000 topographic maps published by the Geographical Survey Institute of Japan, landforms before the incision were reconstructed by completely smoothed contours on the remaining smooth slopes. The contour interval of the maps is generally 10 m and partly 5 m. Using the map enlargement, the altitude of a point can be read with an accuracy of ± 1 m. Next, 25 lattice points with a 125-meter interval were plotted within each cell (Fig. 5b). At each lattice point, the difference in altitude between the present landform and the reconstructed landform was measured. The mean altitude difference for the 25 points (ED) represents the average depth of hillslope erosion since the Late Glacial.

Using the ED values for the selected cells, the average erosion rates of all the mountain slopes in the basins were estimated. The landforms above the fan apex have three major components: the smooth slopes, the incised slopes, and fluvial surfaces (Figs. 2 and 3). Using a digitizer with a resolution of 0.1 mm, the total area of each landform component within the selected cells was measured on the 1:25,000 landform classification maps. Next, the areal percentage of the upper and lower incised slopes to all the mountain slopes was calculated. The percentage, which is referred to as PI, represents the degree of development of the incised slopes. For each of the three regions, the selected cells were categorized according to the values of PI. The range of PI for one category was determined to be 5 % for the Matsumoto region while 10 or 20 % for the Yamagata and Kitakami regions, depending on the total number of available cells. Fig. 6 is a plot of the average ED values for each category versus the modal values of PI. The figure indicates that the two parameters are positively correlated for each region. The ED value should be zero when the PI ratio equals zero, and thus a power function was fitted to the PI–ED relation for each region. The results were:

For the Matsumoto region:

(1) $ED = 0.019\ PI^{1.37}$; $r = 0.95$

For the Yamagata region:

(2) $ED = 0.037\ PI^{1.16}$; $r = 0.96$

For the Kitakami region:

(3) $ED = 0.10\ PI^{0.795}$; $r = 0.95$.

The good correlation of the equations indicates that the average ED can be estimated from the values of PI. OGUCHI (1988b, 1994a, 1994b) had already calculated the mean PI ratio for each basin. The substitution of the PI ratios into Eqs. (1), (2), and (3) yielded mean ED for the basins (Table 1). The ED values were divided by the duration of the Late Glacial and the Holocene (13,000 yrs.), which revealed that the average hillslope erosion rates since the Late Glacial ranged between 0.20 and 0.73 mm/yr. Assuming that the mean specific gravity of the earth crust is about 2.7 t/m³, the erosion rates can be converted into 540 to 1,970 t/km²/yr (Table 1). These values are considerably higher than the average contemporary erosion rates in the world, i.e., 150 or 183 t/km²/yr (HOLEMAN 1968, MILLIMAN & MEADE 1983). However, they are normal for the small mountainous river basins in South Asia and East Asia (MILLIMAN & SYVITSKI 1992).

In discussing the erosion rates of the whole basins, not only hillslope erosion but also river incision due to erosional terrace formation should be considered. The sediment supply from the erosional terraces can be estimated from the terrace classification maps and their longitudinal profiles (OGUCHI 1991). In five basins out of the fifteen basins studied, the amount of sediments supplied from the terraces was estimated to be only about 5 to 10 percent of the total sediment supply from both hillslopes and the terraces (OGUCHI 1991). The smaller sediment supply from the terraces can be confirmed for the other ten basins, because both the relative area of the erosional terraces and the depth of the degradation are as small as those of the five basins. Consequently, the hillslope erosion rates inferred here approximate the erosion rates in a whole basin.

The ED–PI relations of Eqs. (1), (2), and (3) probably reflect regional differences in climatic conditions. Fig. 6 shows that ED for the same PI varies according to the regions, being the largest for the Matsumoto region, intermediate for the Yamagata region, and the smallest for the Kitakami region. In other words, the average depth of the incised slopes tends to increase in a southward direction. This can be explained from a southward rise in storm intensity over the Japanese Islands (MATSUMOTO 1993) because the degree of postglacial bedrock incision has been affected by storm intensity (OGUCHI 1996).

Estimation of contemporary hillslope erosion rates

The contemporary erosion rates of the fifteen basins were estimated using OHMORI's (1978, 1982) empirical equation, which was derived from the sedimentation rates of 30 reservoirs in Japanese mountainous basins:

(4) $ER = 0.35 \times 10^{-6}\ D^{3.2}$; $r = 0.77$

where ER is the present hillslope erosion rate (mm/yr), and D is the standard deviation of altitudes within a unit area of 1 × 1 km. In Japanese mountains, the standard deviation

Late Quaternary hillslope erosion rates 177

Region River basin	Basin area above fan apex (km²)	Mountain slope area above fan apex (km²)	Maximum altitude of basin (m)	Altitude of fan apex (m)	Relief ratio of basin above fan apex	Percentage of area of incised slopes	Hillslope erosion since the Late Glacial			Mean standard deviation of altitude in 1x1 km² cells (m)	Present hillslope erosion rate		Ratio of past hillslope erosion rate to present erosion rate
							erosion depth of hillslopes (m)	hillslope erosion rate (mm/yr)	hillslope erosion rate (t/km²/yr)		based on relief (mm/yr)	based on dam sediments (t/km²/yr)	
Matsumoto													
Chi (1)	38.2	34.4	2647	830	0.22	88	8.8	0.68	1840	116	1.41	—	0.48
Ashima (2)	11.9	10.5	2283	820	0.26	93	9.5	0.73	1970	124	1.75	—	0.42
Nakafusa (3)	56.3	54.0	2922	720	0.21	90	9.0	0.69	1860	128	1.94	—	0.36
Karasu (4)	61.7	56.8	2857	720	0.21	79	7.6	0.58	1570	117	1.45	—	0.40
Kurosawa (5)	14.4	13.8	2051	770	0.29	86	8.5	0.65	1760	120	1.58	—	0.41
Kusari (6)	52.0	48.9	2446	830	0.16	43	3.3	0.25	675	95	0.74	—	0.34
Metoba (7)	32.6	29.0	1935	760	0.13	63	5.5	0.42	1130	85	0.52	—	0.81
Susuki (8)	77.8	68.5	2034	750	0.10	78	7.4	0.57	1540	86	0.54	—	1.06
Yamagata													
Midare (9)	71.8	66.1	1284	200	0.11	69	5.0	0.38	1030	86	0.54	—	0.70
Tachiya (10)	51.6	44.9	1264	200	0.11	86	6.5	0.50	1350	77	0.38	—	1.32
Mamigasaki (11)	70.3	59.8	1840	200	0.13	84	6.3	0.49	1320	73	0.32	1650	1.53 (0.80)
Kitakami													
Dai (12)	22.6	34.3	701	130	0.068	65	2.8	0.22	590	65	0.22	—	1.00
Samusawa (13)	15.0	22.9	968	210	0.090	59	2.6	0.20	540	49	0.09	—	2.22
Shiritaira (14)	28.0	43.6	968	210	0.061	61	2.6	0.20	540	74	0.34	—	0.59
Geto (15)	48.9	73.7	1373	220	0.087	71	3.0	0.23	620	71	0.29	—	0.79
													mean: 0.83

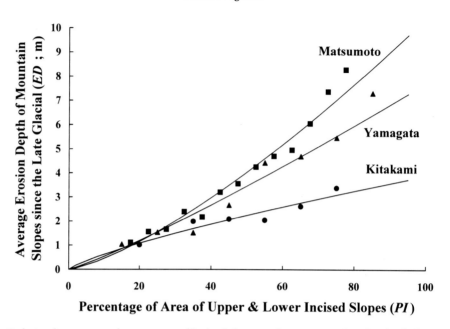

Fig. 6. Relation between areal percentage of incised slopes and average erosion depth of all hillslopes estimated from contour restoration method. Squares: the Matsumoto region, Triangles: the Yamagata region, Circles: the Kitakami region.

of altitudes shows a positive correlation with the local slope (OHMORI & HIRANO 1984). Thus, Eq. (4) expresses the relationship between the erosion rates and the hillslope angle. Using the 125 × 125 m grid systems in Fig. 5b, the standard deviation of the altitudes of the 25 lattice points was calculated for each 500 × 500 m cell. As indicated by EVANS (1972), the standard deviation of altitude changes with the size of the cell used for calculation. To describe this change, OHMORI (1978) derived the following empirical equation for Japanese mountains:

(5) $\log D_s = -0.0374 + 0.9656 \log D; \quad r = 0.9963$

where D_s is the standard deviation of altitudes in a 500 × 500 m square. Because Eq. (5) shows a notably high correlation, the regression line of D on D_s is statistically equivalent to that of D_s on D. Thus, the following equation rearranged from Eq. (5) was used to derive the value of D from the value of D_s:

(6) $\log D = (0.0374 + \log D_s) / 0.9656$

The calculated values of D were averaged for each basin, and the averages were substituted into Eq. (4) to estimate contemporary erosion rates. The estimated rates ranged from 0.09 to 1.94 mm/yr, which are equivalent to 240 and 5,240 t/km²/yr (Table 1).

Among the fifteen basins studied, the Mamigasaki river basin in the Yamagata region has an artificial reservoir in the middle reach. According to AKOJIMA (1983), the sedimentation rate in the reservoir per unit area of the upstream basin is 888 m³/km²/yr. Because the density of the sediments in a Japanese dam is about 1.75 t/m³ (ASHIDA 1971), the sedimentation rate is equivalent to 1,554 t/km²/yr. Since the dissolved loads in the river were estimated to be 105 t/km²/yr (AKOJIMA 1983), the contemporary erosion rate of the Mamigasaki river basin amounts to about 1,650 t/km²/yr.

Discussion

Table 1 shows the ratios of the hillslope erosion rates since the Late Glacial to contemporary ones. When the present erosion rates were estimated from the altitude deviation, the ratios ranged from 0.36 to 2.22 with their average being 0.83. As for the Mamigasaki river basin, the ratio inferred from the sum of dam sediments and the dissolved load was 0.80. These values suggest that, if we deal with the basins as a whole, the present erosion rates and the past ones are almost comparable and certainly of the same order, although they were estimated by entirely different methods.

In Circum-Pacific humid regions such as East Asia, South Asia and New Zealand, contemporary erosion rates of mountains are ranked among the world largest (YOSHIKAWA 1974, LI 1976, ADAMS 1980, OHMORI 1983, MILLIMAN & MEADE 1983). This active erosion results in a large sediment discharge and rapid accumulation of alluvial fans in the lower reaches (CAMPBELL & JOHNSTON 1982, SAITO 1993, OGUCHI & OHMORI 1994). So far two different explanations have been proposed for the high erosion rates in these regions. One attributes them to natural conditions including frequent heavy rains and/or the conspicuous relief of mountains associated with rapid tectonic uplift (HOLEMAN 1968, OHMORI 1978, 1983, ADAMS 1980, OGUCHI 1994c). The other stresses the effects of human factors such as farming, grazing, and deforestation, and thus infers that erosion rates were much lower before widespread human interference since ca. 2,000 yr. B.P. (MILLIMAN et al. 1987, MILLIMAN & SYVITSKI 1992). The results of the present study indicate that the former explanation is more appropriate for Japanese mountains, because the average erosion rates since the Late Glacial are comparable to the present rates.

In Japan, high erosion rates could occur without a disturbance of the vegetation and soil associated with human activities. As noted, the Japanese Islands were affected by the onset of frequent heavy rains during the Pleniglacial–Late Glacial transition. The new climatic regime with high erosive force promoted extensive hillslope erosion by landslides and gullying even in the bedrock, despite the re-establishment of thick forests due to the increase in temperature (OGUCHI 1992).

The rapid erosion associated with natural conditions may also apply to the Southern Alps of New Zealand. According to ADAMS (1980), the average erosion rates of the Southern Alps since the Late Glacial (from ca. 12,000 yr. B.P.) are generally comparable to, or even larger, than the present rates. ADAMS's (1980) calculation of the past erosion rates was based on several assumptions, because it is an indirect estimate from the volume of fjord sediments, fan deposits, valley-fills, and lake sediments. As noted, the estimation using lake sediments is highly convincing because lakes trap sediments with a high efficiency. ADAMS (1980) reported that the lake sediments showed high erosion rates since the Late Glacial, generally exceeding 1,000 t/km²/yr. This observation suggests that the rapid erosion in the contemporary Southern Alps of New Zealand is essentially unrelated to human interference. Thus, it is assumed that the erosional conditions in mountains of the Circum-Pacific regions depend mostly on physiographic factors rather than on human factors. The validity of this hypothesis should be confirmed based on further studies. Moreover, we need to evaluate the accuracy of long-term hillslope erosion rates estimated from morphometry combined with landform classification.

Acknowledgements

The author is grateful to Prof. H. Ohmori, Prof. K. Saito, and Dr. J. Bartman for their useful comments on this work.

References

ADAMS, J. (1980): Contemporary uplift and erosion of the Southern Alps, New Zealand. – Geol. Soc Amer. Bull., Part 2, 91: 1–114.

AKOJIMA, I. (1983): Comparison between the past and the present denudation rates of mountains around the Yamagata Basin, northeast Japan. – Trans. Japanese Geomorph Union 4: 97–106. [in Japanese with English abstract]

ALBRECHT, A., G.F. HERZOG, J. KLEIN, B. DEZFOULY-ARJOMANDY & F. GOFF (1993): Quaternary erosion and cosmic-ray-exposure history derived from ^{10}Be and ^{26}Al produced in situ – an example from Pajarito plateau, Valles caldera region. – Geology 21: 551–554.

ANIYA, M. (1985): Contemporary erosion rate by landsliding in Amahata River basin, Japan. – Z. Geomorph. N.F. 29: 301–314.

ASHIDA, T. (1971): Sedimentation in reservoirs. – In: YANO, K. (ed.): Science on disasters caused by water: 522–540; Gihodo, Tokyo. [in Japanese]

CAMPBELL, I.B. & M.R. JOHNSTON (1982): Nelson and Marlborough. – In: SOONS, J.M. & M.J. SELBY (eds.): Landforms of New Zealand: 285–298; Longman, Auckland.

DETHIER, D.P., C.D. HARRINGTON & M.J. ALDRICH (1988): Late Cenozoic rates of erosion in the western Española basin, New Mexico: Evidence from geologic dating of erosion surfaces. – Geol. Soc. Amer. Bull. 100: 928–937.

EVANS, I.S. (1972): General geomorphometry, derivatives of altitude, and descriptive statistics. – In: CHORLEY, R.J. (ed.): Spatial analysis in geomorphology: 17–90; Methuen, London.

HOLEMAN, J.N. (1968): The sediment yield of major rivers of the world. – Water Resources Res. 4: 737–747.

LI, Y.H. (1976): Denudation of Taiwan Island since the Pliocene Epoch. – Geology 4: 105–107.

MATSUMOTO, J. (1993): Global distribution of daily maximum precipitation. – Bull. Dept. Geogr. Univ. Tokyo 25: 43–48.

MILLIMAN, J.D. & R.H. MEADE (1983): World-wide delivery of river sediment to the oceans. – Jour. Geol. 91: 1–21.

MILLIMAN, D., O. YUN-SHAN, R. MEI-E & Y. SAITO (1987): Man's influence on the erosion and transport of sediment by Asian rivers: the Yellow River (Huanghe) example. – Jour. Geol. 95: 751–762.

MILLIMAN, D. & P.M. SYVITSKI (1992): Geomorphic / Tectonic control of sediment discharge to the ocean: the importance of small mountainous rivers. – Jour. Geol. 100: 525–544.

OGUCHI, T. (1988a): Landform development during the Last Glacial and the Post-Glacial ages in the Matsumoto Basin and its surrounding mountains, central Japan. – The Quat. Res. (Tokyo) 27: 101–124. [in Japanese with English abstract]

– (1988b): Differences in landform development during the Late Glacial and the Post-Glacial ages among drainage basins around the Matsumoto Basin, central Japan. – Geogr. Rev. Japan 61A: 872–893. [in Japanese with English abstract]

OGUCHI, T. (1991): Quantitative study of sediment transport in mountain drainage basins since the Late Glacial. – Trans. Japanese Geomorph. Union 12: 25–39. [in Japanese with English abstract]
- (1992): Responses of drainage basins to the Pleistocene–Holocene climatic change: Japan and other mid-latitude regions. – Bull. Dept. Geogr. Univ. Tokyo 24: 51–73.
- (1994a): Late Quaternary Geomorphic development of mountain river basins based on landform classification: the Kitakami region, Northeast Japan. – Bull. Dept. Geogr. Univ. Tokyo 26: 15–32.
- (1994b): Late Quaternary development of alluvial fan–source basin systems based on landform classification: the Yamagata region, Japan. – Geogr. Rev. Japan 67B: 81–100.
- (1994c): Average erosional conditions of Japanese mountains estimated from the frequency and magnitude of landslides. – Proc. Intern. Symp. Forest Hydrol. Tokyo, 1994: 399–406.
- (1996): Factors affecting the magnitude of post-glacial hillslope incision in Japanese mountains. – Catena 26: 171–186.

OGUCHI, T. & H. OHMORI (1994): Analysis of relationships among alluvial fan area, source basin area, basin slope, and sediment yield. – Z. Geomorph. N.F. 38: 405–420.

OHMORI, H. (1978): Relief structure of the Japanese mountains and their stages in geomorphic development. – Bull. Dept. Geogr. Univ. Tokyo 10: 31–85.
- (1982): Functional relationship between the erosion rate and the relief structure in the Japanese mountains. – Bull. Dept. Geogr. Univ. Tokyo 14: 65–74.
- (1983): Characteristics of the erosion rate in the Japanese mountains from the viewpoint of climatic geomorphology. – Z. Geomorph. N.F., Suppl. Bd. 46: 1–14.

OHMORI, H. & M. HIRANO (1984): Mathematical explanation of some characteristics of altitude distributions of landforms in an equilibrium state. – Trans. Japanese Geomorph. Union 5: 293–310.

POETERAY, F.A., P.A. RIEZEBOS & R.T. SLOTBOOM (1984): Rates of subatlantic surface lowering calculated from mardel-trapped material (Gutland, Luxembourg). – Z. Geomorph. N.F. 28: 467–481.

RENEAU, S.I., W.E. DIETRICH, D.J. DONAHUE, A.J.T. JULL & M. RUBIN (1990): Late Quaternary history of colluvial deposition and erosion in hollows, central California Coast Ranges. – Geol. Soc. Amer. Bull. 102: 969–982.

RUXTON, B.P. & I. MCDOUGALL (1967): Denudation rates in northeast Papua from potassium-argon dating of lavas. – Amer. Jour. Sci. 265: 545–561.

SAITO, K. (1993): Effectiveness of a dynamic equilibrium model for alluvial fans in the Japanese Islands and Taiwan Island. – Jour. Saitama Univ., Fac. Educ. (Humanities and Social Sci.) 42: 33–48.

SCHUMM, S.A. (1956): Evolution of drainage systems and slopes in badlands at Perth Amboy, New Jersey. – Geol. Soc. Amer. Bull. 67: 597–646.

SUGAI, T. (1993): River terrace development by concurrent fluvial processes and climatic changes. – Geomorphology 6: 243–252.

SUZUKI, T. (1969): Rate of erosion in strato-volcanoes in Japan. – Bull. Volcanol. Soc. Japan 14: 133–147. [in Japanese with English abstract]

YOSHIKAWA, T. (1974): Denudation and tectonic movement in contemporary Japan. – Bull. Dept. Geogr. Univ. Tokyo 6: 1–14.

YOSHINAGA, S., K. SAIJO & N. KOIWA (1989): Denudation process of mountain slope during Holocene based on the analysis of talus cone aggradation process. – Trans. Japan. Geomorph. Union 10: 179–193. [in Japanese with English abstract]

Address of the author: Dr. TAKASHI OGUCHI, Department of Geography, Graduate School of Science, The University of Tokyo, 7-3-1, Hongo, Bunkyo-ku, Tokyo, 113, Japan.

A comparison of the interrill infiltration, runoff and erosion characteristics of two contrasting 'badland' areas in southern France

JOHN WAINWRIGHT, London

with 7 figures and 2 tables

Summary. Small plot experiments have been carried out to investigate the interrill response of two badland areas in southern France to intense rainfall episodes. The first area is situated in the "terres noires" of the French pre-alps, and has well-developed channel networks and high relief. The second area is situated on molasse sediments in the Aquitainian basin, and has much lower relief and very poorly defined channels. The experiments measured infiltration, resistance to flow and erosion rates using rainfall simulation techniques. The results show similar responses for infiltration and sediment transport rates, but significant differences between the two areas in their resistance to flow. These responses are related to the surface characteristics of the two areas. Consideration of these results, together with data on frequency and magnitude of rainfall events suggests that both the areas are relatively in equilibrium, and that the difference in form between the two areas is larely a result of climatic controls.

Zusammenfassung. *Vergleich der Zwischenrinneninfiltration, Abflußmenge und Erosionscharakteristika zweier unterschiedlicher ‚Ödlandgebiete' im Süden Frankreichs.* – Experimente mit Hilfe kleiner Versuchsquadranten wurden durchgeführt, um die Reaktion des Bodens zwischen Entwässerungsrinnen in zwei Ödlandgebieten im Süden Frankreichs zu intensiven Niederschlagsereignissen zu untersuchen. Das erste Untersuchungsgebiet liegt in den "terres noires" in den französischen Voralpen und hat ein gut entwickeltes Entwässerungsnetz und ein hohes Relief. Das zweite Gebiet liegt in den Molassesedimenten des Aquitanischen Beckens und hat ein niedrigeres Relief und ein schlecht entwickeltes Entwässerungsnetz. Die Experimente dienten dazu, Infiltration, Fließresistenz und Erosionsraten mit Hilfe von Niederschlagssimulationstechniken zu messen. Die Ergebnisse zeigen ähnliche Reaktionen der Systeme in bezug auf Infiltration und Sedimenttransport, aber erhebliche Unterschiede in bezug auf Fließresistenz in beiden Gebieten. Diese unterschiedlichen Reaktionen können durch unterschiedliche Oberflächencharakteristika in den beiden Gebieten erklärt werden. Interpretierung der Ergebnissse zusammen mit Information über Frequenz und Intensität der Niederschlagsereignisse, erlaubt die Schlußfolgerung, daß sich beide Gebiete in einem Zustand des relativen Gleichgewichts befinden, und daß Formunterschiede zwischen den Gebieten vor allem auf klimatische Bedingungen zurückzuführen sind.

Résumé. *Une comparaison des caractères de l'infiltration, du ruissellement et de l'érosion dans la zone inter-rigole de deux zones de 'badland' dans le Midi de la France.* – On a utilisé des expériences sur petites parcelles pour étudier la réponse inter-rigole aux orages extrêmes de deux régions de "badlands" dans le Midi de la France. Le première région se situe dans les "terres noires" des Prealpes françaises. Elle possède un reseau de chenaux et un relief marqué. La deuxième se trouve dans les sédiments molassiques du bassin d'Aquitaine, et ne présente qu'un relief mineur et des chenaux pauvrement définis. Les expériences ont mesuré l'infiltration, la résistance hydraulique et les taux d'érosion en se servant d'un simulateur de pluie. Les résultats montrent des réponses similaires pour l'infiltration et les taux d'érosion, mais des différences signifiantes de la résistance hydraulique entre les deux régions. Ces réponses sont liées aux caractères surficielles des deux régions. En pondérant ces résultats avec ceux des fréquences et des magnitudes d'évènements orageux, on suppose que toutes les deux régions se trouvent rélativement en équilibre, et que la différence des formes des deux régions est plutôt liée aux facteurs climatiques.

1 Introduction

Rapid erosion and flooding are characteristics of the intense storm events which frequently occur in the Mediterranean region. One event which caused such significant effects occurred in the Vaucluse and Drôme regions of Southern France on 22nd September 1992 (ARNAUD-FASSETTA et al. 1993, FLAGEOLLET et al. 1993, WAINWRIGHT 1996a and b). During this event, up to 300 mm of rain fell over a five hour period (Anon. 1993, BENECH et al. 1993).

One of the reasons why runoff is so rapid in the Mediterranean region is the common occurrence of badland areas (BRYAN & YAIR 1982), which tend to have low infiltration rates (SCOGING 1982, 1989, SCOGING & THORNES 1979). Badlands make up a significant part of the Ouvèze catchment, which was the most intensely affected area in the storm of September 1992. It was therefore decided to study runoff and erosion production at a badland site within the area of the September 1992 event, and compare its response with that from another badland area in Southern France.

The aim of this study was to characterize the interrill response to an extreme storm event as fully as possible. This includes definition of the infiltration rate and therefore runoff production, the resistance of the slope to flows, and the sediment transport. Therefore an experimental design was produced to measure these factors as near simultaneously as practicable. In this way, it is possible to observe the interactions between the different processes as far as possible. Furthermore, by measuring the response in this way, it is possible to parameterize physicaly based models of runoff and erosion.

2 Study areas and methods

Two study areas were chosen (Fig. 1) with differing characteristics. The Propiac study area (44° 16′ 55″ N, 5° 11′ 30″ E) was chosen as being characteristic of the badland areas in the catchment area of the 1992 event. These areas are generally located on the dark-coloured marls ("terres noires") which are common in the Alpine zone of Southern France (e.g. BUFALO & NAHON 1992). At Propiac, the badlands occur in thinly-bedded, blue-grey marls belonging to the bathonian stage of the Jurassic (MONIER & CAVELIER 1992). They are generally steeply sloping (usually > 25°) with sharp divides which are often vegetated or covered with a massive limestone caprock. The surface cover of the slopes is dominated (55 %) by platy stones with a mean size of 18.6 mm formed by the break-up of the marl along bedding. The remainder of the surface cover is more finely broken down marl (< 2 mm: 24 %), vegetation (16 %) and vegetation litter (14 %). Gully systems are usually well developed, and often contain multiple nickpoints. The relief of these badlands is usually between 100 and 200 m, and they cover areas of up to several square kilometres. By comparison, the St. Sernin badlands (43° 55′ 50″ N, 2° 6′ 30″ E) are much more restricted, with relief not exceeding thirty metres, and with areal coverages usually of several hundred square metres. They are restricted to hilltop settings within land other-

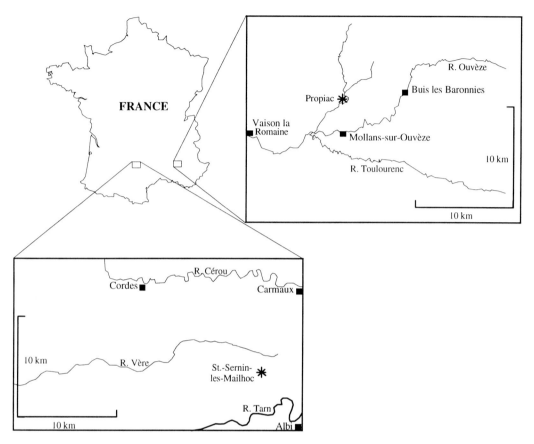

Fig. 1. Location of the two study areas discussed in the text.

wise used for agriculture. Their form is much more rounded. These badlands are lighter coloured, and occur within a sequence of Tertiary fluviatile sediments and limestones (COLLOMB et al. 1989). The badlands are found in the molasse sediments of the sequence, which contain clays, gravels and calareous nodules. The mean slope of 15° is more gentle than at Propiac. The surface cover is again predominantly gravel (52 %) and fine sediment (44 %), with very low vegetation cover (4 %). The mean size of the gravel at the surface is 8.8 mm.

Plot studies at each of the sites were carried out using a rainfall simulator. The simulator was based on the field design suggested by MUNN & HUNTINGTON (1976). Six 61 × 61 cm plots were set up on each site. Rainfall application was at a target rate of 100 mm h^{-1}, although the control method used led to some variation about this value. The experimental design permitted the almost simultaneous measurement of runoff (and therefore infiltration), resistance to flow and erosion rates. It is described in detail in WAINWRIGHT (1996a). In the case of the Propiac study, both overland-flow and splash erosion were measured. Only overland-flow erosion was measured at St. Sernin.

Following each plot experiment, proportional surface cover of vegetation, litter, fine sediment (< 2 mm) or coarse sediment (≥ 2 mm) was determined. The grain size of the coarse sediment was also characterized by measuring the intermediate axis. A shear vane was used to measure soil shear strengths both before and after the experiments, as several authors have shown relationships between this variable and splash and erosion rates (e.g. AL-DURRAH & BRADFORD 1982, RAUWS & GOVERS 1988).

For the Propiac study area, two other methods were used to compare the hydrological characteristics of the soils. Ten undisturbed soil cores (50 mm diameter × 55 mm tall) were sampled and the falling head technique (KLUTE & DIRKESEN 1986) used to measure the saturated hydraulic conductivity (K_{sat}). The figure used for K_{sat} was arrived at as the average of five replicates for each core. Secondly, ponded infiltration rates were calculated by carrying out thirty single cylinder infiltration experiments. Each experiment lasted thirty minutes, for direct comparison with the final infiltration rates of the rainfall simulator plot experiments.

3 Results

3.1 Infiltration rates

The infiltration results at the Propiac study area show a high degree of variability, both within and between each of the three techniques used (Table 1, Fig. 2). The three different techniques produce different distributions of values for final infiltration rate, with the falling head method producing the lowet values and the cylinder infiltrometer method producing the highest values. Comparisons using t tests suggest that the cylinder infiltration results are significantly higher than those for either the rainfall simulation ($p = 0.0003$) or the falling head ($p = 0.0001$) techniques. However, there is no significant difference ($p = 0.32$) between the falling head and the rainfall simulation results, despite the large difference in the mean values.

WATTS (1989) also found a difference of an order of magnitude between the values of K_{sat} obtained using the falling head method and of the final infiltration rate of cylinder infiltrometer experiments, for a series of samples on agricultural terraces in southeast Spain. This difference was explained in terms of the experimental difficulties of using

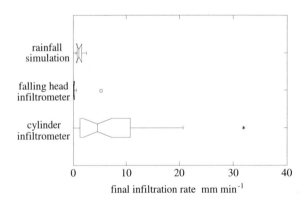

Fig. 2. Comparisons of final infiltration rates at Propiac measured using rainfall simulation, cylinder infiltrometer and falling head techniques.

Table 1. Comparison of infiltration rates measured using three different techniques at Propiac. μ is the mean value and σ the standard deviation.

	final infiltration rate mm h^{-1}		Green and Ampt 'sorptivity' mm	
	μ	σ	μ	σ
rainfall simulation	1.2	0.7	0.8	0.4
ponded	7.0	7.6	7.2	7.6
falling head	0.6	1.6	-	-

cylinder infiltrometers (see also BOUWER 1986). However, POULOVASSILIS et al. (1991) suggest on theoretical grounds that values of final infiltration rate derived for rainfall conditions should be the same as those for ponded conditions. The difference found in the present experiments may be due to the spatial pattern of vegetation. Higher infiltration rates are associated with vegetation cover ($p = 0.013$ for ponded conditions), but the scattered nature of the vegetation means that it has a relatively smaller effect on the measured final infiltration rate due to the larger plot size of the rainfall simulation experiments. This explanation may also be extended to the lower values from the falling head results. As it is impossible to sample for vegetation in the falling head sample cores, beyond the inclusion of some, necessarily truncated, roots, then the values obtained by this method will only reflect the lower infiltration values corresponding to the effects of the soil matrix only.

A further explanation for the difference between the rainfall simulation and ponded infiltration is the rainfall dependence of final infiltration rates. HAWKINS (1982) demonstrated that final infiltration rate increases with rainfall rate, because of the spatial variability of infiltration rates. As the rainfall rate increases, a greater proportion of the plot will have infiltration rates lower than the rainfall rate and therefore contribute runoff. This increase in contributing area will only stop when the rainfall rate is greater than the maximum value of infiltration rate for the plot. Thus, the values for rainfall-simulation infiltration would only be expected to be the same as for cylinder infiltration if the rainfall rate is higher than the maximum infiltration value. This positive correlation between final infiltration (i_{fin} in mm h^{-1}) and rainfall rate (r_f in mm h^{-1}) can be seen in the experimental results (Fig. 3), with the relationship:

(1) $\quad i_{fin} = -28.5 + 409.0\,(1e^{-r_f/409.0})$,

which is a slightly modified version of the form developed by Hawkins, explaining 95.7 % of the variance in the infiltration rate. However, this diagram also shows that the maximum infiltration rate is not reached (the maximum value of infiltration of 31.9 mm min^{-1} from the cylinder infiltrometry suggests this would require a rainfall rate of at least 1914 mm h^{-1}), and therefore the rainfall simulation values would be expected to be less than the cylinder infiltration ones according to this model. The rates measured here are probably of the same order of magnitude as those reported for other *terres noires* by BUFALO & NAHON (1992), who found that little runoff was produced on vegetated surfaces with rainfall at 46 mm h^{-1} for two hours (mean plot vegetation plus litter cover for the present

a.

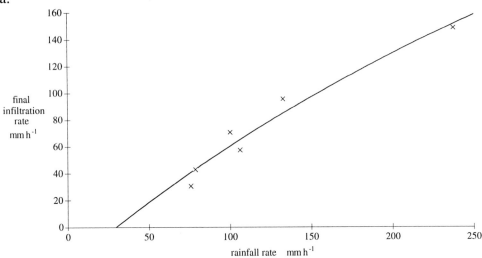

b.

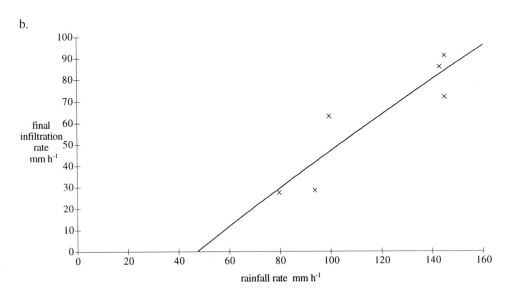

Fig. 3. Relationship between rainfall intensity and final infiltration rate at the Propiac (a) and St. Sernin (b) study areas.

experiments was 17.6 %; by ANTOINE et al. (1995) who present values of 2.1 to 16.7 mm h^{-1}; and by Caris (in ANTOINE et al. 1995) who cites values of 62.5 to 166.7 mm h^{-1}. These relatively high rates contrast with the low values reported for marls in Spain (SCOGING & THORNES 1979). A similar model is found to be a good predictor (adjusted r^2 = 0.828) at St. Sernin:

(2) $i_{fin} = -45.5 + 636.0\,(1e^{-r_f/636.0})$.

The results presented here from Propiac suggest that three commonly used techniques for measuring final infiltration rates can produce results which vary by an order of magnitude. Considerations of both the spatial sampling implied by the techniques, and the processes involved imply that these results are more than the product of experimental errors. Therefore, care must be taken when using infiltration data for surface-hydrology or soil-erosion modelling. Obviously, the most applicable results will be obtained using rainfall simulation. However, this technique is time-consuming and difficult to carry out in more remote locations. In this case, for extensive studies, it may be useful also to carry out cylinder infiltration studies, especially if spatial variability is a concern. It is suggested that K_{sat} values derived from falling head samples will always underestimate the final infiltration rate, and therefore that such K_{sat} values should be restricted to use in subsurface flow phenomena.

Comparison of the two sets of infiltration parameters from the rainfall simulation experiments were carried out using t tests. Neither the final infiltration rate ($p = 0.552$) nor the rate of decline of infiltration rate through time (using the simplified Green and Ampt equation; $p = 0.214$) show any significant difference between the two locations. These results imply that, although the characteristics of the surfaces at Propiac and St. Sernin are markedly different lithologically, their infiltration response to rainfall is similar.

3.2 *Runoff characteristics*

The values of the Darcy-Weisbach friction factor were compared with flow and surface characteristics using stepwise multiple regression. For the Propiac study area, the use of the Reynolds number (Re) explained 66 % of the variance in the friction factor (f: Fig. 4). Addition of the ground surface cover properties (percentage litter cover, L%) increases the explained variance to 70.1 %, producing the following relationship:

(3) $\quad f = 1.83 \times 10^5 \, Re^{-2.571} \, 10^{0.06 L\%}$.

The scatter the St. Sernin data is much greater, with no significant relationship between the friction factor and the Reynolds number. The surface cover variables do however explain 55.5 % of the variance in the friction factor:

(4) $\quad f = 7.85 \times 10^4 \, 10^{(0.234 \, V\% - 0.149 \overline{G})}$.

where V% is the proportion of the surface covered by vegetation and $\overline{G}$ is the mean stone size in mm.

A comparison of the Propiac and St. Sernin results (Fig. 4) suggests a different response to changing flow conditions. Although there is no significant f–Re relationship for the St. Sernin plots, this is largely due to the scatter at low values of Re, and there is a general trend towards lower friction factors with higher Reynolds numbers. For individual experimental plots, this inverse relationship is clearly marked in three cases at Propiac (Fig. 5), whereas the majority of cases at St. Sernin show complex or invariant curves (cf. ABRAHAMS et al. 1994). The surface cover variables providing explanation also differ between the two areas, due to the differing characteristics of the surfaces at the

a.

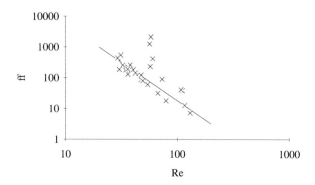

b.

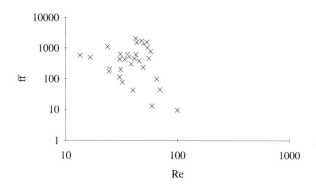

Fig. 4. Comparison of measured friction factor – Reynolds number relationships at the Propiac (a) and St. Sernin (b) study areas. The line in (a) represents the relationship $f_r = 1.67 \times 10^6 \, Re^{-2.484}$ ($r^2 = 0.66$).

two sites. However, the ranges of friction factors are markedly different. The values are significantly lower at Propiac (mean 293.25 compared with 585.25 at St. Sernin, with standard deviations of 436.66 and 547.62, respectively; p = 0.028). Consequently, the Reynolds numbers of the experimental flows at Propiac are significantly higher (mean 55.99, standard deviation 28.71), than those at St. Sernin (mean 42.55, standard deviation 17.27: p = 0.028). This may be because the larger stone size at Propiac is more effective at channeling flows. Thus, given the similar infiltration rates, there is a difference in the runoff behaviour in the two areas, with the Propiac surface offering less resistance, and therefore producing higher velocity flows for the same runoff.

3.3 Sediment transport

The sediment transport rate (q_s, in kg m^{-2} s^{-1}) may be explained as a function of the flow discharge (q in m^3 s^{-1}) and the slope, which is a means of expressing the power of the flow (JULIEN & SIMONS 1985). For the Propiac data, a large proportion of the variance is explained by the discharge and slope (s in m m^{-1}), with no significant addition from surface characteristics:

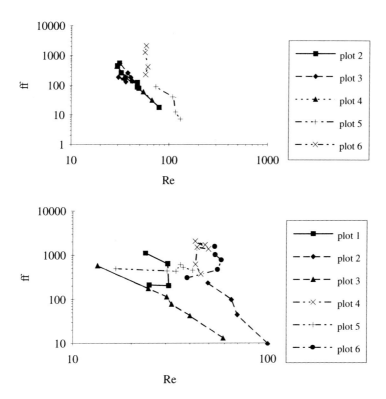

Fig. 5. Comparison of measured friction factor – Reynolds number relationships by experimental plot at the Propiac (a) and St. Sernin (b) study areas.

(5) $\quad q_s = q^{0.827} s^{0.191}$

with a standard error of the estimate of 0.258 log units (Fig. 6).
At St. Sernin, the discharge and slope explain 38.5 % of the variance:

(6) $\quad q_s = 3.38 \times 10^{-5} q^{0.205} s^{0.674}$.

Adding the percentage of the surface cover of fines (F%) increases the explained variance to 51.6%:

(7) $\quad q_s = 2.08 \times 10^{-5} q^{0.861} s^{0.113} F\%^{0.392}$.

EVERAERT (1991) has suggested the use of alternative models for sediment transport rate. The excess shear velocity and effective stream power models were also tested on the data sets here, but only in the case of the St. Sernin data are significant fits found, but in both cases the explained variance is low, with values of 20.3 % and 36.1 % respectively.

The amounts of splash through time at Propiac were calculated as the mean rates over a five-minute period (Fig. 7; all splash rates have been corrected for the surface area of the collector following TORRI & POESEN 1988). Although there is some variability,

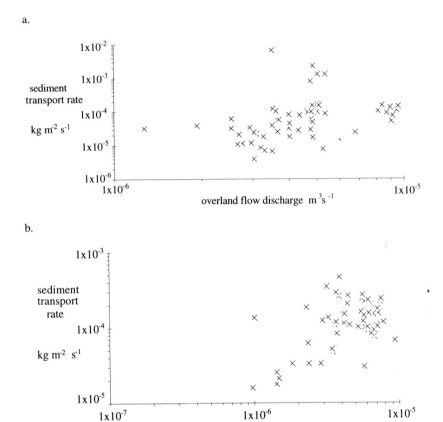

Fig. 6. Measured overland flow sediment discharge at the Propiac (a) and St. Sernin (b) study areas.

the rates are generally high at the start of the experiment (ranging from 2.8 to 24.2 g m^{-2} min^{-1} in the first five minutes of the experiment), and decrease through time, so that the values in the last five minutes of the experiments range from 0.4 to 8.4 g m^{-2} min^{-1}. This decrease through time may either be explained as the result of increasing flow depth reducing the raindrop energy impacting on the soil surface (TORRI et al. 1987), or to exhaustion effects, where the raindrop energy has already detached all the mobile elements at the ground surface (ELLISON 1945). In both these cases, the decrease in the rate of raindrop detachment leads to reduced transport by splash.

Stepwise multiple regression was used to investigate the controlling variables of the splash rate and therefore suggest the extent to which these alternatives are appropriate. Surface characteristics explain 87.4 % of the variability in the splash rates. Of this, 58.2 % is due to the surface covered by fines, probably because splash rates are minimal with particles larger than 2 mm, and zero beyond an upper threshold of 12–20 mm (WAINWRIGHT et al. 1995, POESEN & SAVAT 1981). Other significant surface factors are the mean grain size of the coarse fraction, the shear strength of the fine fraction at saturation

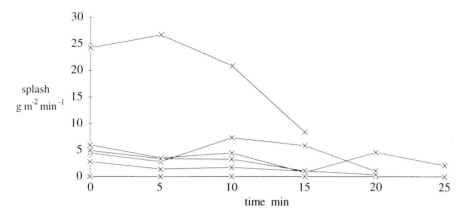

Fig. 7. Splash erosion rates measured at Propiac.

(σ_w in kPa), both reflecting surface resistance, and the percentage cover by vegetation and litter (which afford protection of the surface from raindrop energy). In the final relationship:

(8) $\qquad q_{sp} = 0.003\ F\% - 0.002\overline{G} + 0.007\sigma_w - 0.003\ V\% - 0.002\ L\% - 9.622\ \overline{d}$

only an additional 1.4 % variance in the splash rate is accounted for by the mean flow depth term (d in m). This suggests that the surface depletion effects are far more significant than the surface energy effects. Furthermore, the mean depth of the interrill flows generated was only 2 to 3 mm, and on average 44% (standard deviation 25 %) of the surface remained emergent throughout the whole of the experiments, which suggests that the protection from raindrop impact afforded by the flows was minimal. In contrast with the results presented here, BRADFORD et al. (1986) suggested a negative correlation between the rate of splash and surface shear strength. Another potential problem is the lack of a slope-dependence term, which has been demonstrated by others (e.g. MOSLEY 1973, POESEN & SAVAT 1981, QUANSAH 1981). However, TORRI & POESEN (1992) have shown that surface microtopography effects may counter-balance the slope effect. Thus a straightforward relationship between slope and splash rate may not necessarily be expected in a field situation. Although no direct measurement of microtopography was made at Propiac, surface irregularity is high, due to the presence of stones especially where splash pillars have formed.

The splash rates at Propiac vary from 0.4 g m^{-2} min^{-1} to 26.7 g m^{-2} min^{-1} (mean 5.4 g m^{-2} min^{-1}, standard deviation 6.6 g m^{-2} min^{-1}). MOSLEY (1973) measured splash on sandy badland material in the laboratory, producing values of between 9.5 g m^{-2} min^{-1} and 119.0 g m^{-2} min^{-1} for similar slope angles. The difference of approximately an order of magnitude between clayey and sandy sediments is similar to that found by POESEN & SAVAT (1981), and reflects the higher cohesion of the badland surfaces at Propiac.

The Propiac data also allow observation of the relationship between splash rates and overland-flow sediment discharge. If the supply of sediment to interrill overland flow is predominantly controlled by the amount of sediment detachment by raindrops (PARSONS

Table 2. Comparison of measured annual erosion rates at various badland locations in the Mediterranean region.

location	lithology	measured lowering rate mm a^{-1}	source
Propiac, France	marl	2 - 14	this study
St. Genis, France	marl	10 - 20	Bufalo et al., 1989
Vallcebre, Spain	mudrock	9	Solé et al., 1992
Ebro basin, Spain	shales and Holocene sediments	5 - 9	Benito et al., 1992
Ugijar, Spain	sands, marls and conglomerates	6 - 27	Scoging, 1982
Basilicata, Italy	marine clays:		
	calanchi	5.3 - 13.6	Alexander, 1982
	biancane	22.8 - 39.7	
Zin, Israel	loess	0.75 (over 75 ka)	Yair et al., 1982

et al. 1991, 1994), then by calculating the ratio of sediment transported in the flow to the sediment splashed, the effectiveness of the flow to carry the detached sediment may be estimated. The range of vaues at propiac is from 0.04 % to 14176 %, with a median value of 67.8 %. In eleven out of 28 samples, the ratio is greater than 100 %, indicating either that the splash rate significantly underestimates the true detachment rate (see WAINWRIGHT et al. 1995) or that flow detachment is taking place on these surfaces. The values much greater than 100 % in eight of these eleven cases suggest that the latter explanation is valid at least for part of the time. These results suggest that further work is required to clarify the proportional rôles of raindrop and detachment in the soil erosion process on badland surfaces.

The rates of sediment transport in overland flow at Propiac and St. Sernin are compared in Fig. 6. For the most part, the range of values falls between rates of 1×10^{-5} and 1×10^{-3} kg m^{-2} s^{-1}. The relationships between flow and sediment discharge for the two areas (equations 5–7) indicate that the sediment transport reacts to increases in discharge at approximately the same rate for both sites (the exponent of q is 0.827 for Propiac [equation 5] and 0.861 for St. Sernin [equation 7]). These values are significantly lower than others measured in field (JULIEN & SIMONS 1985) or laboratory (EVERAERT 1991) conditions. Measurements from gully systems in *terres noires* are presented by MATHYS et al. (1989), who report good correlations between total sediment transported and maximum discharge, and erosion rates and rainfall intensity. BUFALO et al. (1989) have measured erosion rates of 185 to 284 t ha^{-1} a^{-1} in gully outflows in *terres noires*. These rates are broadly comparable with those measured at Propiac (see Table 2).

The rôle of extreme events at Propiac may be demonstrated with reference to measurements of annual erosion rates on badlands at Propiac and at similar settings in the

Mediterranean. In the year following the 1992 event (May 1993 to April 1994, which experienced 1079.5 mm of rainfall compared with an annual average of 733 mm), erosion pins were used to measure the annual point erosion rate. Rates varied between 2 and 14 mm These values are closely comparable for other measurements in the Mediterranean region (Table 2). Extrapolation of the erosion rate results measured above suggest that between 4 and 54 mm were lost in the interrill zone in the event of 22nd September 1992. Although this is the equivalent of several times the annual rate, in no way could it really be described as catastrophic. In this sense these particular badlands, although still producing rapid runoff and erosion, could be described as being relatively in equilibrium. This is further suggested by the fact that the return periods of the 179 mm that fell at Vaison-la-Romaine and the 212 mm at Carpentras on 22nd September 1992 have return periods of 87 and 466 years respectively. By comparison, the nearest available rainfall stations to St. Sernin show that high intensity rainfall is much less frequent. Both Albi and Carmaux have return periods of approximately 200 years for a 100 mm event. Thus, although the response to extreme rainfall is similar to the Propiac site, such rainfall is much rarer, and this may explain the fact that the rounded form of the St. Sernin badlands is much more characteristic of splash as a dominant process (MOSLEY 1973).

4 Conclusions

The results show that it is possible to characterize the infiltration, runoff and erosion response of the interrill areas of the badland hillslopes by means of a single, integrated plot study. The experimental design used here is effective in that it enables the simultaneous measurement of the required variables, and thus can assess the interactions between variables.

The infiltration response of the two areas is comparable. The similar ranges of values are most probably related to the infiltration into the clay matrix of the slopes. Once flows have been generated, however, there is a divergence in the response of the slopes in the two areas. The St. Sernin slopes offer more frictional resistance, leading to lower sediment-transport rates. The splash results from Propiac have quantified the relatively high rates of splash from the badland surfaces (cf. HOWARD 1994), as may be adduced by the widespread development of splash pillars on the surface.

The small plot design presented here shows the effective development of a strategy for characterizing slope responses to rainfall events. However, the results presented shows that it also has its disadvantages. The most significant of these is the inability to derive expressions for surface roughness once flow accumulation has taken place. It may be possible to overcome this problem by using longer plots, whilst still maintaining the control on the spatial pattern of the surface characteristics.

Acknowledgements

This research has been supported by grants from NERC (GR9/1058) and the University of Southampton. I wish to thank Laurent and Jean-Michel Carozza for their assistance in the St. Sernin set of experiments. Rainfall data were provided by Météo-France.

References

ABRAHAMS, A.D., A.J. PARSONS & J. WAINWRIGHT (1994): Resistance to overland flow on semiarid grassland and shrubland hillslopes, Walnut Gulch, Southern Arizona. – J. Hydrol. **156**: 431–446.

AL-DURRAH, M.M. & J.M. BRADFORD (1982): Parameters for describing soil detachment due to single waterdrop impact. – Soil Sci. Soc. Amer. J. **46**: 836–840.

ALEXANDER, D. (1982): Difference between "calanchi" and "biancane" badlands in Italy. – In: BRYAN, R.B. & A. YAIR (eds.): Badland Geomorphology and Piping. – 71–85, GeoBooks, Norwich.

Anon. (1993): Echos de l'Ouvèze et du Ventoux. – Les Carnets du Ventoux **14**: 22–25.

ANTOINE, P., A. GIRARD & T. VAN ASCH (1995): Geological and geotechnical properties of the "Terres Noires" in southeastern France. Weathering, erosion, solid transport and instability. – Eng. Geol. **40**: 223–241.

ARNAUD-FASSETTA, G., J.-L. BALLAIS, E. BEGHIN, M. JORDA, J.-C. MEFFRE, M. PROVENSAL, J.-C. RODITIS & S. SUANEZ (1993): La crue de l'Ouvèze à Vaison-la-Romaine (22 septembre 1992). Ses effets morphodynamiques, sa place dans le fonctionnement d'un géosystème anthropisé. – Rev. Géomorph. Dynam. **42**: 34–48.

BENECH, B., H. BRUNET, V. JACO, M. PAYEN, J.-C. RIVRAIN & P. SANTURETTE (1993): La catastrophe de Vaison-La-Romaine et les violentes précipitations de septembre 1992: aspects météorologiques. – La Météorologie **8**: 72–90.

BENITO, G., M. GUTIÉRREZ & C. SANCHO (1992): Erosion rate in badland areas of the Central Ebro basin (NE-Spain). – Catena **19**: 269–286.

BOUWER, H. (1986): Intake rate: cylinder infiltrometer. – In: KLUTE, A. (ed.): Methods of Soil Analysis, Part 1. Physical and Mineralogical Methods. – 825–844, Agronomy Monograph n° 9, Madison.

BRADFORD, J.M., P.A. REMLEY, J.E. FERRIS & J.B. SANTINI (1986): Effect of soil surface sealing on splash from a single waterdrop. – Soil. Sci. Soc. Amer. J. **50**: 1457–1462.

BRYAN, R.B. & A. YAIR (eds.): Badland Geomorphology and Piping. – 408 pp., GeoBooks, Norwich.

BUFALO, M. & D. NAHON (1992): Erosional processes of Mediterranean badlands: a new erosivity index for predicting sediment yield from gully erosion. – Geoderma **52**: 133–147.

BUFALO, M., C. OLIVEROS & R.E. QUÉLENNEC (1989): L'érosion des Terres Noires dans la région du Buëch (Hautes-Alpes). Contribution à l'étude des processus érosifs sur le bassin versant représentatif (BVRE) de Saint-Genis. – La Houille Blanche **1989** (3/4): 193–195.

COLLOMB, P., H. GRAS, M. DURAND-DELGA, B. DELSAHUT, R. CUBAYNES, P., MOULINE & J.P. PARIS (1989): Notice Explicative de la Feuille Albi à 1/50 000. – 56 pp., Éditions du BRGM, Orléans.

ELLISN, W.D. (1945): Some effects of raindrops and surface flow on soil erosion and infiltration. – Transact. Amer. Geophys. Union **26**: 415–429.

EVERAERT, W. (1991): Empirical relations for the sediment transport capacity of interrill flow. – Earth Surf. Proc. Landf. **16**: 513–532.

FLAGEOLLET, J.C., P. DE FRAIPOINT, P. GOURBESVILLE, N. THOLEY & J. TRAUTMAN (1993): La crue de l'Ouvèze à Vaison-la-Romaine de septembre 1992: origines, effets, enseignements. – Rev. Géomorph. Dynam. **42**: 57–72.

HAWKINS, R.H. (1982): Interpretations of source area variability in rainfall-runoff relations. – In: SINGH, V.P. (ed.): Rainfall-Runoff Relationship. – 303–323, Water Resources Publications, Littleton.

HOWARD, A.D. (1994): Badlands. – In: ABRAHAMS, A.D. & A.J. PARSONS (eds.): Geomorphology of Desert Environments. – 213–242, Chapman and Hall, London.

JULIEN, P.Y. & D.B. SIMONS (1985): Sediment transport capacity of overland flow. – Transact. Amer. Soc. Agricul. Eng. **28**: 755–762.

KLUTE, A. & C. DIRKSEN (1986): Hydraulic conductivity and diffusivity laboratory methods. – In: KLUTE, A. (ed.): Methods of Soil Analysis, Part 1. Physical and Mineralogical Methods. – 687–734, Agronomy Monograph n° 9, Madison.
MATHYS, N., N. MEUNIER & C. GUET (1989): Mesure et interprétation du processus d'érosion dans les marnes des Alpes du Sud à l'échelle de la petite ravine. – La Houille Blanche 1989 (3/4): 188–192.
MONIER, P. & C. CAVELIER (1991): Notice Explicative de la Feuille Vaison-la-Romaine à 1/50 000. – 55 pp., Éditions du BRGM, Orléans.
MOSLEY, M.P. (1973): Rainsplash and the convexity of badland divides. – Z. Geomorph. Suppl. 18: 10–25.
MUNN, J.R. & G.L. HUNTINGTON (1976): A portable rainfall simulator for erodibility and infiltration measurements on rough terrain. – Soil. Sci. Soc. Amer. J. 40: 622–624.
PARSONS, A.J., A.D. ABRAHAMS & S.-H. LUK (1991): Size characteristics of sediment in interrill overland flow on a semi-arid hillslope, southern Arizona. – Earth Surf. Proc. Landf. 16: 143–152.
PARSONS, A.J., A.D. ABRAHAMS & J. WAINWRIGHT (1994a): Rainsplash and erosion rates in an interrill area on semi-arid grassland, Southern Arizona. – Catena 22: 215–226.
– – – (1994b): On determining resistance to interrill overland flow. – Water resources Research 30: 3515–3521.
POESEN, J. & J. SAVAT (1981): Detachment and transportation of loose sediments by raindrop splash. Part II: Detachibility and transportibility measurements. – Catena 8: 9–41.
POULOVASSILIS, A., P. KERKIDES, S. ELMAGLOU & I. ARGYROKASTRITIS (1991): An investigation of the relationship between ponded and constant flux rainfall infiltration. – Water Resources Research 27: 1403–1409.
QUANSAH, C. (1981): The effect of soil type, slope, rain intensity and their interactions on splash detachment and transport. – J. Soil Sci. 32: 215–224.
RAUWS, G. & G. GOVERS (1988): Hydraulic and soil mechanical aspects of rill generation on agricultural soils. – J. Soil Sci. 39: 111–124.
SCOGING, H.M. (1982): Spatial variations in infiltration, runoff and erosion on hillslopes in semi-arid Spain. – In: BRYAN, R.B. & A. YAIR (eds.): Badland Geomorphology and Piping. – 89–112, GeoBooks, Norwich.
– (1989): Runoff generation and sediment mobilisation by water. – In: THOMAS, D.S.G. (ed.): Arid Zone Geomorphology. – 87–116, Bellhaven Press, London.
SCOGING, H.M. & J.B. THORNES (1979): Infiltration characteristics in a semiarid environment. – In: The Hydrology of Areas of Low Precipitaiton. – 159–167; IAHS Publication n° 128, Wallingford.
SOLÉ, A., R. JOSA, G. PARDINI, R. ARINGHIERI, F. PLANA & F. GALLART (1992): How mudrock and soil physical properties influence badland formation at Vallcebre (Pre-Pyrenees, NE Spain). – Catena 19: 287–300.
TORRI, D. & J. POESEN (1988): The effect of cup size on splash detachment and transport measurements. Part II: Theoretical Aproach. – In: IMESON, A. & M. SALA (eds.): Geomorphic Processes in Environments with Strong Seasonal Contrasts, Volume 1: Hillslope Processes. – Catena Suppl. 12: 127–137, Braunschweig.
– – (1992): The effect of soil surface slope on raindrop detachment. – Catena 19: 561–578.
TORRI, D., M. SFALANGA & M. DEL SETTE (1987): Splash detachment: runoff depth and soil cohesion. – Catena 14: 149–155.
WAINWRIGHT, J. (1996a): Infiltration, runoff and erosion characteristics of agricultural land in extreme storm events, SE France. – Catena 26: 27–47.
– (1996b): Hillslope response to extreme storm events: the example of the Vaison-La-Romaine event. – In ANDERSON, M.G. & S.M. BROOKS (eds.): Advances in Hillslope Processes. – Vol. 2: 997–1026. John Wiley and Sons, Chichester.
WAINWRIGHT, J., A.J. PARSONS & A.D. ABRAHAMS (1995): A simulation study of the rôle of raindrop erosion in the formation of desert pavements. – Earth Surf. Proc. Landf. 20 277–291.

WATTS, G.P. (1989): Modelling the Subsurface Hydrology of Semi-Arid Agricultural Terraces. – Unpubl. PhD Thesis, Bristol University.

YAIR, A., P. GOLDBERG & B. BRIMER (1982): Long term denudation rates in the Zin-Havarim badlands, northern Negev, Israel. – In: BRYAN, R.B. & A. YAIR (eds.): Badland Geomorphology and Piping. – 279–291, GeoBooks, Norwich.

Address of the author: JOHN WAINWRIGHT, Department of Geography, King's College London, Strand, London WC2R 2LS, UK

Ploughing blocks as evidence of down-slope sediment transport in the English Lake District

ROBERT J. ALLISON, Durham, and KATE C. DAVIES, Southampton

with 12 figures and 3 tables

Summary. Ploughing blocks are often found in upland areas, particularly where the ground is frozen for part of the year. They evolve due to the down-slope movement of regolith and are indicators of sediment transport on slopes. Data presented here elucidate ploughing block characteristics in the English Lake District. Their quantitative examination has seldom been undertaken and their evolution remains obscure. The three morphological components of ploughing blocks, namely the block itself, fronting ridge and rear hollow, were examined by measuring some 26 variables for 150 features. The development of ploughing blocks can be related to key controlling variables such as slope angle but the causal relationships are not always those which might be expected. The location and development of ploughing blocks are useful indicators of variations in mass wasting processes and can be used to define the lower boundary of cryergenic processes as an agent of down-slope sediment movement.

Résumé. Le phénomène de glissements de terrain est souvent observé dans les régions montagneuses, particulièrement où le sol est gelé la majeure partie de l'année. Ils se dévelopent grâce au mouvement des régolithes vers le bas de la pente et sont des indicateurs de transport des sédiments sur les pentes. Les données présentées ici, expliquent les caractéristiques des glissement de terrain dans le Lake District anglais. Leur érude quantitative a rarement été enterprise et leur évolution reste obscure. Les 3 composantes morphologiques des glissements de terrain c'est-à-dire le bloc lui-même, la crête frontale et la depression arrière, ont été examinées en measurant 26 variables pour 150 caractéristiques. Le développement des glissements de terrain peut-être associé des variables de contrôle clefs, telles que l'angle de la pente mais les relations de causes ne sont pas toujours celles qui sone prévues. La localisation et le dévelopment des glissements de terrain sont des indicateurs utiles des variations des mouvements de solifluxion et peuvent être utilisés pour définir les limites les plus basses des procédés cryogéniques comme agent de mouvent des sédiments vers l'aval des pentes.

Introduction

Ploughing blocks are isolated fragments of usually boulder-size rocks, which travel down-slope at a greater velocity than their surrounding material. The result of the differential transport rate is a ridge of soil material immediately in front of the block and a depression to the rear. Optimum conditions seem to exist for the development of ploughing blocks. It appears that regions where the ground is temporarily frozen at certain times of the day of for periods during the year are most conducive to their formation. Ploughing blocks can frequently be found in upland areas and it has been suggested (TUFNELL 1969) that ploughing blocks are amongst the most widespread of the currently developing slope phenomena in periglacial environments.

Fig. 1. A ploughing block with associated mound and depression.

A study of ploughing blocks can give valuable information on slope movements and is also of use when defining the lower boundary of morphologically significant frost action as a process promoting the down-slope transfer of regolith. Ploughing blocks have seldom been examined in detail and there are few quantitative accounts of their morphology and development. This study provides data for ploughing blocks, using the English Lake District as the study area. By characterising the form of ploughing blocks, examining their spatial distribution and discussing the data in the light of other studies, an explanation of this type of slope movement can be made and a model tentatively proposed for ploughing block activity.

Ploughing blocks

REKSTAD (1909) and HOGBOM (1914) were among the first to record ploughing block rock movement. In the UK ploughing blocks have been recorded in the Lake District (HAY 1937, 1942), on the Moor House National Nature Reserve (JOHNSON & DUNHAM 1963, TUFNELL 1972) and in Scotland (GALLOWAY 1961, RAGG & BIBBY 1966, CHATTOPADHYAY 1983). Most previous studies afford no more than a bare outline of ploughing block characteristics. CHATTOPADHYAY (1983) gives quantitative information on the distribution and characteristics of ploughing blocks on the Drumochter Hills of the Scottish Highlands. The occurrence of ploughing blocks is also recorded in the Alps (CAPELLO 1955, PISSART 1964), the Apennines (HOLLERMANN 1964), Norway (CAILLEUX & TAYLOR 1954), Sweden (RUDBERG 1958, 1962), Spain, Greece, the Pyrenees and Massif Cen-

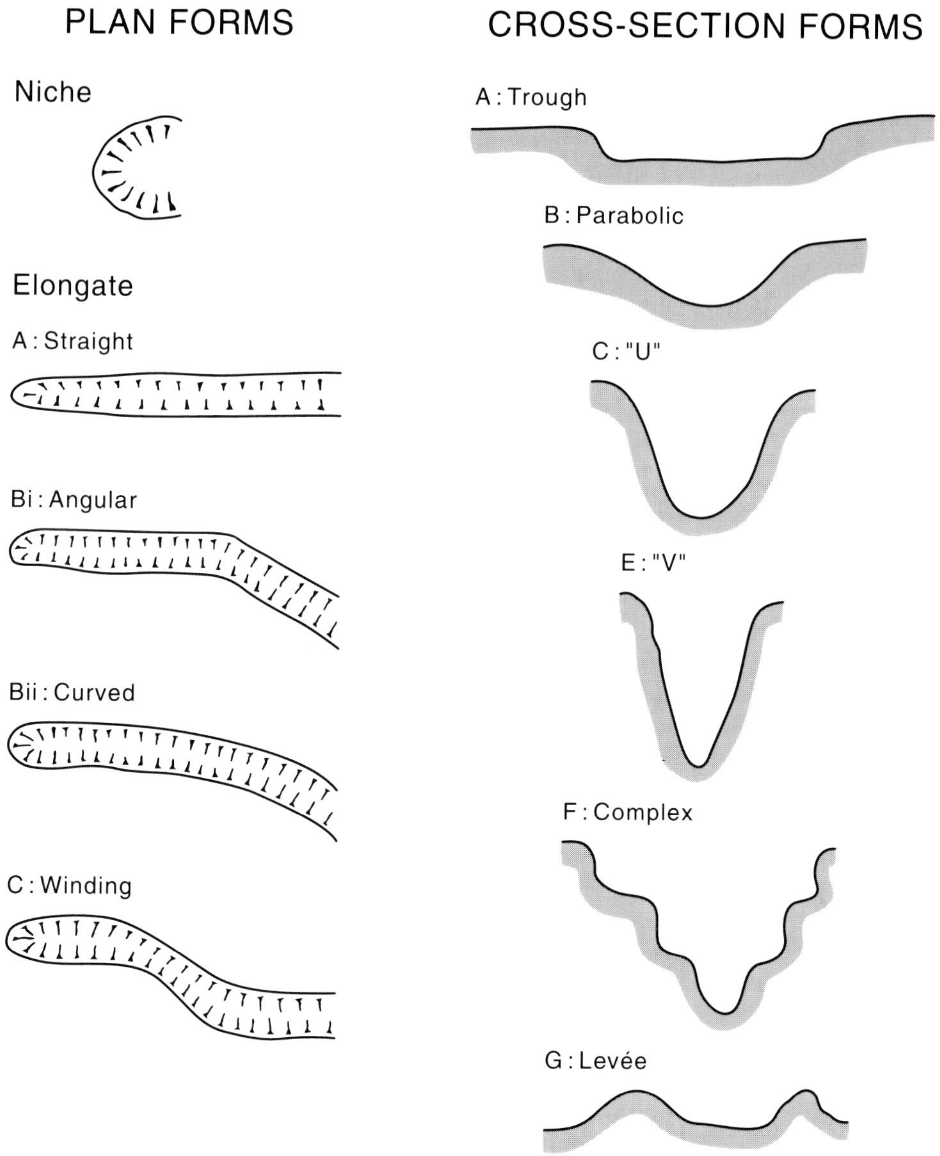

PLAN FORMS

Niche

Elongate

A: Straight

Bi: Angular

Bii: Curved

C: Winding

CROSS-SECTION FORMS

A: Trough

B: Parabolic

C: "U"

E: "V"

F: Complex

G: Levée

Fig. 2. Sketch representation of types of depression associated with ploughing blocks.

tral, central and southern Germany, Rumania and Czechoslovakia (GORBUNOV 1991). Ploughing blocks are documented in the Tien Shan (GORBUNOV 1970, 1980, 1991), in the Himalayas (IWATA 1976), the mountains of south-eastern Iran (KUHLE 1974) and in polar regions (WASHBURN 1979).

TUFNELL (1972) defined the phrase *ploughing block* to denote the blocky components of periglacial slope movements which travel faster than their surroundings, force up the ground before them and leave a depression to the rear. Other terms used include *gliders* or *gliding blocks* (HAY 1937, 1942). In Germany the features are referred to as *wanderblock* an in Russia *wandering, voyaging, creeping* or *floating* rocks, blocks or mounds (GORBUNOV 1991). The term *ploughing block* implies a specific type of movement and landform, including the block itself, the material pushed up in front of it and the depression to the rear (Fig. 1). The movement of ploughing blocks appears to be controlled by four main mechanisms. First, frost heave and temperature related expansion and contraction of the block. Second, the intensity of freeze-thaw activity at the front of the block where the snow melts first, due to block heat storage and conduction. Third, the abundant supply of misture to the soil beneath the block from the depression where snow is late-lying. Finally, a basal frozen layer in the early period of seasonal thaw can promote sliding.

Ploughing blocks have three components. There is the rock block, the up-slope depression and the mound of material pushed up in front of the boulder. Several different types of depression and mound can be defined. Depression length appears to be determined by the speed of block movement and how actively the processes of gelifluction and frost creep operate to obliterate it. Depressions differ in plan and cross-section shape (Fig. 2). The principal factors that determine depression shape are soil texture and moisture content, type of soil and vegetation cover, velocity and nature of ploughing block movement, the block shapes and dimensions and the steepness and micro-relief of the hillsides (GORBUNOV 1991). For example niche depressions are likely to form if the movement of the block only just exceeds that of the surrounding material or if a once moving block becomes virtually immobile and its elongate depression is gradually reduced by other slope processes (TUFNELL 1972). Elongate depressions may be straight, angular, curved or winding. Trough cross-sections are varied (Fig. 2) and some are more common than others. Changes in cross-section along the length of a depression (CHATTOPADHYAY 1983) may be due to the block sinking into the ground as it moves down slope or the effects of other slope processes. Previous investigations (Table 1) suggest that furrow depth remains virtually constant throughout the length of a depression. Depressions are mostly vegetated, modified by other slope processes and grade into the non-deformed hillside. One aspect of this reasearch is to distinguish between different types of depression and determine whether similar patterns of frequency exist for plan and cross-sectional forms.

A mound develops immediately in front of a ploughing block. As with depressions, different types of mound can be defined (Fig. 3). Some are single mounds but they can vary in morphology from one frontal ridge, fronting and lateral ridges or a lateral ridge only. The most common type recorded in other studies (TUFNELL 1872, CHATTOPADHYAY 1983, GORBUNOV 1991) is where the mound surrounds the block both laterally and frontally. Frontal/lateral mounds form where the ploughing block is deeply embedded in the ground, thereby pushing up material. Mounds may be separated from the ploughing block by a small gap, caused by the up-slope settling of a bock. More complex double mounds have two ridges of material, with occasional variations in the size of each ridge and the spacing between them. The average heights of mounds, together with the most dominant mound type reported in previous studies are presented in Table 1.

Table 1. Characteristics of ploughing blocks noted in other studies.

	Lufteilbahn Fiesch - Eggishorn, Switzerland	Moor House Reserve, NE Westmorland, England	Drumochter Hills, Grampian Highlands, Scotland	Zailyskoye Ala Tau, Tien Shan Mountains
Source	Tufnell, 1972	Tufnell, 1972	Chattopadhyay, 1983	Gorbunov, 1991
Sample site	50	500	100	170
Altitude (m)	2200	685 - 840	650 - 910	2400 - 2800
Gradient ($\chi°$)				
range	9 - 32	4 - 40	*	5 - 30
mean	18.4	18.1	*	20 - 30
Block length (cm)				
range	30 - 370	14 - 250	30 - 230	* - 360
mean	146	65	103	100 - 150
Depression length (cm)				
range	30 - 2010	5 - 800	50 - 560	* - 3700
mean	320	76	239	2600 - 2800
Mound height (cm)				
range	*	*	9 - 55	200 - 500
mean	*	*	23	*
Mound type (mode)	B	B	B	B
Depression form (mode)	Parabolic	Trough	Trough	Parabolic / Trough
Long axis alignment to maximum slope ($\chi°$)	0 (83%)	0 (82%)	0 - 15 (52%)	0 (95%)

* data not available

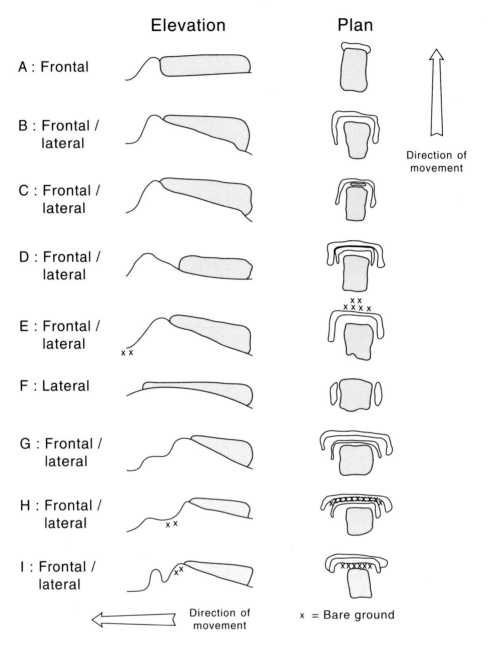

Fig. 3. Sketch representation of types of mound associated with ploughing blocks.

A further characteristic of ploughing blocks is their orientation. Most often, the long axes of blocks are parallel to the slope. On the Moor House Reserve 17% of the blocks had their long axis perpendicular to the steepest slope (TUFNELL 1972); in the Zailiyskoye Ala Tau 95% of the blocks were similarly aligned (GORBUNOV 1991). Blocks aligned away from the maximum slope have often been influenced by factors other than those usually responsible for ploughing block evolution including obstacles and sudden changes in regolith properties.

If the morphological characteristics of ploughing blocks are considered, there must be a set of conditions which are conducive to their development including the correct altitude and latitude for a suitable process environment, the form and dimensions of the block itself and the morphology of the surrounding terrain. The lower altitudinal boundary is determined by the intensity of cryergenic activity. The upper limit results from excessive periglacial action and the development of permafrost.

Past studies suggest a set of optimum block characteristics (TUFNELL 1972). The optimum shape is the rectangular prism. Deviations are likely to slow the rate of movement. The optimum block size varies with the prevailing site conditions. Small blocks may not posses enough weight to form an associated mound and depression. Large blocks need a greater amount of frost action to move them. Ploughing blocks usually occur on grassy slopes where there is sufficient though not necessarily abundant moisture. Vegetation binds slope debris, thus retarding its movement, while larger material which protrudes above the surface can counteract the vegetation effect and move faster than its surroundings. Where vegetation is sparse smaller fragments of rock can move faster than the blocks, leading to a build-up of debris behind the block. Slope angle is another critical factor in the formation of ploughing blocks. The optimum gradient varies with the situation. Very steep slopes encourage the rolling or bouncing of boulders and the risk of slope failure increases. Gentle slopes are unlikely to provide optimum conditions due to the magnitude of processes required for movement. Besides examining the individual characteristics of ploughing blocks an attempt has been made here to explain relationships between the parameters, in order to provide a greater understanding of the conditions that give rise to ploughing block evolution. It is the combination of quantitative data for the English Lake District, comparison with other studies and a statistical analysis of process : form relationships which is significant here.

Field area

The field area lies in the northern region of the Coniston fells of the English Lake District (Fig. 4). The summits of Grey Friar, Little Carrs, Great Carrs and Swirl How are within the area. Altitudes range from just over 800 m on the summit of Swirl How to 400 m. the total ground surface area is approximately 4.5 km^2.

Rocks are of the Borrowdale Volcanic Series and comprise tuffs, rhyolites and andesites. All the outcropping rock types can be broken down, often by gelifraction, into blocks suitable for ploughing. The Devensian glaciation had a considerable impact on the area, leaving a scoured and eroded landscape on which periglacial processes have been free to operate. The relief is varied with steep, scree slopes, rocky outcrops, crags and more gentle grassy slopes. Snow may cover the ground intermittently or continuously during the autumn and winter seasons. The Lake District has been described as having a bitter freeze-and-thaw climate which prevails for long periods over winter, with a high frequency of alternations between cold rain, sleet or driving snow (PEARSALL & PENNINGTON 19773). Even in the summer months warm and dry calm spells are rare. Vege-

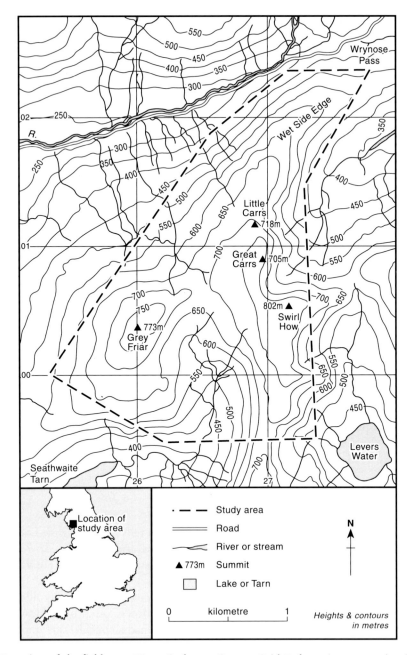

Fig. 4. Location of the field area. Note: Ordnance Survey Grid Referencing system has been used.

tation is sparse and consists mainly of grasses. Plants above shrub height are non existent. Soils are shallow, base deficient, peaty, acidic, frequently waterlogged and therefore generally infertile. The soil types found throughout the area are peaty podzols with a high water content, sandy loam texture and a depth from 40 cm to 60 cm.

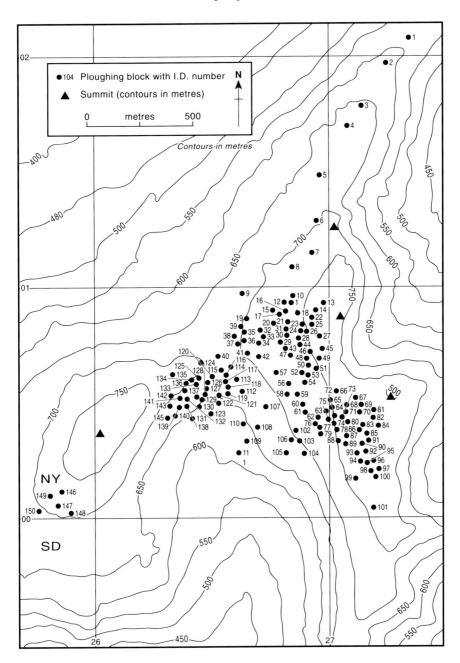

Fig. 5. Ploughing block distribution in the field area. Note: Ordnance Survey Grid Referencing system has been used.

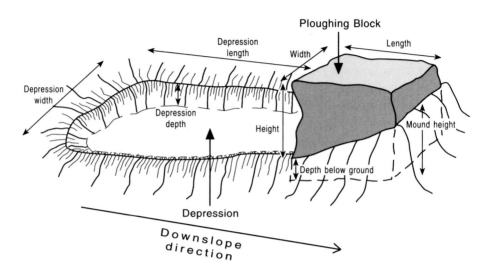

Fig. 6. Ploughing block parameters measured in the field.

Methods

Of the many ploughing blocks in the study area a sample size of 150 was chosen for analysis. Every other ploughing block encountered while traversing the study area was recorded. Markers were placed at the end of each walked transect to prevent repeated measurements. For each block 26 variables were noted. Each block was marked on a 1 : 10 000 scale map of the area and given an identification number (Fig. 5). Eastings, Northings and altitude were derived from the map. The aspect of the slope on which each ploughing block is situated was recorded to the nearest 5°. The gradient of the slope down which a block was moving was measured to the nearest 0.5°. The block, mound and depression parameters measured in the field are illustrated in Fig. 6. Block volume was determined as follows.

$$B_v = B_l \times B_w \times tB_h \qquad (Eq.\ 1)$$

where:
B_v = block volume
B_l = block length
B_w = block width
tB_h = total block height

The circumference of a block was measured to give a further indication of its size, as often the shape was irregular and the length and width were not uniform. Long axis alignment was measured using two metre rules and a protractor (Fig. 7). Seven shapes were defined (Fig. 8).

The depression length was taken as the distance from the rear of the block to the start of the depression. Depression depth is the maximum vertical depth of the hollow. Depression width is defined as the maximum cross-slope measurement of the hollow and is measured at the ground surface. The plan and cross-section form of a depresssion were

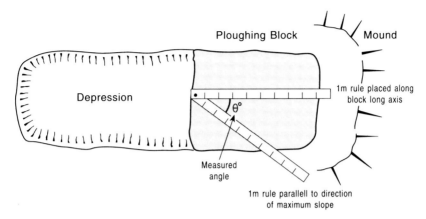

Fig. 7. Technique for measuring ploughing block long axis alignment.

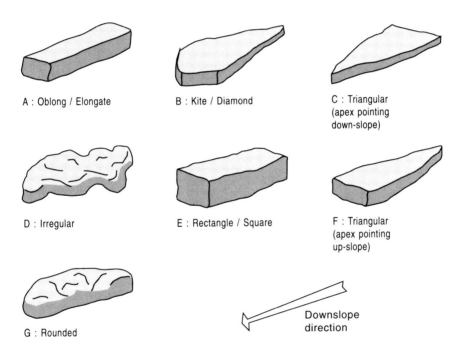

Fig. 8. Ploughing block shape defined as part of this study.

categorised based on the sub-divisions defined by TUFNELL (1972). Mound height is defined as the total maximum mound height and was measured from the top of the mound to the ground surface. In the case of double mounds, the combined height of the two parts was recorded. If there was no frontal mound, as in the case of the single lateral variety, then the height of the lateral mound was recorded. The dominant plant species on each mound were identified following FITTER et al. (1984), as was the vegetation cover

Table 2. Summary statistics for the ploughing blocks.

VARIABLE	MEAN	MEDIAN	RANGE	STANDARD DEVIATION	SKEWNESS
Easting (distance east x10m from grid line NY000)	678.2	684	575-739	31.4	-0.81
Northing (distance north x10m from gridline NY200)	60.3	55	3-208	32.1	+1.74
Altitude (m)	720.3	726	589-794	43.2	-0.79
Gradient ($\chi°$)	13.0	11.5	8-26.7	3.93	+1.24
Block length (cm)	79.7	73	14-220	31.5	+1.63
Block width (cm)	52.2	48	21-149	21.9	+1.68
Block height (cm)	12.3	10	1-74	10.1	+2.71
Block depth (cm)	13.7	12	3-36	5.88	+0.76
Total block height (cm)	25.6	23	9-93	13.0	+2.12
Block circumference (cm)	226	211	102-620	83.4	+1.42
Block volume (m^3)	0.155	0.078	0.01-2.39	0.281	+5.48
Alignment ($\chi°$)	14.4	4	0-90	23.8	+2.20
Depression length (cm)	103.9	97	10-330	59.2	+1.08
Depression depth (cm)	15.2	14	5-37	5.53	+1.12
Depression width (cm)	51.7	49	22-205	19.7	+3.71
Mound height (cm)	16.7	16	3-45	7.47	+0.89

of the area surrounding the block. Patches of ground devoid of vegetation and areas of bog moss (*Sphagnum* spp.) around a block were recorded. The data were analysed using the STATA computer statistics package.

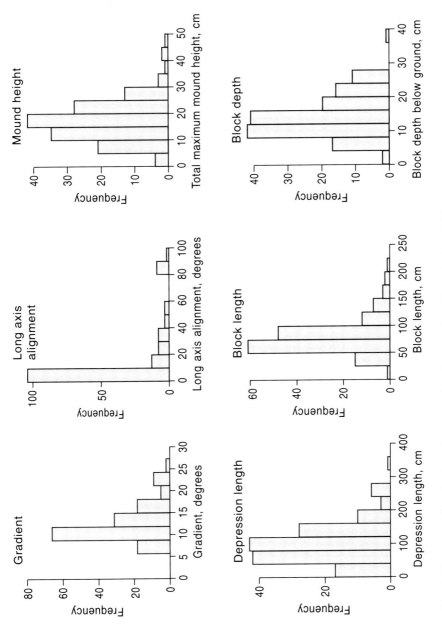

Fig. 9. Frequency distribution characteristics of some ploughing block variables.

Results

Ploughing blocks are not found on the summits of the fells. The slopes to the east of Little Carrs, Great Carrs and Swirl How are steep, unvegetated scree and ploughing blocks are absent. Blocks either move down-slope as talus or move more slowly than the surrounding material. Ploughing blocks are absent below 550 m and above 800 m. Summary statistics are shown in Table 2. The frequency distributions of the gradient, long axis alignment, mound height, depression length, block length and block depth below ground are shown in Fig. 9. The majority of ploughing blocks are found on slope angles between 9° and 15°. Over 70% of blocks have their long axes aligned with 20° of the maximum slope, suggesting that ploughing blocks show a similar alignment to gelifluction deposits. The blocks which are aligned at an angle greater than 15° have been affected by obstacles. Mound hight seldom exceeds 30 cm and all but one block have depression lengths of less than 300 cm. Mean block length is between 50 cm and 100 cm and mean depth below the ground is rarely greater than 30 cm.

The vector mean of aspect is 244.1, with a resultant length of 76.95 and a strength of 51.3%, showing a strong preferred west-south-west direction (Fig. 10). Few blocks are found on north-east facing slopes. The results suggest that aspect has little influence on ploughing block formation. However, the effects of aspect should not be totally ruled out since other factors such as slope angle may have a greater influence in this area. For example, the north-east slope adjacent to Swirl How is far too steep and rocky for the formation of ploughing blocks and periglacial action is too great. Indeed, north-east slopes often include cirques and cirque back walls, as is the case at Calf Cove near the summit of Grey Friar.

Box plots of mound height and aspect, and depression length and aspect are shown in Fig. 11. Depression length and mound height are used here as variables which represent the development of the ploughing block, based on the principle that well formed ploughing blocks will have longer depresssions and larger mounds. As north-east slopes are colder, it might be expected to find blocks with longer depressions and larger mounds. There appears to be no significant difference in mound height between south-east, south-west and north-west facing slopes but north-east slopes do have a higher mean and upper quartile. However depression length is shorter on north-east slopes than on other slope aspects.

Three species of vegetation were found to dominate the study area: *Nardus stricta* (mat grass), *Festuca ovina* (sheep's fescue) and *Juncus squarrosus* (heath rush). Species of moss are present including *Sphagnum* spp. (bog moss) and *Rhacomitrium lanuginosum* (wooly hair moss, sheep's fescue, mat grass and heath rush. The most frequent observations for vegetation are of *Festuca ovina* either on its own or with *Juncus squarrosus*. The vegetation of the adjacent area is not the same as the vegetation on mounds. *Festuca ovina* occurs on its own or with *Sphagnum* spp. on mounds 87 times whereas in the surrounding area the figure is 61. Similarly, there are 74 observations of *Juncus squarrosus* around blocks either on its own or in combination, compared with 34 on mounds. The differences are likely to be caused by the drier conditions on the mound due to better drainage because of its elevated position and better soil texture. The presence or absence of moss *Sphagnum* spp. was recorded as a separate variable. Of the recorded ploughing blocks, 20% had moss present in the surrounding area compared to 15.3% with moss on the

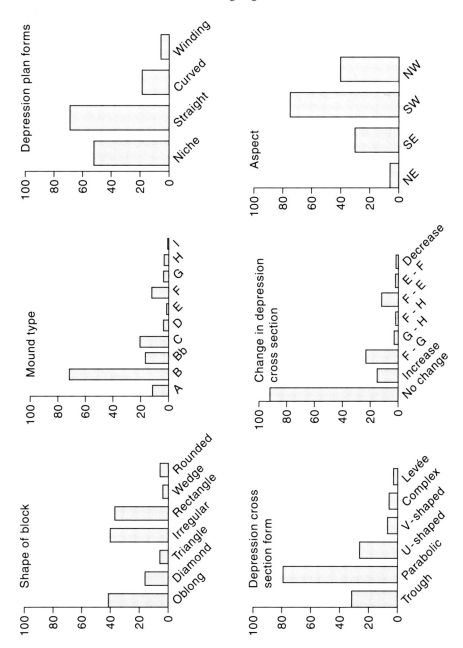

Fig. 10. Summary characteristics of the ploughing blocks.

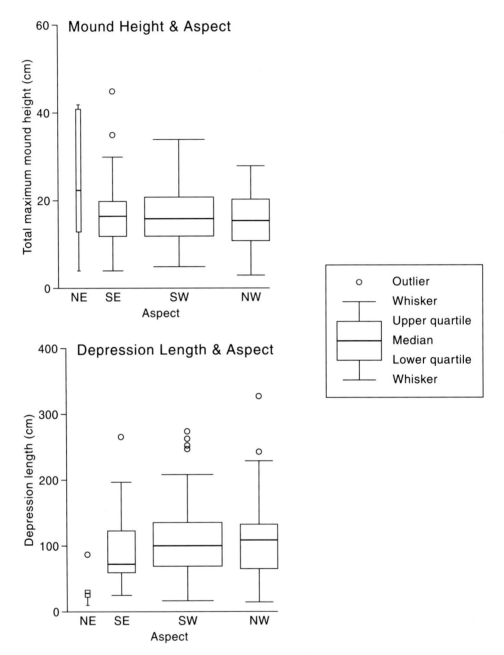

Fig. 11. Box plots of some measured ploughing block variables.

mound. Moss is less abundant on the mounds than in the area adjacent to the block. Drainage is better on the slightly higher ground, making conditions less suitable for the growth of *Sphagnum*. Of all the ploughing blocks measured, 8.7% had patches devoid of vegetation surrounding the ploughing block. This compares with 16% of blocks with

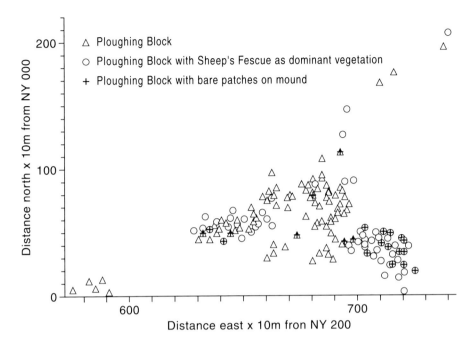

Fig. 12. The distribution of ploughing blocks, mounds with patches devoid of vegetation and Sheep's Fescue as the dominant species.

bare patches found on the mound. Blocks with areas devoid of vegetation in the area surrounding the block are also more likely to have mounds with bare patches. A block with a bare patch on the mound is not necessarily going to have bare patches adjacent to it.

The frequency distribution of block shapes is shown in Fig. 10. The most common shapes of the ploughing blocks are angular oblongs (27%), rectangles (26%) or irregular (25%). These three shapes appear to be the most conducive to mound and depression formation. Maximum mound heights are found with irregular and rectangular shaped blocks. Single lateral mounds, type-F (Fig. 3), are often found with diamonds. Type-F mounds are not found exclusively with triangle shaped blocks and triangles can also give rise to other types of mound. Wedge shape affects mound type. The frequency of mound types (Fig. 3) recorded in the study area is shown in Fig. 10. At the data analysis stage a tenth mound form, type-Bb, was added to the classification, representing single, frontal-lateral mounds which, instead of having bare patches close to their down-slope border as in type-E mounds, have them on top of the mound. Similarly, double mounds were found with bare ground on top of the upper section of the feature as well as mounds with bare patches just down-slope of the smaller, upper part. The mounds were classed together with type-H mounds. Type-B is the most common mound form (48%), followed by type-C (14%). Some 11% of the mounds were of the type-Bb. Only 6% of all blocks recorded were double mounds.

Bar graphs show the frequency distributions for the depression plan form, cross-section form and the change in depression cross-section (Fig. 10). Straight plan forms were the most common (47%), followed by nich forms (35%). Some 51% of ploughing blocks had a parabolic cross-section and 21% were trough shaped. There are eight different types of change in the depression cross-sectional form. First, the depression may show no change. A change recorded as an increase indicates the existence of an increase in depression depth down-slope. Similarly, a decrease indicates a distinct decrease in depression depth down-slope. There is no significant change throughout the majority (61%) of cross-section forms. Some of the depressions do increase in depth down-slope (10%). The change is reasonable as the depression will gradually grade into the above slope and, in so doing, become shallower or the block may sink as it moves down-slope. A significant number of depressions, 15%, change from parabolic to U-shaped forms down-slope. Again, the explanation for this is the erosion of the initial shape of the depression, grading it into the slope. Generally, depressions grade into the slope and, in so doing, their cross-section may change.

Fig. 12 illustrates aspects of the distribution of ploughing blocks in the study area. The 41% of blocks with *Festuca ovina* as the dominant vegetation adjacent to the block are marked by a circle and the 16% of blocks with bare patches on the mounds are shown by a cross. The blocks with Fescue are clustered to the south-west of the summit of Swirl How and around the south-east of Grey Friar. They are generally the higher altitude areas where drainage is better. Here the vegetation is short and the shape and dimensions of the mound and depressions are more easily identified than in the longer and tuftier areas of *Juncus* and *Nardus*. It can be seen that bare patches on mounds are most commonly found in areas with *Festuca* dominant. Only 5% of blocks have bare patches on the mound where Fescue is not the dominant vegetation, probably due to the thinner vegetation cover of the ground which therefore results in areas with a higher susceptibility to erosion.

PEARSON's product correlation coefficients (r) for the size and site characteristics of the blocks are shown in Table 3. With altitude very low correlations emerge for block, depression, mound size and long-axis alignment. There is a relatively high linear association with gradient, suggesting that at higher altitudes the gradient decreases as is the case on hill summits. Depression length may therefore be affected by the inverse relationship existing between gradient and altitude and so associations between altitude and depression length should not be totally ruled out if gradient is also taken into account. There appears to be no distinct relationship between mound type and altitude. However, type-A mounds seem to have slightly shorter depressions than other mound types. The mean depression length for type-A mounds is only 65.4 cm compared to 103.9 cm for the whole data set, confirming the proposal made by TUFNELL (1972) that blocks which have moved a short distance have only frontal mounds.

Conclusions

The data collected in this study indicate that ploughing blocks are found on slopes whose angle is between 4° and 40° and consequently mass wasting processes are active in such places. The nature of terrain on the Coniston fells confirms that ploughing blocks evolve

Table 3. Correlation matrix of ploughing block variables measured in this study.

		1	2	3	4	5	6	7	8	9	10	11	12	13	14
1	Altitude	1.0													
2	Gradient	-0.5	1.0												
3	Block length	-0.06	0.19	1.0											
4	Block width	-0.1	-0.02	0.68	1.0										
5	Block height	-0.25	0.04	0.62	0.62	1.0									
6	Block depth	0.06	-0.2	0.32	0.21	0.27	1.0								
7	Block circumference	-0.13	-0.12	0.85	0.75	0.68	0.22	1.0							
8	Total block height	-0.16	-0.06	0.63	0.57	0.9	0.66	0.63	1.0						
9	Block volume	-0.06	-0.09	0.77	0.71	0.83	0.36	0.72	0.81	1.0					
10	Alignment	-0.2	0.1	0.13	0.27	0.27	0.03	0.2	0.22	0.28	1.0				
11	Depression length	0.23	-0.03	0.15	0.06	0.0	0.08	0.12	0.03	0.07	-0.14	1.0			
12	Depression depth	0.08	-0.09	0.32	0.24	0.12	0.52	0.27	0.33	0.2	-0.02	-0.09	1.0		
13	Depression width	-0.1	-0.02	0.51	0.64	0.66	0.23	0.64	0.62	0.76	0.34	-0.01	0.23	1.0	
14	Mound height	-0.21	0.05	0.19	0.29	0.26	0.29	0.2	0.33	0.25	0.14	0.07	0.19	0.21	1.0

on grassy slopes where there is sufficient moisture and few rock outcrops. Data for ploughing block rock size recorded in this study are consistent with those reported for the Moor House Nature Reserve (TUFNELL 1972) and on the Drumochter Hills (CHATTOPADHYAY 1983). Previous studies suggest that regular prism blocks are the optimum shape for ploughing. Regular prism shaped blocks are common in the field area but there is little difference between their mounds and depressions and those for less regularly shaped blocks. It can also be confirmed that triangular or diamond shaped blocks may hamper mound formation. The length and height of mounds recorded in the study are somewhat larger than those reported elsewhere. If depression parameters are considered, width seldom approximates to the width of the block which has produced the depression. Many of the depression widths are smaller. One or two are larger. The suggestion is that mounds and depressions on moist slopes, in cool, temperate environments are less well preserved than depressions in some other areas. Depression dimensions are frequently reduced following ploughing block movement due to the operation of other slope erosion processes. Parabolic and trough shaped depressions were common, with the most regularly recurring mound being a type-B. The frequency distribution of the depression long profile indicates that straight and nich-shaped forms are the most common, although there is often a decrease in the depth of the depression up-slope or a change in cross-sectional form.

For ploughing block formation specific conditions are necessary including a gradient which is neither too steep nor too flat, moist soil conditions, regolith which contains boulders of a suitable size, certain climatic conditions and grassy sopes. All the required conditions are met in the field area examined here. The different types of vegetation cover recorded do not affect the formation of ploughing blocks although they may influence subsequent erosion and the visibility of the associated features. The lower boundary of cryergenic activity may be delimited directly from the distribution of the blocks in the field area.

The study highlights the need for similar, quantitative data on ploughing blocks as well as more detailed studies of their formation and movement. There are many problems with the analysis and comparison of ploughing block data. The correlation of results is fraught with difficulties because there are many interrelating factors involved in their evolution. Despite such limitations, this study provides an insight into a common but much neglected feature of many landscapes which gives an indication of the general nature of geomorphological activity and specific details of some of the slope processses operating in places where ploughing blocks are found.

Acknowledgements

The authors are grateful to Dr N.J. Cox and Dr I.S. Evans for guidance and advice on the statistical analysis. The diagrams were prepared with dexterity by the staff of the Drawing Office, Department of Geography, University of Durham.

References

CAILLEUX, A. & G. TAYLOR (1954): Cryopédologie études des sols gelés: Expéditions Polaires Françaises Missions Paul-Emile IV. – 218 pp., Hermann & Cie, Paris.
CAPELLO, C.F. (1955): I'massi contornati. – Natura Riv. sci. Naturali 46: 109–119.
CHATTOPADHYAY, G.P. (1983): Ploughing blocks on the Drumochter hills in the Grampian Highlands. – Geogr. J. 149: 211–215.
EVANS, I.S. (1981): General Geomorphometry. – In: GOUDIE, A.S. (ed.): Geomorphological Techniques. – p. 31–37, Allen & Unwin, London.
FITTER, R., A. FITTER, & A. FARRER (1984): Guide to the Grasses, sedges, Rushes and Ferns of britain and Northern Europe. – 256 pp., Collins, London.
GALLOWAY, R.W. (1961): Solifluction in Scotland. – Scott. Geogr. Mag. 77: 75–87.
GOURBUNOV, A.P. (1970): Merzolynuye Yavlenia Tyan shanya. – 265 pp., M. Gidrometeoizdat, Moscow.
– (1980): Borozdyatshiye Valouny. – In: Merzlotnye Issledovaniya V Osvaivaemykh Rayonakh SSSR. – p. 160–167, Novosibrsk, Nauka.
– (1991): Ploughing blocks of the Tien Shan. – Permafrost and Periglacial Processes 2: 237–243.
HAY, T. (1937): Physiographical notes on the Ullswater area. – Geogr. J. 90: 426–445.
– (1942): Physiographical notes from Lakeland. – Geogr. J. 100: 165–173.
HOGBOM, B. (1914): Über die geologische Bedeutung des Frostes. – Uppsala University Geol. Inst. Bull. 12: 257–389.
HOLLERMANN, P.W. (1964): Rezente Verwitterung Abtragung und Formenschatz in den Zentralalpen am Beispiel des oberen Suldentales (Ortlergruppe): – Z. Geomorph. N.F., Suppl.-Bd. 4: 257 pp.
IWATA, S. (1976): Some periglacial morphology in the Sagarmatha (Everest) region, Khumbu Himal. – J. Jap. Soc. Snow and Ice (SEPPYO) 38: 115–119.
JOHNSON, G.A.L. & K.C. DUNHAM (1963): The Geology of Moor House. – Nature Conservancy Monograph. 2: 182 pp.

KUHLE, M. (1974): Vorläufige Ausführungen morphologischer Feldarbeitsergebnisse aus dem S/E-Iranischen Hochgebirge am Beispiel des Kuh-I-Jupar. – Z. Geomorph. N.F. **18**: 472–483.
O'BRIEN, L. (1992): Introducing quantitative Geography. – 356 pp., Routledge, London.
PEARSALL, W.H. & W. PENNINGTON (1973): The Lake District – a Landscape History. – 320 pp., Collins, London.
PISSART, A. (1964): Vitesse des mouvements du sol au Chambeyron (Basses Alpes). – Buil. Peryglac. **14**: 303–309.
RAGG, J.M. & J.S. BIBBY (1966): Frost weathering and solifluction deposits in southern Scotland. – Geogr. Ann. **48A**: 12–23.
REKSTAD, J. (1909): Geologiske iagttagelser fra stroket mellern Songefjord, Eksingedal og Vossestranden. – Norges Geol. Unders: 1–47.
RUDBERG, S. (1958): Some observations concerning mass movements on slopes in Sweden. – Geol. Fören Stockholm **80**: 114–125.
– (1962): A report on some field observations concerning periglacial geomorphology and mass movement on slopes in Sweden. – Biul. Peryglac. **11**: 311–323.
SILK, J. (1979): Statistical Concepts in Geography. – 276 pp., Allen & Unwin, London.
TUFNELL, L. (1966): Some little studied British landforms. – Proc. Cumberland Geol. Soc. **2**: 50–56.
– (1969): The range of periglacial phenomena in northern England. – Biul. Peryglac. **19**: 291–323.
– (1972): Ploughing blocks with special reference to north-west England. – Biul. Peryglac. **21**: 237–270.
– (1976): Ploughing block movements on the Moor House Reserve. – Bull. Periglac. **26**: 311–317.
WASHBURN, A.L. 81979): Geocryology: a survey of periglacial processes and environments. – 2nd edition, 406 pp., Edward Arnold, London.

Addresses of the authors: Dr. ROBERT J. ALLISON, Department of Geography, University of Durham, Science Laboratories, South Road, Durham, DH1 3LE, UK. KATE C. DAVIES, Ordinance Survey, Romsey Road, Maybush, Southampton, SO16 4GU, UK.

Estimating sediment transport and reservoir sedimentation in an EIA framework

(a case study applied to the Pequenos Libombos Dam in Southern Mozambique)

EBENIZÁRIO CHONGUIÇA, Uppsala

with 7 figures and 2 tables

Summary. Sustainable development calls for appropriate tools and approaches to assess environmental impacts, monitor environmental changes for an adequate resource planning and management process. The present paper aims at illustrating some key issues in data generation and impact assessment. The Pequenos Libombos dam has been used as a case study to assess the level of effectiveness of a set of methods used to assess sediment transport and reservoir sedimentation in an EIA framework. Factors determining uncertainty levels on predicted impacts are discussed. It is argued that an EIA approach that applies more than one methods for the assessment purpose, allows for a better mutual calibration of the obtained results, reducing consequently the levels of uncertainty. With adequate auditing and monitoring programmes a more effective basis for the continuous planning and management process can be established.

1 Introduction

The Pequenos Libombos Dam, is a water development scheme implemented within the Umbeluzi basin (Fig. 1), some 35 km west of Maputo City in Southern Mozambique. The 5700 km² drainage basin is shared by Mozambique, Swaziland and South Africa. Concluded in 1987, the major objectives were: I) water supply, II) irrigation, III) flood control and IV) energy generation (cf. CHONGUIÇA and STRÖMQUIST 1992).

From past experiences with similar projects throughout the world, siltation problems with the associated implications over the reduction of reservoir storage capacity, are considered to be but just one of the major drawbacks of such undertakings (cf. RAPP et al. 1972, ROOSEBOOM 1992).

The present paper summarises an attempt to assess current rates of reservoir siltation and the level of effectiveness of the methods used to estimate sediment transport and reservoir sedimentation in an EIA (environmental assessment impact) framework (cf. CHONGUIÇA 1995).

2 Methods

2.1 Sediment transport studies

The adopted sediment sampling approach was based on the depth-integrating sampling technique developed by NILSSON (1969). The approach is based on the derivation of a

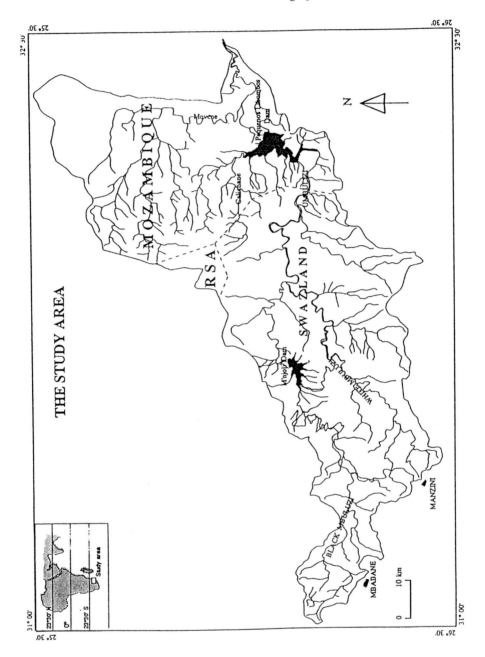

Fig. 1. The Umbeluzi drainage basin.

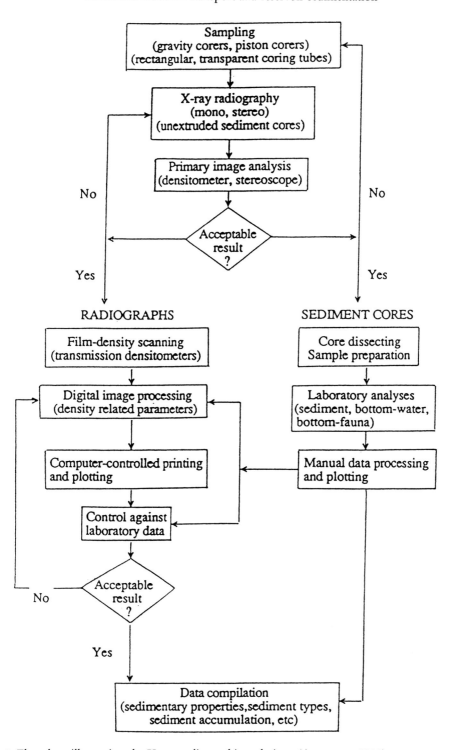

Fig. 2. Flowchart illustrating the X-ray radiographic technique (AXELSSON 1992).

regression model portraying an empirical relationship between water discharge and sediment concentration synchronously measured. The basic equipment is somewhat similar to the USDH-48 sampler, which has the ability to fill the sampling bottle at the same speed as the streaming water surrounding the sampler (cf. CHOW 1964, GRAF 1971, SIMONS & STENTURK 1977).

2.2 Reservoir sedimentation studies

Reservoir sedimentation studies were performed using two different approaches: a) sedimentary properties analysis based on X-ray radiographic technique and b) sedimentation modelling.

2.2.1 X-ray radiographic technique

This technique basically consists of a non-destructive scanning and recording procedure which simplifies the determination of sedimentary properties and facilitates the calculation of reservoir sedimentation rates (cf. AXELSSON 1983, 1992, BODBACKA FÄLTMAN 1993). Major steps of the technique are as illustrated in Fig. 2.

2.2.2 Sedimentation modelling

Reservoir sedimentation modelling was based on the technique developed by SUNDBORG (1992). Its two basic parameters used to calculate sedimentation rates are: a) The trap efficiency of the reservoir and b) Amounts of sediment concentration remaining in suspension.

a) The trap efficiency of the reservoir

(equ. 1) $T = 100 \cdot \{1 - e^{[w \cdot \phi(w) \cdot Aflow(D)]/Q}\}$

where:
T = percentage of materials with a given settling velocity (w) that is deposited in the rservoir – Trap efficiency of the dam
Q = volume of water flowing through the reservoir per unit time
w = settling velocity of sediment particles
$\phi(w)$ = ratio between suspended sediment concentration of particles with a settling velocity (w) close to the bed and mean concentration in the flow
$A\,flow(D)$ = Area of active flow – indicative of the total horizontal area actively contributing to the water flow through the reservoir upstream of the dam under any given conditions
Application of this equation integrating over all water discharges and settling velocities (w) gives the total trapping of materials in the reservoir.

b) Amounts of sediment concentration remaining in suspension

(equ. 2) $S_{wm}(L_{i+1}) = S_{wm}(L_i) \cdot e^{-e\{[w \cdot \phi(w) \cdot Bi+1/2 \cdot \Delta L]/Q\}}$

where:

Q = water discharge
w = settling velocity of sediment particles
$S_{wm}(L_i)$ = mean sediment concentration of particles with settling velocity (w) within a unit volume of the flow at section L_i
$S_{wm}(L_{i+1})$ = mean sediment concentration at downstream section L_{i+1}
ΔL = distance in the flowing direction between sections L_i and L_{i+1}
$B_{i+1/2}$ = mean width of the prevailing active flow between L_i and L_{i+1}
ø(w) = ratio between suspended sediment concentration of particles with a settling velocity (w) close to the bed and mean concentration in the flow, normally ≥1.

As the model requires input data regarding representative grain size distribution of inflowing suspended sediment at different discharge conditions, such data was extrapolated from grains size analysis of deposited materials along the sampling points utilised for reservoir sedimentation studies based on X-ray radiographic technique.

3 Results

3.1 Sediment load

Sediment load estimates through the main Umbeluzi river (station E-10) during the research period (1990–94) were based on derived regression equations differentiated in terms of rising and falling limb of the hydrograph with the following structure:

(equ. 3) $\log Sc = 1.113 \log Q + 1.178$ ($r^2 = 0.904$) Rising limb
(equ. 4) $\log Sc = 1.244 \log Q + 1.084$ ($r^2 = 0.764$) Falling limb

Total suspended sediment transport over the observation period was estimated at 238053 tonnes. Assuming that the same regression equations also applies to the period before the sampling programme, total sediment transport throughout the whole period of reservoir existence, (1987–94) amounts to 363400 tonnes. This corresponds to a mean annual suspended sediment inflow into the reservoir, of 60500 tonnes/yr. The temporal variability of this amount is illustrated in Fig. 3.

The total sediment load into the reservoir should consider the contribution from other tributaries entering the reservoir as well as the sediment input fraction directly flowing from the adequate nearshore areas. As no reliable measurements are available, the additional sediment contribution from those sources is not being taken into account in the current study. The bed load is another sediment input to be considered. It can be determined through theoretical bed load formulae or through direct field measurements, both methods being constrained by a set of limiting factors (cf. EINSTEIN 1950, 1964, SUNDBORG 1992). Existing bed load formulae are flume-derived and do assume an overall relationship between stream power and the bed load transport. This assumption is in fact not fulfilled in rivers with heterogenous materials. Assessment based on field measurements, on the other hand, is hampered by the levels of sophistication required for the field equipment. Given this, estimates of probable magnitudes can be based on empirical data on sediment characteristics in given environmental settings. Using the classification proposed by MADDOCK (cf. LANE & BROLAND 1951), 25% of the suspended load could be applied for the Umbeluzi channel, considering the specificities of its channel material

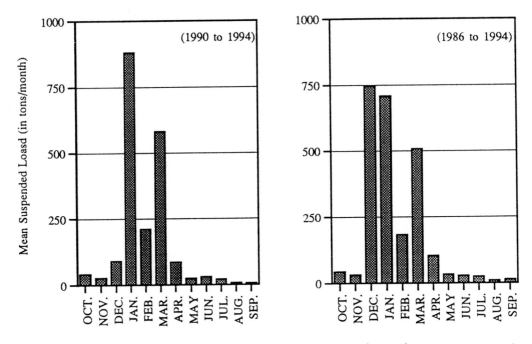

Fig. 3. Calculated mean monthly sediment load at E-10 for the research period (1990–91 to 1993–94) and dam operation period (1986–87 to 1993–94).

(cf. SUNDBORG 1964, 1992, JANSSON 1985). A certain degree of uncertainty to this figure can, however, be anticipated given the wide range of the probable percentage given in Maddock's table.

Considering the annual suspended load of 60500 tonnes and assuming as acceptable the bed load fraction of 25% on the basis of Maddock's tables, the total partial load from the main Umbeluzi river would be estimated at 75625 tonnes per year. This corresponds to an annual sediment yield from the upper catchment of 24 tonnes/km²/yr.

3.2 *Rates of reservoir sedimentation*

3.2.1 *Based on the X-ray radiographic technique*

After the identification of the probable pre-reservoir bottom sediment level through a careful analysis of the X-ray radiographs on core samples in combination with the water and organic content variation with core depth as well as grain size, sedimentation rates have been determined from that starting level upward. This was accomplished considering the X-ray densitometrically calculated values of accumulation of solids, specific to each core. The results are given in terms of weight per unit area and time as well as in thickness per unit time. Spatial distribution of such sedimentation rates are as illustrated in Fig. 4. It should be noted that, taking into account the compression curves, the rates of sedimentation given in terms of mm of thickness per unit time may vary over time. On the basis of measured annual sedimentation rates at each sampling point, average mean annual sediment accumulation for the whole reservoir is estimated at 142202 tonnes. For a reservoir trap efficiency of 70%, the total annual sediment inflow is estimated at 184863 tonnes. The sediment yield corresponding to the Pequenos Libombos upstream catchment, excluding the Black Mbuluzi portion draining into M'njoli Dam, is therefore of 58.2 tonnes/km²/yr.

3.2.2 Based on modelling technique

I. Background data

Considering the theoretical and methodological principles of the sedimentation modelling technique developed by SUNDBORG (1992), the model structure integrates the following basic components: I) reservoir volume and geometric shape; II) reservoir flow conditions, including the influence of density currents; III) water discharge conditions in the main tributaries; IV) sediment inflow characteristics and V) water quality.

Reservoir morphology – The model being applied is based on a concept of the physical process of deposition of particles related to the corresponding settling rates and flow conditions (cf. SUNDBORG 1992). If an attempt is made to apply the model, given the influence the morphological characteristics of the reservoir on the model structure, a differential analysis of the Pequenos Libombos reservoir is then required in terms of the flow and settling conditions specific to the Umbeluzi and Calichane sub-sections of the reservoir. The reservoir area circumscribed to each of these main sub-sections, contain various morphological features such as irregular shorelines, small islands, vegetation and several other physical barriers which may disturb the flow. Therefore, the total horizontal area that actively contributes to the flow throughout the reservoir, under any given condition, has to be reduced because of such potential obstacles (cf. SUNDBORG 1992). This reduction process has yielded three categories of areas of active flow characterised as follows:

- *Maximum area of active flow* – based on the assumption that the undisturbed flow also takes place along areas beyond the main stream channel and its adjacent flood plain zone, as long as the topographic conditions are favourable. Given this assumption, the cutting of the reservoir areas that were assumed to be obstacles to the active flow was reduced to a minimum;
- *Minimum area of active flow* – based on the assumption that undisturbed flow exclusively follows the old river channel and the immediate adjacent areas within the flood plain zone. Reduction of areas assumed to have an obstacle role to the flow were maximised.
- *Mean area of active flow* – corresponds to the mean value area derived from the previous categories.

Water and sediment discharge – Besides the reservoir morphology, the flow regime throughout the dam is strongly constrained by the magnitudes of water entering the reservoir. Such water discharge magnitudes are also decisive in terms of the amounts of sediment being supplied to the reservoir. Therefore, the analysis of duration and frequency of occurrence of various water discharge magnitudes is important for the modelling technique. The mean daily discharges during the research period (1990/94) were converted

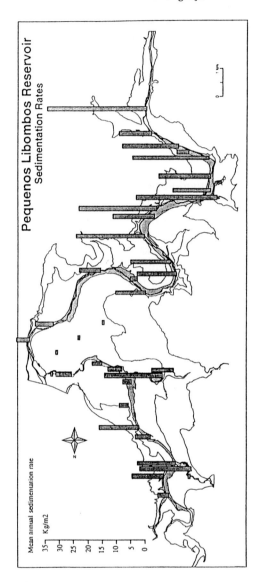

Fig. 4. Reservoir sampling stations and measured sedimentation rates. The highest sedimentation rates occur within the rivers entry points into the reservoir.

into flow duration. From this conversion it is found that the low discharges have prevailed during the research period at gauging station E-10. During 90% of the time, daily mean flows have not exceeded 1 m³/s. Converting duration curves from 1986–87 to 1993–94 into frequencies of water and sediment discharge in percent time (cf. SUNDBORG 1992), it was observed that low discharges have a high frequency with low sediment loads. Conversely, high sediment loads occur at high discharges of low frequencies.

As for Calichane, no daily water discharge data is available, water and sediment discharge from Umbeluzi were hypothetically used simply to check for the response of the adopted grain size distribution models and then allow for a global estimate of the reservoir trap efficiency according to the technique being applied.

Considering that water and sediment discharge data are restricted to the period corresponding to the years 1987 to 1993, it should be noted that, depending on the discharge pattern adopted, different sediment load curves could therefore be obtained. This situation calls for caution in interpreting the results of the modelling exercise, as the data input may not be representative of medium-long term flow conditions of the river system under analysis.

Grain size characteristics – The grain size distribution of the suspended sediments were inferred from the grain-size distribution obtained from the analysis of already de-

Table 1. Major Umbeluzi Grain Size Distribution Groups.

Statistics	Group 1	Group 2	Group 3	Group 4
		Median Grain Size (in Phi units)		
Minimum	8.8	6.3	2.8	0.9
Maximum	12.0	10.8	5.8	2.3
Count	108	16	5	6
		Percent Sand		
Minimum	0.0%	0.0%	43.4%	77.2%
Maximum	11.2%	32.6%	57.7%	91.0%
Mean	0.7%	13.3%	49.4%	83.4%
Median	0.0%	10.9%	46.6%	80.7%
Std Deviation	1.9	11.0	5.4	5.1
Variance	3.8	121.2	29.2	26.2
Count	108	16	5	6
		Percent Silt		
Minimum	3.5%	8.1%	7.7%	1.4%
Maximum	51.0%	69.5%	30.4%	9.5%
Mean	21.6%	42.8%	17.5%	6.5%
Median	20.0%	41.6%	8.0%	6.8%
Std Deviation	9.8	19.4	10.5	2.8
Variance	96.5	377.1	109.9	8.1
Count	108	16	5	6
		Percent Clay		
Minimum	49.0%	30.5%	26.2%	3.0%
Maximum	96.5%	64.5%	40.2%	21.4%
Mean	77.7%	43.8%	32.8%	10.1%
Median	79.0%	40.4%	26.8%	7.7%
Std Deviation	9.9	11.4	6.1	6.9
Variance	99.2	131.1	37.5	47.2
Count	108	16	5	6

posited reservoir sediments. In the sampled cores, an alternate pattern (finer vs. coarser) is generally found in the grain-size distribution along the core layers from top to bottom.

The alternate mode of appearance of finer and coarser material is related to the hydrological conditions that prevailed during deposition as well as to the sediment transport history and sediment source areas (cf. EINSTEIN 1964).

Plotting all grain size distributions from all cores at all depths, peculiar patterns on the graphic displays with relatively similar tendencies of the median particle size and relative proportions of wide range sizes (cf. EINSTEIN 1964) of sand, silt and clay materials could be identified. Based on these similarities (median particle size, relative proportions of sand, silt and clay), the grouping of core sub-samples was made possible as illustrated in Table 1. Assuming that these groups may represent the major grain size distribution patterns of suspended materials transported under various hydrodynamic conditions (group one = low discharges; group two medium to high discharges; group three = high discharges and group four = bed load), they were used for the determination of major grain size distribution models both for Umbeluzi and Calichane. Calculating the mean percentage for each grain size within the samples falling in the same group, four grain size distribution models have been determined for Umbeluzi and three for Calichane. Pre-reservoir bottom or bed material load are represented by the fourth and third models for Umbeluzi and Calichane sub-sections respectively.

II. Computation results

The trap efficiency R was calculated on the basis of equation (2). Computations were applied separately for the two inflowing sub-sections (Umbeluzi and Calichane). Adopted settling velocities for suspended particles ranging from 0.002 mm to 0.2 mm are as presented in Table 2. The function Ø(w) in equation (2) was assumed to be 1 in all conditions. Discharge classes used are specific to the flow characteristics of the two river systems covering the wide spectrum of its variability so as to reflect the corresponding high magnitude/low frequency events.

Trap efficiency was calculated covering three scenarios on the basis of active flow areas for both Umbeluzi and Calichane sub-sections. Scenario 1 corresponds to maximum active flow area measured, scenario 2 to the minimum active flow area and scenario 3 to the mean between the two previous area measurements. All area measurements were

Table 2. Settling velocities for spherical particles of 2.65 g/cm^{-3} at water temperature of +20 °C.

Particle Size		Settling Velocity	
(in mm)	(in ø)	(cm/s)	(10^{-6} m/s)
0.002	9.0	0.00036	3.6
0.004	8.0	0.00145	14.5
0.006	6.5	0.0032	32
0.01	6.0	0.008	80
0.02	5.5	0.036	360
0.04	4.5	0.145	1450
0.06	4.0	0.32	3200
0.1	3.0	0.8	8000
0.2	2.2	2.4	24000

Source: SUNDBORG 1992

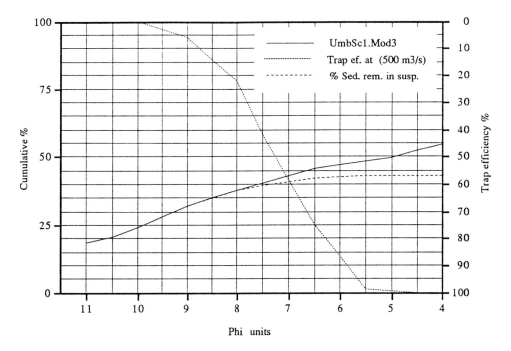

Fig. 5. Calculation of total amount of sediment remaining in suspension at a water discharge of 500 m³/s.

performed considering a reservoir level of 42 m, the average value observed during the research period.

Percentage of sediment through flow – Based upon calculated trap efficiency of the dam for specific grain size and water discharge, the actual percentage of sediments remaining in suspension for a specific grain size distribution model and discharge magnitude was determined as portrayed in Fig. 5. This basically corresponds to the graphical application of equation (2). For Umbeluzi grain size distribution model 3, under trap efficiency scenario 1 at 500 m³/s, the accumulated percentage of material remaining in suspension is 43%. Similar procedures have been applied for all grain size distribution models previously presented and at all trap efficiency scenarios and various water discharge magnitudes observed during the dam operation period. Obtained results are plotted in Fig. 6 for Umbeluzi sub-section.

Reservoir sedimentation – Using the curve of total sediment load over the dam operation period (1987–1994) as previously presented, reservoir sedimentation is calculated for different grain size distribution models and trap efficiency scenarios. Multiplying the calculated percentage of materials remaining in suspension (for a specific grain size distribution model at a given water discharge and trap efficiency scenario) by the corresponding amount of sediment load, sediment outflow is determined for specific water discharge. By integrating such values over all discharge intervals, total sediment outflow over the period is obtained.

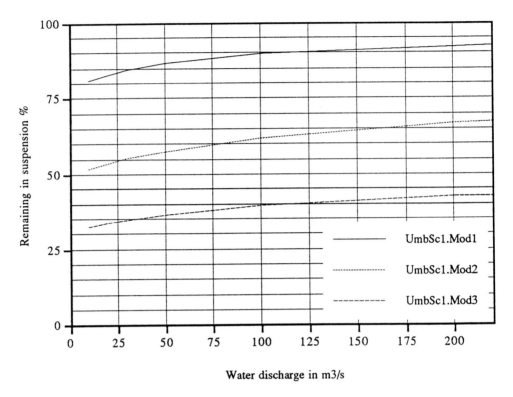

Fig. 6. Percent of inflowing suspended sediments remaining in suspension at different water discharges and grain size distribution models (Umbeluzi).

For Umbeluzi sub-section, scenario 1 (max. active flow area of 8.5 km²) and the grain size distribution model 1, total sediment load over the dam operation period of 363395 tonnes yields a total amount of sediment through flow of 277918 tones. This results in a trap efficiency of 24%.

Using grain size distribution model 2, the total sediment through flow amounts to 166694 tonnes, corresponding to a trap efficiency of 54%. Grain size distribution model 3 generates a total through flow of 93701 tonnes, which corresponds to a trap efficiency of 74% (Fig. 7).

For Calichane sub-section, considering the adopted grain size distribution models and assuming hypothetical water and sediment discharge conditions similar to those from Umbeluzi, scenario 1 (max. active flow area of 4.6 km²) yields a trap efficiency of 39% and 64% for grain models one and two, respectively. Scenario 2 (min. active flow area of 1.6 km²) yields a trap efficiency of 31% and 60%, respectively. Assuming that the most representative area of active flow and grain size distribution models for Umbeluzi corresponds to scenario 1 model 3, and for Calichane, to scenario 1 model 2, the total reservoir trap efficiency is estimated at 70%.

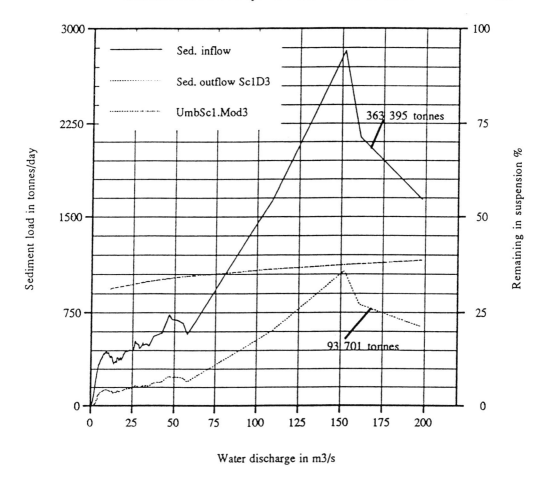

Fig. 7. Sediment outflow from the reservoir deduced from the percentage remaining in suspension at different water discharge levels (Scenario 1 – Model 3).

4 Discussion and conclusions

4.1 Sediment transport

Sediment yield determined on the basis of sediment transport when compared with estimates based on X-ray technique, seems to be an underestimate by about 100%. This may be attributed to:

I)   Estimate according to X-ray technique reflects better the total sediment inflow to the reservoir over the whole period of its existence;
II) Estimate based on sediment transport reflects better what was captured during the observation period;
III) Other streams entering the reservoir were not accounted for;
IV) The observation period was dominated by a prolonged dry spell.

Taking the high variability of the sediment transport into account as well as the occurrence of high magnitude-low frequency events and the sediment availability within the basin, it seems reasonable to expect a much higher load from the main Umbeluzi from a medium to long term perspective. It should, however, be noted that a more precise determination of the potential sediment transport impact on the reservoir needs to consider other set of factors related to the sediment deposition process within the reservoir itself (e.g. grain size distribution of inflowing materials, reservoir hydrodynamic conditions and so forth).

Summary of findings: Given the high level of fluctuation of sediment concentrations versus water discharge, regression models often yield a considerable level of uncertain results (cf. WALLING 1977, 1984, JANSSON 1982, FREGUSON 1986, GRETENER 1994). Level of underestimation that usually occur, is due to the sampling difficulties especially during periods of high discharge and along all other tributaries. Despite all those limitations, it should be noted that sediment transport estimates seem to be one of the most significant prediction techniques to be applied in EIA's related to water development schemes. It is therefore believed that if a major effort is made so as to overcome these types of sampling difficulties, the adopted sampling approach can provide significantly improved results.

4.2 *Reservoir sedimentation*

– *X-ray radiographic technique* – Based on this technique, estimated annual sedimentation rate amounts to about 142 200 tonnes. Such figure cannot, however, be assumed to be representative for a medium to long term time span. The levels of sediment availability within the basin constitutes one reason for such a statement. Factors influencing the deposition process within the reservoir should also be taken into account. Such factors are especially related to the grain size distribution of the inflowing sediments, variability of inflowing water discharge, reservoir morphological conditions (especially the canyon geometry of the old river channel) and the magnitude and frequency of turbidity currents.

The occurrence of turbidity currents has been observed during the observation period and their relevance is related to their capability to activate erosion, re-suspension and further transport of the already deposited material. From the assessment of spatial distribution of accumulation rates within the reservoir, it was found that the higher rates occur within the river entry sections into the reservoir, while the lower accumulation rates occur in the deeper parts. Similar trend was observed with regard to the grain size distribution of the already deposited sediments. Coarser particles accumulate at the entry sections while finer materials settle within the deeper portions of the reservoir. This is further substantiated by the spatial distribution pattern of void ratio. Under such condition, possibilities for erosion and re-suspension are more likely to occur within the deeper parts of the reservoir, meaning that portions of already deposited materials can be flushed out of the reservoir. Therefore, in assessing the spatial distribution of sediment accumulation rates, the hypothesis of having zones with negative deposition rate cannot be completely discarded.

Summary of findings – This technique seems to have great potential to provide accurate information on the amounts of sediment presently accumulated in the reservoir.

It also has the capacity to reveal qualitative information regarding the character of the deposited materials such as the grain size, mineral composition, organic content, water content and density related parameters. This information is provided not only in quantitative terms but also with regard to the spatial distribution. This allows for identification of the potential sediment source areas despite the possibilities to clarify the nature of the probable sub-aquatic environments being formed in the lake (cf. PASSEGA 1977, WEAVER & STONE 1980). From an environmental impact assessment perspective, those capabilities open ways for analysis of a much wider range of environmental changes covering both the quantitative and qualitative water attributes. Given the fact that such measurements can only be performed after reservoir filling, this technique cannot be used for pre-project prediction purposes. Its auditing capabilities are, however, unique. As the sedimentary properties of the collected sediment core samples are recorded on the X-ray radiograph, this technique also holds a considerable monitoring power. The provided results can also be used to support the formulation of appropriate reservoir management strategies.

– *Modelling technique* – The major objective was the testing and provision of a methodological framework for an adequate determination of reservoir trap efficiency. It has, however, a logical consequence, namely the improvement of estimates of reservoir sedimentation rates as long as adequate suspended sediment transport estimates and grain size distribution models are provided. From the testing done, however, it became clear that previous estimates of the Pequenos Libombos reservoir trap efficiency by Montreal Engineering Company (1979), of 90%, can be questioned. Considering the prevailing grain size distribution of deposited materials, reservoir morphology, hydrodynamic conditions and probable influence of turbidity currents, the estimated trap efficiency of the reservoir in the present study is of 70%. The exercise indicated that a high sediment load occurs during high discharges of very low frequency. It should be observed that if, on one hand, those short periods of high water discharge provide the highest sediment inflow into the dam, on the other hand, it is also during those periods that an increase of the amounts of solids remaining in suspension peaks up. What remains to be deposited on the reservoir bottom would correspond to the net difference between those two values. Considering, however, that during these periods turbidity currents are more likely to occur, and also taking into account the grain size distribution of inflowing sediments, real determination of how much material remains in suspension, as well as of how much settles, seems to be more complex than indicated by the simple substraction. Besides the set of factors mentioned above concerning grain size distribution and occurrence of turbidity currents, level of aggregation of suspended materials as well as the possibilities of flocculation should be taken into account.

The study indicated that among the set of factors influencing trap efficiency, grain size distribution of inflowing sediments assumes considerable relevance. This was reflected by the range of variation in the amounts of sediments remaining in suspension when grain size distribution patterns are changed in the model, while other interfering factors (e.g. water discharge, area of active flow, sediment load) are kept constant. Concerning the role of reservoir area of active flow, the model displays a very low level of sensitivity to the variations of this parameter. Changes in area of active flow while holding constant all other variables interacting in the model resulted in a very low variation in estimates of the trap efficiency.

Summary of findings – The model for reservoir sedimentation, as any model, is conditioned to the quality of the data input. Critical input data are: the sediment discharge, the grain size characteristics of the suspended sediments and the water discharge magnitudes. Its major drawback is related to its limitation in incorporating an accurate mathematical description of a three-dimensional flow field as well as an adequate analysis of the mechanics of sediment transport and deposition processes. Requirements of a steady and uniform flow with constant depth and velocities are not fulfilled by the model. However, by careful application, it simplifies the reproduction of the depositional process in a reasonable manner (cf. SUNDBORG 1992). The model provides a more effective estimate of the trap efficiency when compared to other techniques such as the Brune curves (cf. BRUNE 1953). It portrays in an adequate manner the dynamic nature of the trap efficiency concept. With adequate input data, the technique seems to be very suitable for prediction purposes in the context of an EIA applied to this type of water development project. The character of results it provides are also suitable for formulation of managment strategies, especially with respect to the determination of adequate periods for sediment sluicing and flushing, if necessary.

4.3 *Concluding remarks*

From the study results it can be concluded that an EIA should be viewed as a procedural tool of using current scientific knowledge to a specific planning and managment context (cf. STRÖMQUIST & TATHAM 1992).

Predictive methods in EIA studies on river basin development projects, yield results with a higher degree of uncertainty. This is due to: a) the dynamic nature of drainage basin systems, b) scale of analysis and c) the limitations inherent to the analytical instruments used in the assessment process. Under such circumstances it seems reasonable to always utilise more than one method for the assessment purpose to allow for mutual calibration of obtained results. Given this, adequate auditing and monitoring programmes need to be effectively implemented after an EIA exercise.

Finally, there is a need for the planners to have an adequate understanding that planning and managment should be viewed continuous but an informed process of correction and adjustment.

References

AXELSSON, V. (1992): X-ray Radiograph Techniques in studying Sedimentary Properties of Reservoir Sedimentation – A Manual. – In: JANSSON, M.B. & A. RODRIGUES (eds.) (1992): Sedimentological studies in the Cachi Reservoir, Costa Rica. Sediment inflow, reservoir sedimentation and effects of flushing. – UNGI Rapport 81: 195–217.

BODBACKA FÄLTMAN, L. (1993): Sedimentary Structures and Sediment Accumulation in the Lakes Lilla Ullfjärden and Stora Ullfjärden. – UNGI Rapport 83: 113 p.

BRUNE, G.M. (1953): Trap Efficiency of Reservoirs. – Trans. Am. Geoph. Union: 1920–1965.

CHONGUIÇA, E. (1995): Environmental Impact Assessment of the Pequenos Libombos Dam – an evaluation of methods for terrain analysis, sediment transport and reservoir sedimentation in an EIA framework. – UNGI Rapport 90: 213 p.

CHONGUIÇA, E. & L. STRÖMQUIST (1992): The Environmental Impacts of the Pequenos Libombos Reservoir, Mozambique – an outline of a research project. – In: STRÖMQUIST, L. (ed.) (1992):

Environment, Development and Environmental Impact Assessment: Notes on Applied Research. – UNGI Rapport **82**: 67–85.
CHOW, V.T. (1964): Handbook of Applied Hydrology: A compendium of water-resources technology. – McGraw-Hill Book Company, New York.
EINSTEIN, H.A. (1950): The Bed Load Function for sediment Transport in Open Channel flows. – US Department of Agriculture Soil Conservation Service. techn. Bull. **1026**: 70 p.
– (1964): River sedimentation. – In: CHOW, V.T. (ed.) (1964): Handbook of Applied Hydrology. – pp 17.35–17.67, McGraw-Hill Book Company, New York.
FREGUSON, R.I. (1986): River Loads Underestimated by Rating Curves. – Water Resources Res. **22**: 74–76.
GRAF, W.H. (1971): Hydraulics of sediment transport. – 509 p., Series in Water Resources and Environmental Engineering, McGraw-Hill.
GRETENER, B. (1994): The River Fyris. A study of fluvial transportation. – UNGI Rapport **87**: 241 p.
JANSSON, M.B. (1982): Land Erosion by Water in Different Climates. – UNGI Rapport **57**: 151 p.
LANE, E.W. & W.M. BORLAND (1951): Estimating Bed Load. – Trans. Am. Geoph. Union Vol. 32, Washington.
NILSSON, B. (1969): Development of a Depth-Integrating Water sampler. – UNGI Rapport **2**: 17 p.
PASSEGA, R. (1977): Significance of CM Diagrams of Sediments Deposited by Suspensions. – Sedimentology **24**: 723–733.
RAPP, A., L. BERRY & P. TEMPLE (eds.) (1972): Studies of Soil Erosion and Sedimentation in Tanzania. – Geogr. Ann. **54A**: 105–379.
ROOSEBOOM, A. (1992): Sediment Transport in Rivers and Reservoirs – A Southern African Perspective. – Report to the Water Research Commission. WRC Report **297/1/92**: 160 p.
SIMONS, D.B. & F. SENTURK (1977): Sediment Transport Technology. – Water Resources Publication, 807 p., Fort Collins, Colorado.
SUNDBORG, Å. (1992): Reservoir Sedimentation – A Manual. – In: JANSSON, M.B. & A. RODRIGUES, (eds.) (1992): Sedimentological Studies in the Cachi Reservoir, Costa Rica. Sediment inflow, reservoir sedimentation and effects of flushing. – UNGI Rapport **81**: 171–193, Uppsala.
WALLING, D.E. (1977): Limitations of the rating Curve Technique for Estimating Suspended Sediment Loads with Particular Reference to british Rivers. – In: Erosion and Solid Matter Transport in Inland Waters. – Proceedings of the Paris Symposium. IAHS Publ. **122**: 34–48.
– (1984): The Sediment Yield of African Rivers. – In: Challenges in African hydrology and water resources. – Proceedings of the Harare symposium. IAHS Publ. **144**: 265–283.
WEAVER, A.V.B. & A.W. STONE (1980): Distinguishing Sub-environments of Sediment Deposition in Reservoirs Using Particle Size Distribution Parameters. – J. Limn. Soc. South. Afr. **6** (1): 59–65.

Addresses of the author: EBENIZÁRIO CHONGUIÇA, Eduardo Mondlane University, Department of Geography, P.O. Box 257, Maputo – Mozambique, and Uppsala University, Institute of Earth Sciences, Programme on Applied Environmental Impact Assessment, Norbyvägen, 18 B, S–753 22 Uppsala, Sweden.

Morphologic characteristics of debris flow fans in Xiaojiang Valley of southwestern China

XILIN LIU, Chengdu, P. R. China

with 14 figures and 1 table

Summary. Both static and dynamic characteristics of debris flow fans are important topics in alluvial fan morphology. The features of morphometry and sedimentology of 54 debris flow fans in semiarid subtrophics of southwestern China are summarized here based on field surveys and interpretations of aerial photographs and measurements of topographic maps. The influence of depositional processes on the morphologic characteristics of debris flow fans is studied through depositional model experiments. These experiments were conducted near a debris flow gully just after debris flows took place. Debris flow fan morphometry in Xiaojiang Valley of southwestern China differs from that of debris flow fans in Canadian Rocky Mountains and Eastern Italien Alps. Planimetric form and longitudinal profiles of debris flow fans are sensitive to variations in depositional activity at the base of the slope and debris flow unit weight. As deposition at the base of the slope increases and debris flow unit weight decreases, the plan form changes from semi-circular to nearly circular to elliptical, and the longitudinal profiles display a transition from simple straight slopes to complex curved slopes.

1 Introduction

The alluvial fan problem has recently obtained a new recognition and attention (LECCE 1990). Debris flow fans, as a specific category of alluvial fan, are greatly influenced by mountain exploitation and development. Greater understanding of their behaviour is urgently needed.

From a survey in central Virginia, USA, KOCHEL & JOHNSON (1984) provided a generalized physical description of alluvial fans. Moreover, KOCHEL (1990) investigated the depositional processes of alluvial fans in Appalachian Mountains of eastern Unitd States. BEATY (1990) described debris flow fans of White Mountains in California and Nevada, USA. MARCHI & TECCA (1995) showed the morphometry and depositional features of debris flow fans in eastern Italien Alps. TAKEI & MIZUHARA (1982), SUWA & OKUDA (1982) studied the sedimentary characteristics of debris flow deposits on alluvial fans. TAKAHASHI (1980, 1982) researched the formative and erosional processes of debris flow fans. LIU (1991) analysed the relationships between debris flow fans and the river-bed morphology of receiving rivers taking Xiaojiang Valley of southwestern China as an example. TANG et al. (1991) approached the spatial distribution and combination of debris flow fan aprons in the same valley. However, in the international bibliography of alluvial fan, there is almost none from China (LECCE 1990). This incompleteness of site selection, may hinder the construction of a general theory for explanation of alluvial fan devellpment internationally (BEATY 1970, WASSON 1977, LECCE 1990). Here presentation is a studied case of debris flow fans in semiarid subtropical valley of southwestern China.

Debris flow fans are the final products of debris flow processes. Debris flow processes involve various mechanisms such as initiation, transportation and deposition. Unfortunately, these mechanisms are still poorly understood (COSTA 1984). Debris flow researches still face two problems. One is the lack of an integrated theory of debris flow itself, so there are not universally recognized methodologies in this field. Another is the lack of sufficient original data. Although some ideas from hydraulics, physics and mathematics may be borrowed, because of the scarceness of actual data, making a substantial breakthrough in debris flow research still requires an unremitting effort. Therefore, a "black-box" approach (i.e., a statistical method which is interested in the "input" and the "output", while little concerned about intrinsic laws) is still a feasible way to deal with debris flows.

Because of the rare chance to observe debris flows in the field and the danger of testing debris flows in the gully, model experiments and indoor tests have been useful (VAN STEIJN et al. 1989, ZIMMERMANN 1991, LIU et al. 1992, 1993, MAINALI et al. 1994), but the similarities of debris flow mass and of experimental models have been unsolved, and how to utilize the experimental data, remains a problem.

In this paper, the author has carried out both statistical analysis and an experimental study on the morphologic characteristics of debris flow fans. The understanding of the development of debris flow fans and the processes of debris flow deposition are the ultimate objective of the research.

2 Study area

Xiaojiang Valley is located in an old and long-term active structural fault formed in Proterozoic era and composed of several proximate faults. The main fault surface has a steep dip angle of approximately 55°. Along the main fault, there is an extensively fractured zone with width ranging from hundreds to thousands of meters. The dominant rock types are Proterozoic slate, sandstone and shale, Cambrian dolomite and mudstone, Permian basalt and pelitic limestone and Triassic sandstone. The highest peak is 4344 m above sea level (a.s.l.) and most of the other peaks are also more than 4000 m a.s.l., while the local erosion base of rivers is 1100 m a.s.l., and the lowest point is only 695 m a.s.l. The relief is pronounced with high mountains and deep canyons.

Xiaojiang Valley is situated between 25° 54' N and 26° 33' N. Below 1600 m a.s.l., debris flow fans are extensively developed. The climate is typically semiarid subtropical, with annual rainfall of about 700 mm, and potential evapo-transpiration of about 3700 mm. The annual average temperature is about 20 °C, with the highest temperature over 40 °C and the lowest temperature down to –2 °C. The months from April to September are summer; the other months are spring and autumn; there is no winter.

Vegetation cover of Xiaojiang Valley is very low, only 9% in 1983, and vegetation types are mainly scattered trees with bushes and grasses, including *Gossampinus malabarica, Melia azedarach, Morus alba, Opuntia dillenii, Plantago asiatica* etc.

Xiaojiang River has a length of 144 km and a drainage basin area of 2950 km^2 (Fig. 1). There are numerous debris flow gullies in Xiaojiang Valley. Every year debris flows

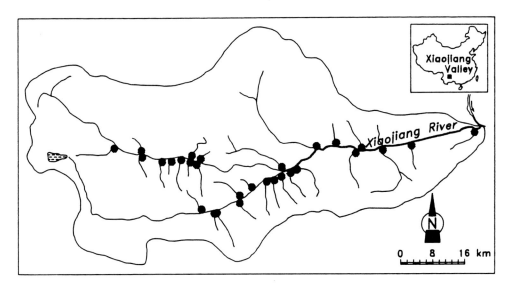

Fig. 1. Location map of Xiaojiang Valley in southwestern China.

occur. Either many gullies simultaneously produce debris flows or a single gully produces several debris flows. In Jiangjia Gully of the valley, a permanent Debris Flow Observation and Research Station of the Chinese Academy of Sciences has been established, so various rainstorm induced debris flows with high-frequency and large-magnitude may be witnessed there. The dense distribution and frequent activity of debris flows, make Xiaojing Valley an excellent site for scientific research on debris flow not only in China but also in the world. "Natural Museum of Debris Flow" another name of Xiaojiang Valley, really matches its reality.

3 Statistical analysis on the morphology of debris flow fans

3.1 Morphometry of debris flow fans

Based on long-term field surveys and in situ measurements, a recognized description of debris flow fans has been formed. They are: always convex cross-sections, mostly concave longitudinal profiles (BLISSENBACH 1954, BULL 1977, ZHAO 1986), but sometimes convex profiles too, and not always smooth exponential curves (TANG 1989, LECCE 1990). The plan forms of debris flow fans vary in their different stages. In the initial stage, they are mainly semi-circular and are called "cones" (Fig. 2). In the developmental stage, they are of irregular shapes and asymmetrical expansion. During this stage, depending on the activity of the receiving river, the fan may elongate quickly or slowly. The lengths of the two sides of the fan may be different, and the ratio of the two sides sometimes reaches up to 2 or more (Fig. 3). In the mature stage, the fan shapes are mostly circular sectors, which are commonly called "fans", and sometimes may display a series of fans without necessary tectonic uplift (Fig. 4). Because of undercutting of the receiving river, the front rim of the fan may become very steep. The upper longitudinal slopes of the fan are

Fig. 2. An initial debris flow fan and its developing drainage basin.

Fig. 3. An unsymmetrical debris flow fan showing its elongated right side (lower direction of receiving river).

Fig. 4. A small debris flow fan born of the large debris flow fan.

Fig. 5. A mature debris flow fan showing its stiff rim cut by receiving river.

Table 1. Morphometric characteristics of debris flow fans in Xiaojiang Valley of southwestern China

Characteristic value	Min. value	Max. value	Average value	Std. Dev.
Fan area S (km²)	0.10	5.74	0.75	0.92
Fan max. length L (km)	0.20	4.20	1.02	0.79
Fan max. width B (km)	0.20	3.20	0.93	0.60
Fan max. thickness T (m)	12	293	68	47
Fan average slope S_f (°)	2	9	4.7	1.7
Range of angles subtended at fan vertex R (°)	2.5	100	53	15
Drainage basin area A (km²)	0.45	93.20	11.23	17.63
Drainage basin relief H (km)	0.46	3.02	1.28	0.45
Main channel length D (km)	0.70	18.00	4.67	3.79
Main channel gradient I (%)	7.0	46.6	23.7	9.0
Melton ruggedness number*	0.23	1.04	0.55	0.19

*Melton ruggedness number: Mel. = $H/A^{0.50}$

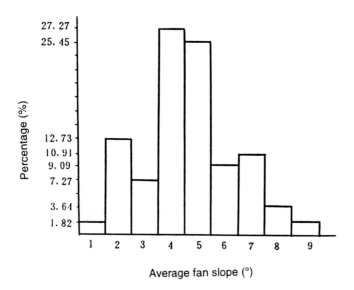

Fig. 6. Distribution of debris flow fan average slope along the bisector.

usually greater than 10° and the lower longitudinal slopes of the fan are gentle with mostly less than 5° (Fig. 5).

Through interpretation of aerial photographs on the scales of 1 : 38000–42000 and measurements of topographic maps on the scale of 1 : 50000, complemented with some purposive field surveys, the morphometric characteristics of 54 debris flow fans in Xiaojiang valley are shown in Table 1. The average longitudinal fan slopes and range of fan angles (i.e., angles subtended at fan vertex) in this valley are respectively shown in Figs. 6 and 7.

3.2 Sedimentology of debris flow fans

Debris flow depositional process is a rapid dynamic geomorphological process. The sediments of debris flow fans, besides fewer fine materials owing to the latter scour of water

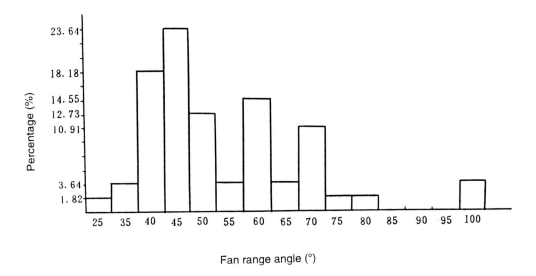

Fig. 7. Distribution of debris flow angles subtended at fan vertex.

flow, are similar to those of debris flow mass. In Jiangjia Gully of this valley, grain size of debris flow fan is clay 5%, and 20% and gravel 75%, and that of debris flow mass averages clay 7.8%, sand 22.6% and gravel 69.6%. Grain size distributions are quite similar. In the whole valley, average grain size (by weight) of 25 sediment samples shows that, debris flow fans have clay (d < 0.005 mm) ranging from 2% to 12%, mean 5%; sand (d = 0.005–2 mm) ranging from 18% to 55%, mean 37.5%; gravel (d > 2 mm) ranging from 40% to 79%, mean 57.5%. The average ratios of clay to sand to gravel are 1 : 7.5 : 11.5. Proportion of clay to sand to gravel in viscous debris flow mass is generally 1 : 8 : 11 in Southwest China (LIU 1986).

4 Experimental study on the morphology of debris flow fans

4.1 Materials and device

Model experiments of debris flow depositional processes were conducted in Donghcuan Debris Flow Observation and Research Station of the Chinese Academy of Sciences, located in Jiangjia Gully of Xiaojiang Valley. In the small hours of 1st September, 1991, a large debris flow took place in Jiangjia Gully, lasting nearly 6 hours. Three days later, perfectly preserved debris flow mass was used as the experimental material (Sample I). At 22–24 o'clock on 6th September, another medium size debris flow occurred again. Next morning, fresh debris flow mass was collected as the experimental material (Sample II). The unit weight and mean grain size of Sample II were 2.12 g · cm^{-3} and 0.8 mm. The unit weight and mean grain size of Sample II were 2.12 g · cm^{-3} and 1.7 mm. Grain size distributions of the two samples are shown in Fig. 8.

The small experimental device was made up through three components.

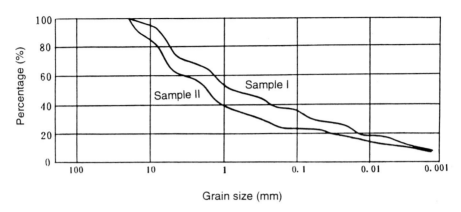

Fig. 8. Distribution of grain sizes of debris flow samples used in the experiments.

(1) *Debris flow supply cubic box*

With each side length of 46 cm and a maximum volume of 97336 cm³, may be loaded debris flow mass up to 200 kg. There was a hole with a diameter of 5 cm in the box bottom, which may be controlled by a lock board, so as to control debris flow supply volume. The box was supported with a fixed prop.

(2) *Debris flow rectangle flume*

With interior width of 16 cm, interior height of 18 cm and effective flow length of 150 cm. The flume was supported with a movable prop, which can be moved forward and backward so as to adjust flume slope ranging from 10° to 34°.

(3) *Debris flow deposition board*

With width of 150 cm and length of 200 cm, which was divided into 300 grids with each area of 10 × 10 cm. The board was directly placed in the experimental field, and its slope is adjustable.

The board and flume were smoothly connected. In order to give the debris flow mass more energy when it flowed out of the box, the flume and the hole did not touch each other. They were apart approximately from 30 cm to 10 cm when the flume slope varied from 10° to 34°.

4.2 Influences of deposition board slope and debris flow unit weight on the morphometry of debris flow fans

4.2.1 Influence of deposition board slope

Fixed parameters were flume slope of 15°, debris flow unit weight of 2.03 g · cm⁻³ and debris flow supply volume of 6318 cm³, variable parameter was deposition board slope ranging from 0° to 10°. The experiment included 11 runs. When each depositional process was entirely stopped, the maximum length and the maximum width were recorded; thicknesses from the fan apex to toe were equidistantly measured, the plan form was described. After each run, the deposits were washed out with water, and the next run was initiated.

The experimental results are shown in Figs. 9 and 10.

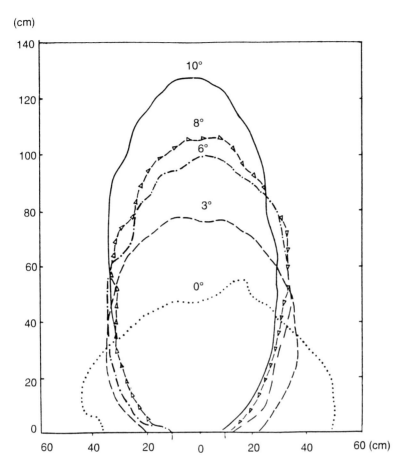

Fig. 9. Influence of deposition base slope on plan shapes of debris flow fan.

4.2.2 Influence of debris flow unit weight

Fixed parameters were flume slope of 15°, deposition board slope of 5° and debris flow supply volume of 7700 cm^3; variable parameter was debris flow unit weight ranging from 2.20 g · cm^{-3} to 1.62 g · cm^{-3} through dilution of debris flow mass with water. The experiment included 10 runs. Operational procedures were the same as the above.

The experimental results are shown in Figs. 11 and 12.

5 Discussions

5.1 Debris flow fan morphology

One of the most characteristic parameters is the fan average slope, which is considered as the most representative index of fan development (BULL 1961). KOSTASCHUK et al. (1986) quantitativel expressed debris flow fan average slope (S_f) as a power function of

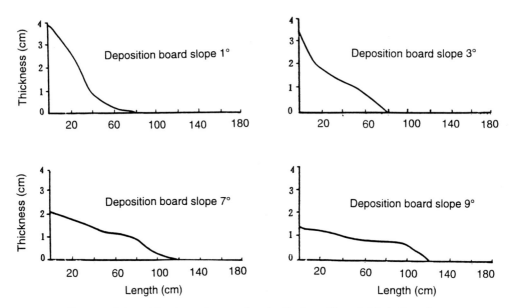

Fig. 10. Influence of deposition base slope on longitudinal profiles of debris flow fan.

Melton ruggedness number (Mel.): $S_f = aMel.^b$; $b > 1$; in Canadian Rocky Mountains $S_f = 0.19$ Mel.$^{2.33}$ (12 samples) (KOSTASCHUK et al. 1986). This kind of positive allometric characteristic means that the fan average slope is increasing more rapidly than the ruggedness of the basin. MARCHI & TECCA (1995) also confirmed this characteristic in eastern Italian Alps using 60 debris flow fans ($S_f = 0.22$ Mel.$^{1.33}$). JACKSON et al. (1987) developed a method to distinguish debris flow fans from alluvial fans using bivariate criteria: $S_f > 4°$ and Mel. > 0.25 in Canadian Rocky Mountains, and inferred that it should have wide applicability for continuously graded basins in unglacierized regions. This means that the fan average slope greater than 4° and Melton ruggedness number greater than 0.25 are the two exclusive characteristics of debris flow fans. MARCHI et al. (1993) followed a similar approach to that developed by JACKSON et al. (1987). They proposed a similar method with Mel. > 0.5 and a similar fan average slope. Using geomorphic and sedimentologic evidences to identify debris flow fans, and using historical data, they confirmed the suitability of the method.

Debris flow fans in Xiaojiang Valley of southwestern China, however, seem not to have these characteristics. The differences are: (1) the fan slope – Melton's number power function is $S_f = 0.1022$ Mel.$^{0.47}$ (54 samples, $r = 0.37 > r_{0.01} = 0.35$). This kind of negative allometric characteristic means that the fan average slope is increasing more slowly than the ruggedness of the basin, which is different from the result obtained by KOSTASCHUK et al. (1986) in the Canadian Rocky Mountains. (2) If we use the criteria developed by JACKSON et al. (1987) to examine these debris flow fans, 26 of 54 debris flow fans do not qualify as debris flow fans. (3) If we use the method proposed by MARCHI et al. (1993), 25 of 54 debris flow fans are not eligible. The two methods can not therefore be used in southwestern China, because of the different climatic, geologic and tectonic en-

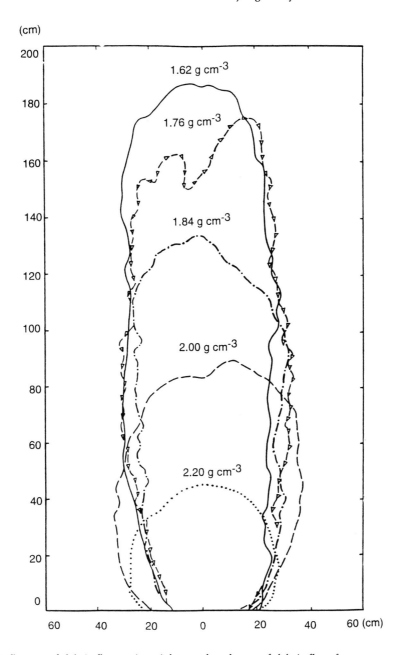

Fig. 11. Influence of debris flow unit weight on plan shapes of debris flow fan.

vironments for the fan and the basin development. In fact, Chinese researchers usually pay more attention to the drainage basin and use a comprehensive assessment method of basin elements to discriminate whether or not it is a debris flow dominated basin, although these methods still face many challenges (TAN 1988, JIANG 1994).

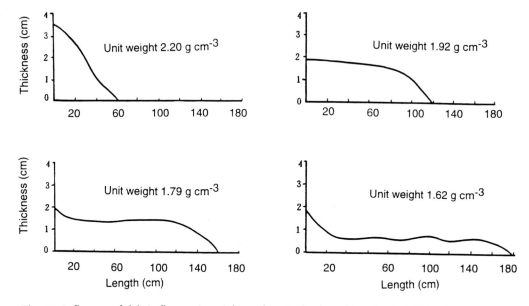

Fig. 12. Influence of debris flow unit weight on longitudinal profiles of debris flow fan.

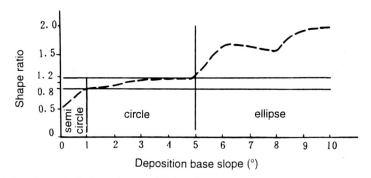

Fig. 13. Variational trend of plane shapes of debris flow fan with variations of deposition base slope.

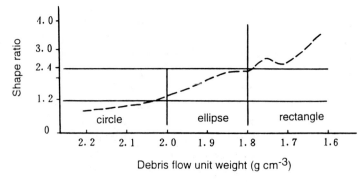

Fig. 14. Variational trend of plane shapes of debris flow fan with variations of debris flow unit weight.

Table 1 has summarized the primary morphologic characteristics of debris flow fans in Xiaojiang Valley.

5.2 *Experimental studies*

Experimental studies on the morphology of debris flow fans were also carried out. Earlier depositional model experiments have demonstrated that, debris flow supply volume affects fan size but not fan shape, and flume slope affects flow velocity but also not fan shape, nor fan size if debris flow supply volume is a constant (LIU et al. 1992, 1993). Therefore, deposition board slope and debris flow unit weight need to be examined. Experimental results show that plan forms and longitudinal profiles of debris flow fans are very sensitive to the variations of these two parameters. As one can see from Fig. 9, 10, 11, and 12, when the deposition board slope increases, the fan shape ratio, defined as the maximum fan length to the maximum fan width, increases but the fan maximum thickness decreases. When debris flow unit weight increases, the fan shape ratio decreases but the fan maximum thickness increases. Thus, other things being equal, the more gentle deposition base slope and more viscous debris flow, the shorter and thicker debris flow fan; the steeper deposition base slope and less viscous debris flow, the longer and thinner debris flow fan.

The plan forms of viscous debris flow fans differ by stage. In Fig. 13 there are two thresholds of slope at 1° and 5° where fan plan form changes from semi-circular to elliptical. The deposition base slopes of < 1°, 1°–5°, and > 5° could perhaps represent different stages of fan development.

In Fig. 14 of debris flow unit weight versus the fan shape ratio, there may be seen two thresholds of the unit weigth at 2.0 g · cm^{-3} and 1.8 g · cm^{-3} where fan plan form changes from circular to elliptical to rectangular. According to ZHANG's (1993) classification of debris flows in China: viscous debris flows have unit weights of > 2.0 g · cm^{-3}, transitional debris flows have unit weights of 1.8–2.0 g · cm^{-3} and low-viscous debris flows have unit weights of 1.5–1.8 g · cm^{-3}. For viscous debris flow, the fan plan forms should therefore be circular, transitional debris flows elliptical and low-viscous debris flows rectangular.

Longitudinal profiles of debris flow fans in the experiments are more complicated than their ideal simple concave profile. Our experiments show that the profiles usually contain several segments such as the straight, the concave and the convex. A complex of multiple slopes usually becomes one of the basic characteristics of debris flow fans.

Influences of deposition base slope and debris flow unit weight on the longitudinal profiles of debris flow fans are interesting. When deposition base slope increases, the fan thickness decreases but the fan length increases. Hence the fan profile should vary from the simple to the complex. Deposition base slope of 5° is a threshold. When deposition base slope is less than 5°, the fan profile is mainly straight or convex; when the slope is greater than 5°, the fan profile is mainly a complex of upper convex – middle concave – lower convex.

When debris flow unit weight decreases, the fan thickness decreases but the fan length increases. This also implies that the fan profile varies from simple to complex. When the unit weight is greater than 1.8 g · cm^{-3}, the fan profile is mainly straight or convex; when the unit weight is less than 1.8 g · cm^{-3}, the fan profile is mainly a complex of upper concave–lower convex, or upper concave – middle straight – lower convex.

6 Conclusions

Xiaojiang Valley is a typical region of semiarid subtropics and a representative area of debris flows in China. This paper has presented the basic morphologic characteristics of debris flow fans in this valley. This will enrich the international bibliography on alluvial fans.

The depositional model experiments make it possible to study quantitatively the morphologic characteristics of debris flow fans under different depositional criteria. The following conclusions may be drawn from this research.

(1) The more gentle deposition base slope and more viscous debris flow, the shorter and thicker debris flow fans; the steeper deposition base slope and less viscous debris flow, the longer and thinner debris flow fans.

(2) The fan plan forms of viscous and transitional debris flows are further affected by deposition base slope. When the slope < 1°, the shapes are mostly semi-circular; when the slope is from 1° to 5°, the shapes are mostly nearly circular; when the slope is greater than 5°, the shapes are mostly elliptical. The fan plan forms of low-viscous debris flows have a weak relation to deposition base slope. They often display nearly rectangular forms.

(3) The fan longitudinal profiles of viscous and transitional debris flows are also further affected by deposition base slope. when the slope is less than 5°, the profiles are mostly straight or convex; when the slope is greater than 5°, the profiles are mostly three-segment complex slopes, and the intermediate base slopes are usually associated with concave profiles. The fan longitudinal profiles of low-viscous debris flow also have a weak relation to deposition base slope. They often display two-segment complex slopes, with the upper concave section most obvious.

Acknowledgements

This paper was written when the author was a Senior Visiting Scholar in CNR–IRPI in Padova (Italy). The research work was funded by National Natural Science Foundation of China (Grant No. 49000011). The author is grateful to his colleagues Dr. Tang Chuan and engineer Zhang Songlin for their collaboration in the model experiments, and also grateful to Drs. L. Marchi and P.R. Tecca for his and her helpful discussions. Many thanks go to Prof. Olav Slaymaker for his initial review on this manuscript.

References

BEATY, C.B. (1970): Age and estimated rate of accumulation of an alluvial fan, White Mountains, California, U.S.A. – Amer. J. Sci. **268**: 50–70.
– (1990): Anatomy of a White Mountain debris-flow – making of an alluvial fan. – In: RACHOKI, A.H. & M. CHURCH (eds.): Alluvial Fans: A Field Approach: 69–89; John Wiley & Sons Ltd., Chichester.

BLISSENBACH, E. (1954): Geology of alluvial fans in semiarid regions. – Geol. Soc. Amer. Bull. **65**: 175–190.
BULL, W.B. (1961): Tectonic significance of radial profiles of alluvial fans in Western Fresno County, California. – U. S. Geol. Surv. Prof. Pap. **424B**: 181–184.
– (1977): The alluvial fan environment. – Progr. Phys. Geogr. **1**: 222–270.
COSTA, J.E. (1984): Physical geomorphology of debris flow. – In: COSTA, J.E. & P.J. FLEISHER (eds.): Developments and Applications of Geomorphology: 268–317; Springer-Verlag, Berlin.
JACKSON, L.E. Jr., R.A. KOSTASCHUK & G.M. MACDONALD (1987): Identification of debris flow hazard on alluvial fans in the Canadian Rocky Mountains. – Geol. Soc. Amer. Rev. Eng. Geol. **7**: 115–124.
JIANG, ZHONGXIN (1994): A simple discriminant method of rainstorm induced debris flow valley in southwest mountainous areas of China. – J. Natur. Disasters **3**: 75–83. [Chinese]
KOCHEL, R.C. (1990): Humid fans of the Appalachian Mountains. – In: RACHOCKI, A.H. & M. CHURCH (eds.): Alluvial Fans: A Field Approach: 109–129; John Wiley & Sons Ltd., Chichester.
KOCHEL, R.C. & R.A. JOHNSON (1984): Geomorphology and sedimentology of humid-temperate alluvial fans, central Virginia. – In: KOSTER, E. & R. STEEL (eds.): Gravel and Conglomerates. – Canad. Soc. Petrol. Geol. Mem. **10**: 109–122.
KOSTASCHUK, R.A., G.M. MACDONALD & P.E. PUTNAM (1986): Depositional process and alluvial fan-drainage basin morphometric relationships near Banff, Alberta, canada. – Earth surf. Proc. Landf. **11**: 471–484.
LECCE, S.A. (1990): The alluvial fan problem. – In: RACHOCKI, A.H. & M. CHURCH (eds.): Alluvial Fans: A Field Approach: 3–24; John Wiley & Sons Ltd., Chichester.
LIU, XILIN (1986): Components and properties of debris flows. – Potential Science **7**: 25–26. [Chinese]
– (1991): Influences of debris flow dynamic processes on the morphometry of receiving river. – In: TANG, B. & S. CHENG (eds.): Proceedings of the 2nd National Symposium on Debris Flow. – pp. 281–289; Science Press, Beijing. [Chinese]
LIU, XILIN, C. TANG, S. ZHANG, & M. CHENG (1992): The geomorphological characteristics of debris flow fans and their model experiments. – The Chinese Journal of Geological Hazard and Control **3**: 368–377. [Chinese].
LIU, XILIN, S. ZHANG, C. TANG & M. CHENG 81993): Model experiments on risk range of debris flow fan. – Geogr. Res. **12**: 173–181. [Chinese]
MAINALI, A. & N. RAJARATNAM (1994): Experimental study of debris flows. – J. Hydr. Engin. **120**: 104–123.
MARCHI, L. & P.R. TECCA (1995): Alluvial fans of the Eastern Italian Alps: morphometry and depositional processes. – Geodinamica Acta **8**: 20–27.
SUWA, H. & S. OKUDA (1982): Sedimentary structure of debris flow deposits, at kamikamihori fan of Mt. Yakedake. – Ann. Disaster prevent. Res. Inst. Kyoto Univ., No. **25B–1**: 307–321. [Japanese]
VAN STEIJN, H. & P.J. COUTARD (1989): Laboratory experiments with small debris flows: physical properties related to sedimentary characteristics. – Earth Surf. Proc. Landf. **14**: 587–596.
TAKAHASHI, T. (1980): Study on the deposition of debris flows (2): process of formation of debris fan. – Ann. Disaster Prevent. Res. Inst. Kyoto Univ., No. **23Br–2**: 443–456. [Japanese]
– (1982): Study on deposition of debris flows (3): Erosion of debris fan. – Ann. Disaster Prevent. Res. Inst. Kyoto Univ., No. **25B–2**: 327–348. [Japanese]
TAKEI, A. & K. MIZUHARA (1982): Sedimentation of debris flow on alluvial fan. – Study Report on Prevention and Mitigation of the sediment disaster on Alluvial fans, pp. 15–25. [Japanese]
TAN, BINGYAN (1988): The regional synthetical systems for evaluating the extent of debris flow activities in mountain areas. – Proceedings of Interpraevent 1988 – Graz, **2**: 243–252, Tagungspublikation.
TANG, CHUAN, J. ZHU, J. DUAN & R. DU (1991): Research on debris flow fans in Xiaojiang Valley, Yunnan Province. – Mountain Research **9**: 179–184. [Chinese]

WASSON, N.J. (1977): Catchment processes and the evolution of alluvial fans in the lower Derwent Valley, Tasmania. – Ann. Geomorph. **21**: 147–168.
ZHANG, SHUCHENG (1993): A comprehensive approach to the observation and prevention of debris flow in China. – Natural Hazards **7**: 1–23.
ZHAO, SHANGXUE (1986): Depositional landforms of viscous debris flow. – In. LI, H. & S. ZHENG (eds.): Proceedings of Debris Flow Symposium in Lanzhou: 171–174; Sichuan Science & Technology Press, Chengdu. [Chinese]
ZIMMERMANN, M. (1991): Formation of debris flow cones: results from model tests. – Proceedings of Japan – U.S. Workshop on Snow Avalanche, Landslide, Debris Flow Prediction and Control: 463–470.

Address of the author: Prof. XILIN LIU, Institute of Mountain Disasters and Environment, Chinese Academy of Science, P.O. Box 417, Chengdu, Sichuan 610041, P.R. China.

A simple portable device for the measurement of ground loss and surface changes

A. W. WELLS and M. R. BENNETT, Greenwich

with 5 figures and 1 table

Summary. The four most widely used techniques for measuring soil erosion are reviewed and evaluated in order to assess whether they fulfil the needs of a postgraduate research project. These include erosion pins, erosion frames, sediment traps and coloured stones. These are then discussed in relation to a newly developed profiling device that is equally accurate at measuring erosion but is cheaper and easier to construct.

Zusammenfassung. *Ein einfaches, tragbares Gerät zur Messung von Bodenverlust und Oberflächenveränderungen.* – Die vier am häufigsten verwendeten Methoden der Bodenerosionsmessung werden besprochen und beurteilt, um aufzudecken, ob sie den Bedürfnissen eines Forschungsprojektes entsprechen. Die vier besprochenen Methoden umfassen Erosionsstäbe, Erosionsrahmen, Sedimentfallen und farbmarkierte Steine. Diese Methoden werden mit einem neuentwickelten Gerät für die Messung des Oberflächenprofils verglichen, das nicht nur eben so genau bei der Messung, sondern auch billiger und leichter zu bauen ist.

Résumé. *Un procédé simple et portable pour mésurer les changements de surface et l'érosion du sol.* – Les quatre techniques les plus utilisées de mesure de l'érosion des sols sont examinées es évaluées afin de vérifier si elles correspondent aux besoins d'un projet de recherche de troisième cycle. Celles-ci comprennent des épingles d'érosion, des cadres dérosion, des pièges à sediments et des pierres colorées. Celles-ci sont ensuite comparées à un procédé de profilage récemment élaboré qui est aussi précis pour mesurer l'érosion mais moins cher et plus facile à construire.

1 Introduction

This paper describes a simple device designed specifically to fulfil the needs of a postgraduate research project to determine the rate of erosion and its variation within a semi-arid region of southern Spain. The majority of techniques presently used to measure erosion were considered to be unsuitable because they either do not produce data of high enough quality or they are difficult to construct and use. A new technique and measuring device was required which: (1) would provide large quantities of accurate erosion data; (2) would allow rapid data collection; (3) would be portable, given the remote nature of many of the sites to be studied; (4) would not require constant monitoring and upkeep, and; (5) should be cheap and easy to construct. These requirements are not unique to this project but are relevant to a wide range of other studies, especially as there is a growing interest in the degradation of semi-arid regions (e.g. SCHUMM 1964, KIRKBY & KIRKBY 1974a, CAMPBELL 1977, BRYAN & CAMPBELL 1982, LE ROUX 1986, YAIR & ENZEL 1987, ROMERO-DÍAZ et al. 1988).

This paper first reviews and evaluates the techniques and measuring devices used by other workers and then goes on to describe the design and construction of a simple measuring device and its use in southern Spain.

2 Evaluation of techniques

The majority of erosion studies use one or more of the following approaches to measuring surface erosion on soft substrates. These are: (1) erosion pins, (2) erosion frames, (3) sediment traps and (4) coloured stones/sand tracers. Two of these techniques, the use of pins and frames, provide a direct measure of erosion, whilst the other two techniques provide an indirect measure by quantifying total sediment loss (GOUDIE 1990). The advantages and disadvantages of each of the four techniques are summarised in Table 1.

3 Discussion

None of the four approaches, in their unmodified form, met the five requirements set out in the introduction of this study. Neither the sediment trap nor the coloured stones approach could provide erosion data of sufficient accuracy nor could they be monitored as frequently as would be necessary to provide reliable results. Erosion pins are a cheap, easy to use option, but they do affect natural slope processes. In particular, they can disturb any surficial crusts and interfer with sheet flow and hence sediment movement. Erosion frames present the best method for rapid collection of large quantities of accurate data. They are, however, difficult to construct and their weight and shape make them awkward to carry and place in position without disturbing the ground. Further, the data collected by any rectangular frame will suffer if any of the four pins are lost. Nevertheless, the frames that stand on two pins did provide the basis for the final design. They are considerably lighter, easier to construct and can enhance the measurements taken from erosion pins with little effort, both in terms of construction and finances.

4 Design and operation

The final design is shown in Fig. 1–3. A horizontal upper bar is supported at either end by two adjustable legs. The adjustable legs allow the upper bar to be fixed horizontally with reference to a spirit level placed on it. Six plumb lines hang from this upper bar and once lowered to the ground the length of each line can be measured, hence profiling the ground surface.

The upper bar had to be constructed of a light, durable material that was relatively cheap. Wood is light, readily available and easy to work, but may bend or warp under the harsh climatic conditions found in semi-arid areas. An aluminium bar would not warp, but would be considerably heavier and would have involved great, expense to construct. The final design utilised a light, workable plastic called perspex, the dimensions of which are shown in Fig. 2. Perspex is a strong, hard polymer and is one of the trade names for polymethyl methacrylate. It was decided that sufficient measurements could be made from a bar 1 metre long and a piece of perspex this length would remain rigid

Table 1. Evaluation of various approaches used to measure erosion.

	Erosion Pins	Erosion Frames	Coloured stones/sand	Sediment traps	Profiler (This paper)
Accuracy	Accurate to about ±2 mm but complicated by micro-relief of ground at base of pin and interaction with slope processes.	CAMPBELL (1970) quotes an accuracy of ± 0.5 mm. Whilst BENITO et al, (1992) quotes ±2 mm.	Complicated by difficulty of locating samples. Therefore, may be considerable error. Can be solved by use of fluorescent tracers.	Very accurate measurement of sediment moved if slope is enclosed. Only produces indirect measurement of slope profile changes.	Accurate to about ±1-2 mm if devices is levelled correctly.
Expense	Cost of pins.	Cost of pins, construction, materials (aluminium) and transportation.	Very little - paint. Fluorescent dyes more expensive.	Cost of equipment to capture sediment and to enclose the slope.	Limited cost for pins, legs and perspex upper bar.
Ease of Construction	None.	If aluminium is used a workshop may be necessary.		Trap construction dependent on geology! Enclosing the slope may be very difficult.	Simple, with few tools and relatively quickly.
Ease of use	Simple, untrained operators required.	Difficult to hammer in 4 pins and line them up correctly to be able to support frame. May also be quite heavy and therefore difficult for one person to place on pins.	Easy to install but can be difficult to locate once they have been moved.	Only requires sediment collection and weighing once installed.	One person can easily install pins, level and use the device even on steep slopes.
Flexibility	Can be used anywhere.	Requires 4 firm pins. Steep slopes may be a problem.	Can be placed almost anywhere, except maybe steep slopes. Particles size should be similar to sediment.	Very difficult on steep slopes.	Used anywhere, across rills/gullies, steep slopes narrow confined channels.
Effects on slope processes	Surface disrupted when hammered in. Pins influence movement of water and sediment.	Only in terms of pins.	Paints cause cohesion of sediment and alters their heat expansion characteristics.	Drastic when slope is enclosed.	Only in terms of pins.
Portability	None once pins are installed.	CAMPBELL (1970) weighs 8.6 kg and large frame.	Only measuring device.	Only scales need be carried.	Weighs less than 2 kg and very easy to carry.
Summary of Limitations	Disruption of surficial crust. Influence water and sediment movement in their vicinity.	Can be heavy and cumbersome to use. Difficult and expensive to construct.	Only provides measure of distance sediment is transported	Require regular maintenance and monitoring. Enclosing site with barrier disrupts natural flow of water/sediment	Device must be levelled accurately. Pins must be hammered in vertically.
Examples	SCHUMM (1956), GERSON (1977), GOUDIE (1990), HAIGH (1977), SALA (1988), IMESON & VERSTRATEN (1988)	CAMPBELL (1970), LAM (1977), BENITO et al. (1990) DISECKER & RICHARDSON (1961, 1962), CURTIS & COLE (1972), MOSLEY (1975)	KIRKBY and KIRKBY (1974), TELEKI (1966)	GERLACH (1967), YAIR et al. (1980), LEWIS (1988), GOUDIE (1990)	This paper.

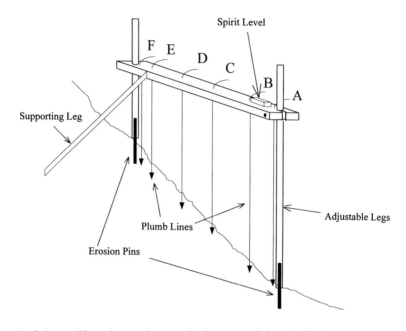

Fig. 1. Sketch of the profiling device showing the location of the plumb lines on the upper bar.

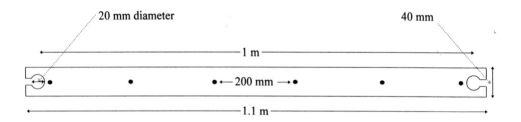

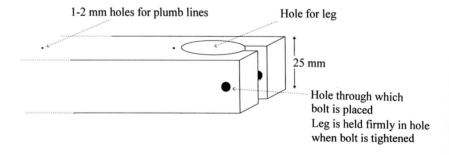

Fig. 2. Sketch of the upper bar showing its dimensions (above) and the detail of one end (below).

Fig. 3. View of profiling device on a slope in southern Spain.

and would not be too heavy. Six very small holes were drilled in the upper bar though which the plumb lines hang.

The legs of the device are simply hollow aluminium tubes, approximately 20 mm in diameter, that can be bought from any hardware store. They are attached to the upper bar through these holes, but can be locked in position by tightening the wing nuts at either end of the bar (Fig. 2). A horizontal cut is made exactly 100 mm from the bottom of each leg into which a flat piece of plastic is glued. This ensures that the legs sit on the pins at the same height each time. A third, supporting leg was designed that would ensure the device did not lean to the right of left (Fig. 2).

Plumb lines were chosen as opposed to narrow rods because with only 25 mm of perspex in the upper bar to guide the rods, they could not consistently return to the smae point on the ground. This effect could have been reduced by either increasing the depth of the perspex bar or by having two separate bars, one above the other, which would guide the rods more effectively. Not only does this add to the weight of the device, but after trials it was realised that some of the rods would have to be almost 2 metres long to be able to measure the steeper slopes, creating further transportation problems. The plumb lines hang from the upper bar through the small holes drilled along its length (labelled A, B, C, D, E & F in Figs. 1 and 2) and are secured on the upper side of the bar using small clips. The small plumb weights themselves can be purchased from most hardware stores.

Field tests were carried out in Almería, southern Spain, where 30 sites had been set up in an area roughly 72 kms by 60 kms. Each site consisted of a line of 5 or more pins, that were hammered vertically into the ground. The lines were orientated in an up-slope direction. When using the device, each leg was placed over an erosion pin, then the upper bar was slided up or down the legs until perfectly horizontal, then locked into position using the two wing nuts. The flexibility of the device allowed profiles to be made across rills and gullies or on the steep walls of narrow badland gullies. Such settings could have proved difficult for a square erosion frame which would need four firmly installed pins. If any of the 5 pins was disturbed, the majority of the measurements could still be made. Had a square frame been used, no further measurements could have been taken until a new set of pins were installed. At each site the six plumb lines were lowered to the ground, secured in that position with the clips and then measured. The height of each end of the bar above the ground was also recorded, so that the device could be easily repositioned and further profiles made. The device can profile the gound between a set of five pins in approximately 20 minutes, which compares favourably with CAMPBELL's (1974) frame that takes 30 minutes to measure 25 points in a 1 metre square. Once the sites readings have been taken, the device can be partially dismantled because the longer lower leg can be separated into two halves and strapped to the upper bar. The device weights less than two kilograms and can therefore be carried easily to the next site or back to the car.

5 Field experiment

In an attempt to examine the efficiency and accuracy of the device, a series of experiments were conducted. On a small plot, approximately 2 m by 1.5 m, of a slope in southern Spain, four steel pins were hammered in, two of which were positioned so as to support the device. Also a small sediment trap, 1 metre by 250 mm by 250 mm deep was dug at the base of the slope. To simulate an erosive period 50 litres of water, in ten stages each of 5 litres, were sprinkled onto the test plot through a simple suspended container. After each stage a number of measurements were taken; the protruding height of each pin was measured; the device was placed onto its pins and all the plumb lines measured, and, a photograph of the test plot was taken. The results of this preliminary experiment are shown in Figs. 4 and 5.

After the ten stages 2.1 kg of fine sediment had been deposited in the trap, the majority of which was in suspension and clearly showed that erosion had occurred. Fig. 4, however, showing the change in height of the ground measured by the erosion pins themselves, suggests that no erosion had taken place and there was instead a net growth in the height of the profile. Although fine sediment was clearly being eroded from the slope, the erosion pin measurements indicate that deposition had occurred. This illustrates very well that erosion pins do influence the movement of sediment around their base.

Fig. 5 which shows the change in slope profile measured by the device, suggests two distinct points. Firstly, the measurements made in the immediate vicinity of the pins (A and F) show more accumulation than erosion. This further emphasises the problems associated with using traditional erosion pins that are described above. It also suggests that the pins' sphere of influence is not confined to the ground in their immediate vicinity, but may in fact cover an area of at least 3 mm around each pin. Secondly, those measurements made further from the pins (B, C, D and E) show a net decrease in the profile.

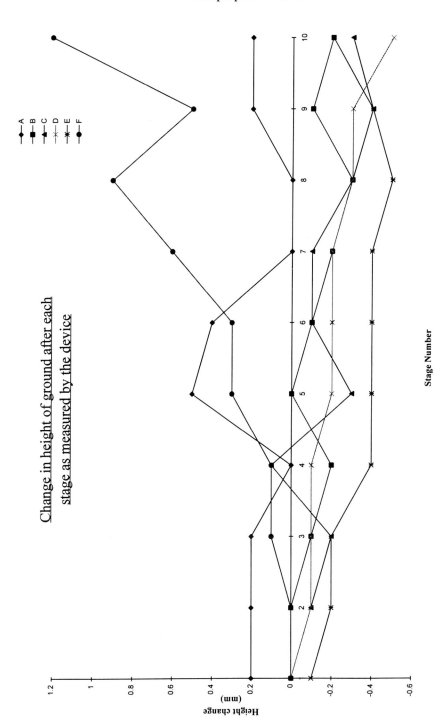

Fig. 4. Graph showing the change in height of the ground measured by the erosion pins. The numbers in the legend refer to pin numbers whose location on the slope can be seen in the sketch.

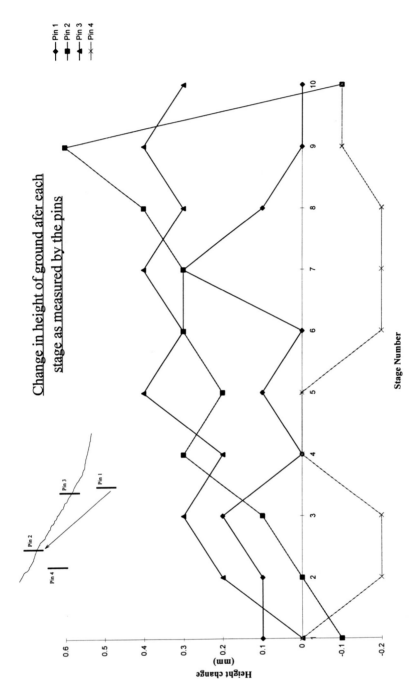

Fig. 5. Graph showing the change in height of the ground measured by the device. The letters in the legend refer to the plumb lines labels (see Fig. 1 and 2).

They do not, however, show a constant loss of profile height, but this is to be expected since sediment moving down-slope will be deposited intermittently behind coarse particles or in micro-depressions. This will naturally depend on the supply of water, but will nevertheless produce a micro-profile that in the short term varies considerably and, in the long term, will show a net decrease.

This preliminary experiment demonstrates that there is more variation in the measurements taken from the single erosion pins due to their influence on sediment movement. The device can effectively measure small changes in the profile of a slope, since the measurements are made away from the influence of the pins.

6   Initial difficulties

During the field tests and preliminary experiment a few problems were identified. Firstly and most importantly, the device had to be levelled accurately every time. This might seem unnecessary, especially as the height of each end of the bar above the ground is recorded, but it does help to identify any movement of the pins themselves. Secondly, at first it was difficult to align the two pins, so that the device sat comfortably on them. This problem was overcome by roughly levelling the device as it sat on the first pin. The other leg would then indicate where the second pin had to be located. Finally, the process of measuring the length of each plumb line was very important and had to be carried out accurately each time. This is of particular importance since minute changes in profile height may be obscrured by measurement errors.

7   Conclusion

The profiling device described here does meet the five requirements outlined at the start of the paper.

(1) It can produce a large quantity of erosion data without compromising its accuracy.
(2) It can do so rapidly.
(3) It weights relatively little and so can be carried easily whether in a car or on foot.
(4) The need to continually monitor the site is removed since only the pins remain at the sites and they only need to be visited periodically.
(5) Finally, the construction of the device was simple; it required no unusual facilities nor materials and was overall very cheap to build.

Although some devices do exist that can masure erosion more accurately and with greater efficiency than this device, such a simple device still has its place. It entirely fulfils the needs of projects, such as those carried out by undergraduates or postgraduates, where accurate data is required but finances are constrained. However, considering the many difficulties associated with measuring erosion in semi-arid regions, it would seem applicable to any study, not only those carried out by students, that focuses on the degradation of these regions.

Acknowledgements

Alistair Wells would like to acknowledge the advice and assistance of the Natural Resource Institute Site Services and the hospitality of Lindy Walsh at Cortijo Urra, Sorbas. This research was carried out during tenure of a postgraduate bursary to Alistair Wells from the University of Greenwich. We would also like to thank Pat Brown for his help with the photos and express our grateful thanks to Joanne Rosso, Peter Doyle, Fiona Cox and Anne Mather for their many helpful comments.

References

BENITO, G., M. GUTIÉRREZ & C. SANCHO (1992): Erosion rates in badland areas of the Central Ebro Basin (NE-Spain). – Catena 19: 269–286.
BRYAN, R.B. & I.A. CAMPBELL (1982): Surface flow and erosion processes in semi-arid micro-scale channels and drainage basins. – I.A.S.H. Hydr. Publ. 137: 123–133.
CAMPBELL, I.A. (1970): Micro-relief measurements on unvegetated shale slope. – Prof. Geogr. 22: 215–220.
– (1977): Sediment origin and sediment load in a semi-arid drainage basin. – In: DOEHERING, D.O.: Geomorphology in arid lands, pp. 168–185.
CURTIS, W.R. & W.D. COLE (1972): Microtopographic gauge. – Agri. Eng. 53: 17.
DISECKER, E.G. & E.C. RICHARDSON (1961): Roadside sediments production and control. – Trans. Am. Soc. Agri. Eng. 4: 62–67.
– – (1962): Erosion rates and control methods on highway cuts. – Trans. Am. Soc. Agri. Eng. 5: 153–155.
GERLACH, T. (1967): Hillslope troughs for measuring sediment movement. – Rev. Géomorphol. Dyn. 17: 173.
GERSON, R. (1977): Sediment transport for desert watersheds in erodible materials – Earth Surf. Proc. 2: 343–361.
GOUDIE, A. (ed.) (1990): Geomorphological Techniques. – 570 p., Unwin Hyman, London.
HAIGH, M.J. (1977): The use of erosion pins in the study of slope evolution. – B.G.R.G. Tech. Bull. 18: 31–49.
IMESON, A.C. & J.M. VERSTRATEN (1988): Rills on badland slopes: A physio-chemically controlled phenomena. – In: IMESON, A.C. & M. SALA (ed.): Geomorphic processes in environments with strong seasonal contrasts. Vol. 1: Hillslope processes. – Catena Suppl. 12: 139–150.
KIRKBY, A.V.T. & M.J. KIRKBY (1974a): Surface wash at the semi-arid break of slope. – Z. Geomorph. N.F. 21: 151–176.
– – (1974b): The implications of geomorphic processes for archaeological reconnaissance survey of semi-arid areas. – In: SHACKLEY, M.L. (ed.): Geoarchaeology. – Duckworth, London.
LAM, KIN-CHE (1977): Patterns and rates of slopewash on the badlands of Hong Kong. – Earth Surf. Proc. 2: 319–332.
LE ROUX, J.S. & Z.N. ROOS (1986): Wash erosion on a debris-covered slope in a semi-arid climate. – Z. Geomorph. N.F. 30: 477–483.
LEWIS, L.A. (1988): Measurement and assessment of soil loss in Rwanda. – In: IMESON, A.C. & M. SALA (ed.): Geomorphic processes in environments with strong seasonal constrasts. Vol. 1: Hillslope processes. – Catena Suppl. 12: 151–165.
MOSLEY, M.P. (1975): A device for the accurate survey of small scale slopes. – B.G.R.G. Techn. Bull. 17: 3–6.
REMLEY, P.A. & J.M. BRADFORD (1989): Relationship of soil crust morphology to inter-rill erosion parameters. – Soil Sci. Soc. Amer. J. 53: 1215–1221.

ROMERO-DÍAZ, M.A., F. LÓPEZ-BERMÚDEZ, J.B. THORNEs, C.F. FRANCIS & G.C. FISHER (1988): Variability of overland flow erosion rates in a semi-arid Mediterranean environment under matorral cover Murcia, Spain. – Catena Suppl. 13: 1–11.
SALA, M. (1988): Slope runoff and sediment production in two Mediterranean mountain environments. – In: IMESON, A.C. & M. SALA (ed.): Geomorphic processes in environments with strong seasonal contrasts. Vol. 1: Hillslope processes. – Catena Suppl. 12: 13–29.
SAUNDERS, I. & A YOUNG (1983): Rates of surface processes on slope, slope retreat and denudation. – Earth Surf. Processes Landforms 8: 473–501.
SCHUMM, S.A. (1956): The evolution of drainage systems and slopes in badlands at Perth Amboy, New Jersey. – Geol. Soc. Am. Bull. 67: 597–646.
– (1964): Seasonal variations of erosion rates and processes on hillslopes in western Colorado. – Z. Geomorph. N.F. 5: 215–238.
TELEKI, P.G. (1966): Fluorescent sand tracers. – J. sedim. Petrol. 36: 468–485.
YAIR, A. & Y. ENZEL (1987): The relationship between annual rainfall and sediment yield in arid and semi-arid areas. – Catena 10: 121–135.
YAIR, A., H. LAVEE, R.B. BRYAN & E. ADAR (1980): Runoff and erosion processes and rates in the Zin Valley badlands, Northern Negev, Israel. – Earth Surf. Proc. 5: 205–255.

Address of the authors: ALISTAIR WELLS and Dr. MATTHEW, R. BENNETT, School of Earth Sciences, University of Greenwich, Medway Towns' Campus, Pembroke, Chatham Maritime, Kent, ME4 4AW.

L'évolution hydrogéomorphologique du delta du Danube Étape Pleistocène – Holocène inférieur

GHEORGHE ROMANESCU, Iași, Romania

avec 22 figures et 1 tableau

Résumé. En analysant les séries des transgressions et des régressions qui ont influencé le territoire actuel du delta du Danube on a essayé de reconstituer les paléodeltas et les paléorivières qui se sont formés pendant l'intervalle compris etre le Pléistocène inférieur et le Holocène.

On remarque l'existence de deux hauteurs prédeltaïques (Letea et Caraorman) qui ont fonctionné comme barrière pour les sédiments transportés par le Danube. En même temps ces deux témoins d'érosion constituaient une ligne de littoral qui séparait une zone occidentale lagunaire et une autre orientale, marine.

Cette ligne du littoral située entre Jibrieni et le promontoire Dunavăț se présentait sous la forme d'un cordon littoral de type flèche.

En analysant plus de 100 forages qui ont été réalisés dans le cadre du delta du Danube nous avons mis en évidence deux couches de tourbe: la première située à des profondeurs comprises entre 25 et 35 m, seulement le long du bras Sf. Gheorge; la deuxième, située à des profondeurs comprises entre 3–10 m, en étant présente au long du bras Sf. Gheorghe et aussi au long du bras Sulina.

Zusammenfassung. *Die Entwicklung des Donaudeltas in bezug auf die Hydrogeomorphologie.* – Mit Hilfe einer Untersuchungsreihe von Transgressionen und Regressionen, die das heutige Gebiet des Donaudeltas beeinflußt haben und beeinflussen, wurde der Versuch unternommen, einen Aufbau der Paläodeltas und der Paläotäler im Zeitabschnitt zwischen Unterpleistozän und Holozän zu rekonstruieren. Dabei wurde die Existenz von zwei vordeltaischen Höhen festgestellt (von zwei Reliefformen – Letea und Caraorman), die ein Hindernis für die von der Donau mitgeführten Sedimente dargestellt haben. Diese Höhen markieren auch eine Küstenlinie, die die westliche lagunäre-sumpfige Zone von der östlichen Seezone trennte. Diese Küstenlinie erstreckte sich von Jibrieni bis zum Dunavăț Kap in Form eines Pfeils.

Anhand der Untersuchung von über 100 Bohrungen im Gebiet des Donaudeltas konnte das Vorhandensein zweier Torfschichten, die ununterbrochen verlaufen, festgestellt werden. Eine der Schichten liegt in einer Tiefe von 25–35 m, die andere in einer Tiefe von 3–10 m. Man findet sie sowohl entlang des Sfîntu Gheorge Arms als auch entlang des Sulina Arms.

Summary. *Hydrogeomorphological Evolution of the Danube Delta.* – *Pleistocene-inferieur Holocene Stage.* – Studying a series of transgressions and regressions which influenced the present territory of the Danube Delta, we have tried to reconstitute the paleodeltas and paleo-valleys which developed in the interval between the Inferior Pleistocene and Holocene.

There has also been pointed the existence of two predelta heights (Letea and Caraorman) which functioned as an obstacle in the way of the sediments brought by the Danube. At the same time, these heights constituted a littoral-line which separated a swampy occidental lagoon sector from a maritime oriental sector.

The above mentioned littoral line argues the existence of a beach, situated between Jibrieni and Dunavăț cape and also the existence of a offshore bars of arrow-type.

Tableau 1. Évolution quaternaire du territoire deltaïque sous l'influence des mouvements eustatiques de la mer Noire.

Années	Périodes		EUROPE	RUSSIE	CLIMAT DE LA ROUMANIE	MER NOIRE ET LE DELTA DU DANUBE		DÉPÔTS DELTAÏQUES
+1000	HOLOCÈNE SUP.		SUBATLANTIQUE		continental-tempéré	Phase Mer Noire actuelle; Transgr. Actuelle (Transgr. Valaque - niveau 0 m actuel) Transgr. Nimphéene; Transgr. Historique. Regres. Fanagorienne - niveau de la mer -4m ou -1m.	Complexe aleuritique	Delta Sinoe Delta St. Gheorghe II et Chilia
0					tempéré froid et humide			
-1000								
-3000	HOLOCÈNE MOYEN		SUBORÉAL		tempéré chaud et sec	Phase Mer Noire Nouvelle; Transgr. Flandrienne; Transgr. Néolitique - niveau de la mer +5m.		Delta Sulina
-5000			ATLANTIQUE		tempéré chaud et humide			
-7000			BORÉAL		tempéré chaud et sec			
	HOLOCÈNE INF.		PRÉBORÉAL		tempéré froid et humide	Phase Mer Noire Ancienne - niveau de la mer -4 m.		Delta St. Gheorghe I
			DRYAS RÉCENT					
-10 000			DRYAS ANCIEN IV				Formation du "cordon initial"	Complexe psamitique (psamitique - aleuritique)
-20 000	PLÉISTOCÈNE SUP.	WÜRM	III IV	OSTASCOVO	tempéré froid et humide	Phase Mer Noire Ancienne - niveau de la mer -35m jusqu'à la position actuelle.		
-30 000			III	VALDAI MOLOGA-SEKSNA KALININ				
-35 000			II III		tempéré chaud			
-40 000			II		froid sec			
-50 000			I II		tempéré froid humide			
-75 000			I		froid sec	Phase Néoeouxinique - niveau de la mer -80 m (-35 m).		Complexe psamo-pélitique
			RISS-WÜRM	MICULINA	tempéré relativement doux	Phase Karangat - niveau de la mer 0 m ou +15 m.		
-100 000	PLÉISTOCÈNE MOYEN		RISS 1-2 2 1	MOSCOVA ODINTOVO NIPRU	froid, plus humide à des	Phase Euxine Moyen - regres. jusqu'au -15 m.		Complexe psamitique moyen
-130 000								
-370 000			MINDEL-RISS	LIHVIN	tempéré doux	Phase Uzunlar - transgr.; niveau de la mer +15 m		
-500 000			MINDEL	OKA	tempéré froid		Phase Paléoeuxine	Complexe psamitique ancien
-700 000	PLÉISTOCÈNE INF.		GÜNZ-MINDEL	PLATOV	chaud sec	Étape lacustre; la ligne du littoral à l'est de la position actuelle	Phase Postceauda	
			GÜNZ SAINT PRESTIAN		continental-tempéré		Phase Ceauda - niveau de la mer tout près de la position actuelle.	Complexe psephitique
			DONAU-GÜNZ	VILLAFRANCHIENNE SUP.	tempéré chaud et sec		Phase Gourianne - lac dans la zone centrale de la Mer Noire.	
-1 800 000			DONAU		tempéré froid			

Analysis of over 100 geological drillings executed in the Danube Delta, made possible the discovering of two strata of peat: the first, situated at depths between 25 and 35 m, present only along the Sf. Gheorghe arm and the second, situated at depths between 3 and 10 m, present either along the Sf. Georghe arm, as well as along the Sulina arm.

Quoique le delta du Danube est une formation deltaique récente que la plupart des auteurs le considére comme ayant un âge de 10000 ans (la limite Pléistocène-Holocène), cependent

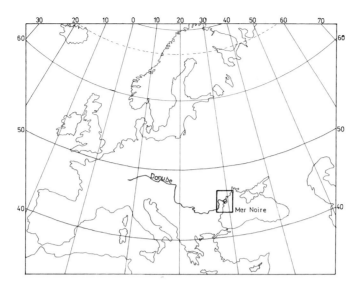

Fig. 1. La position géographique du delta du Danube dans l'Europe.

son cadre naturel de mise en évidence et de sa formation a commencé s'esquisser dépuis longtemps et peut-être élargi jusqu'à la fin de Pleistocène (Romanian Supérieur) et début Pléistocène (Saint Prestian), fait pour lequel son évolution chronologique a été partagé en deux périodes distinctes:

– l'étape paléogéomorphologique (ou du delta Pléistocène) pendant laquelle le territoire du delta a subi des mouvements verticales, dans la plupart une subsidence active surtout dans sa partie du nord, mais initialement aussi une rupture sur sa côté du sud (par rapport du nord de Dobroudja) et une autre sur sa côté du nord (par rapport au Bugeac) et ces subsidences corroborées avec les périodes glaciaires et interglaciaires ont favorisé des transgressions et des régressions repétées qui ont envahi ou non relief pré-deltaïque existent ou en y accumulant des materiaux d'origine fluviale et marine;

– l'étape néogéomorphologique (ou du delta Holocène) a commencé après le parachèvement du bourrelet de plage (du cordon) initial déjàclassique, Jibrieni – Letea – Caraorman, qui représentait en fait une zone de limite existente aux embouchures actuels et qui séparait en réalité deux milieux différents: l'un occidental lagunaire – marécageux et un autre oriental, exclusivement marin.

Pour l'interpretation de l'étape Pléistocène et Holocène on utilisera aussi une série de forages géologiques (en plus de 100 forages), efféctués à des profondeurs situés entre 2–3 m jusqu'à ceux qui dépassent 400–500 m (Fig. 4).

Étape paléogéomorphologique (ou du delta Pleistocène-Holocène)

Pratiquement cette étape se déroule pendant une longue période (Pléistocène inférieur, 1,8 millions ans et Holocène moyen ou même vers la fin de celui-ci) de sort que, en tenant compte du fait que sur le paléoterritoire deltaïque se sont passés des évenements

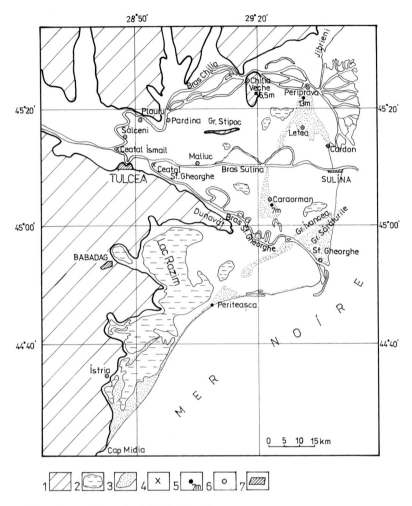

Fig. 2. La position géographique du delta du Danube.
1. Territoire prédeltaïque; 2. Lacs; 3. Levées marins; 4. Bifurcations; 5. Altitudes; 6. Villages; 7. Villes.

très importants pour l'évolution ultérieure du delta du Danube actuel on peut la partager à son tour en plusieurs phases:

- phase Villafranchienne (Gurianā)
- phase Ceauda (Saint Prestian) – le complèxe pséphitique
- phase Uzunlar – Karangat – le complèxe psamitique moyen
- phase Néoeuxine – le complèxe psamito – pélitique (psamitique supérieur)
- phase la Vieille Mer Noire – le complèxe psamito – (aleuritique), (Fig. 6).

Toutes ces phases se détachent par des traits singulières parfois à des caractères distincts, autrefois estompés mais d'une manière assez précise pour pouvoir mettre en évidence les traits des vielles unites deltaïque (paléodeltas). Le complèxe supérieur sédimen-

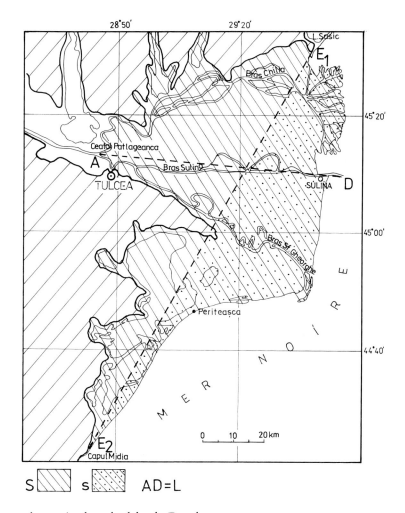

Fig. 3. Géomorphometrie plane du delta du Danube.
A = apex du delta; D = point distal du delta; E_1 et E_2 = extremités du littoral deltaïque; L = AD- longueur du delta; S = surface du delta; s = aire saillante du delta.

taires aleuritique qui représente le delta active fait l'objet d'étude de létape Holocène Historique.

Phase Villafranchienne (Gurianã)

Cette phase représente proprement-dit la période de transition qui établi une corrélation entre la fondation du delta et l'installation des premières formations à caractère deltaïque (Fig. 7). Le fait que la partie supérieure du fondement est couverte d'une couche des dépôts villafranchiennes de type "tarra rosa" mets en évidence que pendant cette période

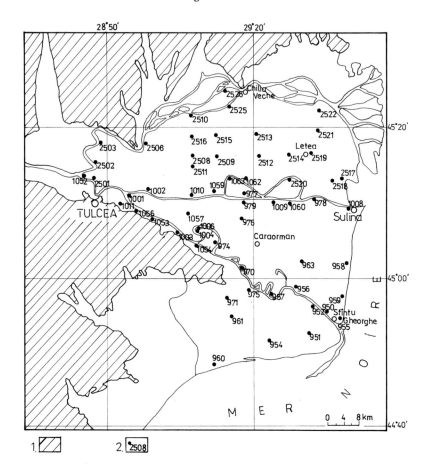

Fig. 4. Emplacement des principaux forages du delta du Danube.
1. Terre prédeltaïque; 2. Numero du forage.

existait un climat totalment différent par rapport à ceux qui ont suivi et implicitement à L'égard de celui actuel. C'était en fait un climat spécifique à la Dépression Pré-Dobroudja et au sud du Bugeac, caracterisé par des températures élevées et des précipitations réduites (POSEA et collab. 1973), qui présentait une faune des mamifères de steppe (PFANNENSTIEL 1947).

La plus importante étape dans l'évolution et le parachèvement du relief prédeltaïque qui représentait la phase Guriană de la Mer Noire se déroulait dans la deuxième partie du Villafranchien. Dans la zone de la Mer Noire pendant la phase déjà mentionnée existait en fait un lac avec Unio procumb. (PFANNENSTIEL 1947) qui couvrait la partie centrale et orientale.

La ligne du littoral en fance de notre pays se trouvait à 150 km vers est par rapport à la position actuelle (POPESCU-VOITEȘTI 1938). C'est la phase avec la surface aquatique la plus réduite et avec le niveau le plus baissé, determiné probablement par les submerssion du fondement (MURATOV 1952), par un bilan hydrologique négatif (POSEA et collab.

1969) à cause du climat sec mais torrentiel ou par la manifestation des climats froids (phases glaciaires) au nord et périglaciaires au centre et à l'est de l'Europe.

Le paléoterritoire deltaïque fonctionnait comme une zone basse, de submersion, avec un prolongement jusqu'à la partie centrale du bassin Pontique. Comme on peut remarquer dans la Fig. 8, certainement sur le territoire Villafranchien du delta du Danube, dans sa partie orientale, respectivement dans la zone actuelle du bourrelet de plage (du cordon) initial Jibrieni – Letea – Caraorman se trouvaient deux témoins d'érosion comme les collins Letea et Caraorman qui présentaient par rapport aux zones environnants des altitudes élevées et des pentes modérées (en faisant exception seulement le nord-ouest de l'élévation (Letea), en se prolongeant de l'hautereur de l'actuelle localité Periprava jusqu'aux environs du bout extrêmement orientale du promontoire Dunavăț. L'altitude relative de ceux deux témoins d'érosion était +65 m pour Letea et +45 m pour Caraorman, en s'enfonçant en même temps –35 m et respectivement –55 m par comparaison à la surface topographique du delta actuel.

Pour consolider l'idée énoncée on mentionne aussi que les bourrelets, comme on semble, sont une construction superficielle dans la morphologie du delta et que leur présence comme tel ne pourrait pas se resentir à une profondeur de –50 m ou plus bas. Cependant, du point de vue hydrochimique leur existence est confirmée à ces profondeurs. Le fait que la présence du complèxe aquifère de profondeur est signalisée partout, dans le sous-sol du delta, c'est absolument normal mais, le fait que les eaux de la profondeur ont une autre composition chimique et appartient aux autres classe d'eaux, sur les bourrelets par comparaison aux dépressions, ça peut nous indiquer que l'existence des principaux bourrelets du delta pourrait être suivi jusqu'aux époque plus vieilles du Quaternaire, ou, que les dépôts composants ont sur des grades épaisseurs un autre caractère lithologique que les dépressions. De toute manière on semble que les principaux bourrelets du delta ne sont pas des simples formations de surface, mais, comme j'ai déjà souligné, on semble qu'il y a des implications beaucoup plus profondes, surtout qu'elles représentent des propolgements des vieilles paléoformations.

Ces collins – comme nous verrons plus tard, par leur position médiane ont favorisé l'apparition et le développement du bourrelet de plage initial qui pratiquement s'est moulé sur leurs hauteurs.

C'est intéressants à remarquer que dans le sous-sol d'actuel bourrelet Sărăturile les forages n'ont pas entrecroisé des hauteurs pareilles, une construction qui nous dorige directement à la conclusion que ce bourrelet a en totalité une autre structure et origine que les autres deux.

Le fait que ce lac Gurian occupait une petite surface dans la partie centrale du bassin de la Mer Noire a permis au paléoDanube et implicitement aux paléorivières qui traversaient le territoire prédeltaïque d'avoir une pente accentuée et de transporter la plupart des sediments dans la cuvette centrale. Par les pentes accentuées des ces paléorivières évidement que leur force d'érosion et de transport était beaucoup plus grande, en favorisant en fait le transport et le dépôt d'un matériel grossier de type pséphitique (gravier et du sable grossier).

Après le tracement des isobathes des dépôts Quaternaires (prédeltaïques) on peut reconstituer et suivre assez facilement des vallées qui traversaient ce territoire avec des analigies et une continuité sur la plate-forme continentale (shelf) jusqu'au lac Gurian. Le paléocours du Danube suivait un tracé sudique dû au fait qu'il y existait aussi un enfon-

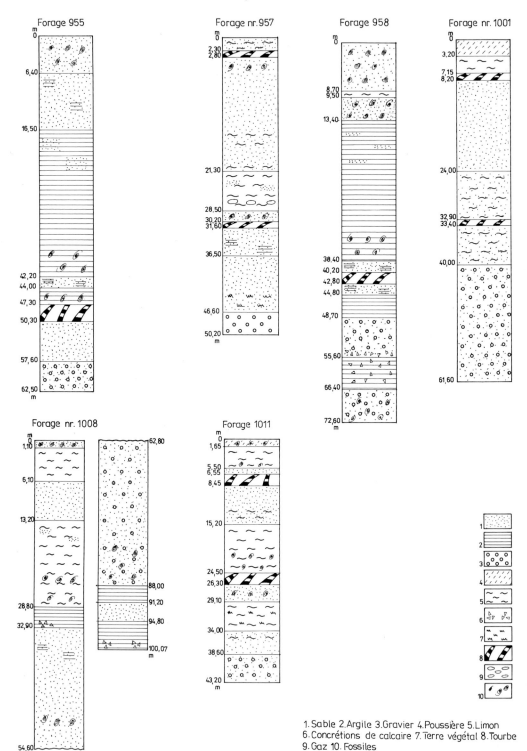

1. Sable 2. Argile 3. Gravier 4. Poussière 5. Limon 6. Concrétions de calcaire 7. Terre végétal 8. Tourbe 9. Gaz 10. Fossiles

L'évolution hydrogéomorphologique du delta du Danube 275

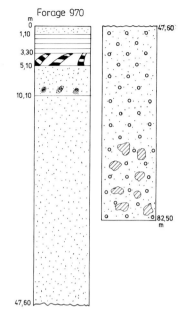

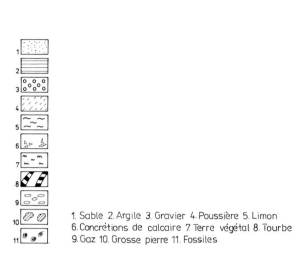

1. Sable 2. Argile 3. Gravier 4. Poussière 5. Limon
6. Concrétions de calcaire 7. Terre végétal 8. Tourbe
9. Gaz 10. Grosse pierre 11. Fossiles

Fig. 5. Forages dans le delta du Danube.
1. Sable; 2. Argile; 3. Gravier; 4. Poussière; 5. Limon; 6. Concrétions des calcaire; 7. Terre végétale; 8. Tourbe; 9. Gaz; 10. Grosse pierre; 11. Fossiles.

cement plus accentué dans ce secteur du delta par rapport à celui nordique, tracé qui s'insinuait parmi les promontoires du Dunavăț et l'hauteur de Caraorman, avec un prolongement jusque vers Sf. Gheorghe ou même au sud vers la zone Zătoanele. Dans la même mésure on peut reconstituer assez précis les paléprivières Jalpug, Katalpug et Kitai qui apportaient à leur tour de la zone de Bugeac des importantes quantités des matériaux (PFANNENSTIEL 1943, mentionne dans la zone de la Mer Noire des importantes quantités des graviers de Nistru, Nipru, Prut, Botna et aussi des vestiges d'Elephas meridionalis) qui étaient déposés surtout derrière les ahuteurs Letea – Caraorman qui fonctionnaient pratiquement comme une zone de barriere autant pour les influences d'ouest que pour celles d'est pendant qui'une autre quantité de matériel soit transporter beaucoup vers l'extérieur du delta.

En même temps jusqu'à la retrait de la mer sous la forme d'un lac, le paléoterritoire deltaïque fonctionnait dans des autres conditions géomorphologiques déjà énoncées, la surface du delta a été desséchée et transformée dans une plaine collinaire avec des vallées profondes qui ont trouvé des conditions favorables de s'enfoncer dans les argiles Villafranchiennes, soumisses à une rapide érosion dans un climat sec.

Phase Ceauda (Saint Prestian) – le complèxe pséphitique

Pour la phase lacustre de la Mer Noire (ȘELARIU 1974), une étape aussi importante, avec une grande puissance d'érosion et d'accumulation grossière est représentées par la phase Ceauda avec un déroulement pendant la période Saint Prestian, ce qui correspondrait au stade glaciair Gunz quand sur le territoire de notre pays aux altitudes réduites s'était installé un climat témpéré-continental qui à son tour a détérminé pratiquement une augmentation du niveau du lac central Pontique jusqu'à proximité de la position actuelle (POSEA et collab. 1969). C'est la période favorable de sédimentation transgressive quand la surface complèxe pséphitique (LITEANU & PRICĂJAN 1963).

Comme j'ai déjà mentionné, les hauteurs Letea et Caraorman par leur grande extension spatiale et par leur rôle de "barrière" entre le secteur oriental et occidental du paléoterritoire deltaïque ont favorisé une accumulation pséphitique accentuée dans leur partie occidentale et en même temps dans des couches assez épaiss mais à une altitude plus baissé par rapport à la situation actuelle dans la partie orientale (Fig. 7, 8, 9, 10, 11, 12, 13).

L'épaisseur de ce complèxe varie entre 10 m et 20 m dans la partie centrale du secteur fluvial du delta jusqu'à 70–80 m en face de la localité Sf. Gheorghe, lieu qui pourrait indiquer l'embouchure d'une rivière et en même temps la cote d'élévation de la transgression.

Au derrière de ces hauteurs (Fig. 9) à moitié distance entre la terre ferme continental et la ligne Letea – Caraorman, les dépôts pséphitiques présentent des épaisseurs plus grandes surtout par suite de la contribution du matériel provenu par l'intermediaire des paléorivières Bassarabiennes et des torrents nord de la Dobroudja qui sont disposés aux confluences sous la forme des cônes de déjection. D'ailleurs l'entière structure du paquet deltaïque est dans la plupart croisée.

La paléoDanube suivi dans la plupart le même trancé de la période Villafranchienne, pendant que le paléoJalpug et paléoKatalpug par conséquence à l'alluvionnement aux points de jonction supporte un léger déplacement vers ouest. Le paléoKitai présente en

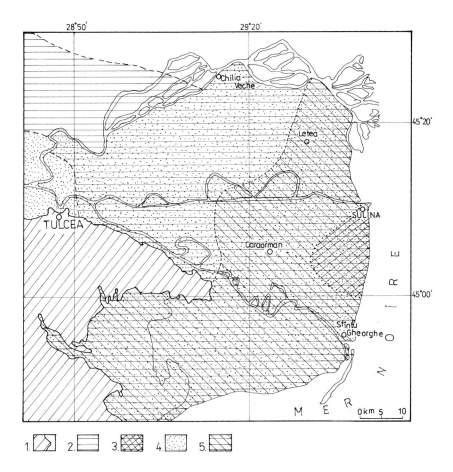

Fig. 6. Carte des transgressions Quaternaires de la Mer Noire sur le territoire actuel du delta du Danube.
1. Terre prédeltaïque; 2. Transgression Paléoeuxine; 3. Transgression Karangat; 4. Transgression Néoeuxine; 5. Transgression le stade la Vieille Mer Noire. D'après LITEANU & PRICAJAN (1961).

même temps un léger déplacement vers sud et ça c'est à la cause de remplissage de la vallée avec des matériaux fluviaux du sud de l'hauteur Letea. La composition variée de ces graviers avec du matériel carpatique et balkanique est un indice que dans l'époque correspondante a existé un réseau hydrographique très élargi du golfe qui a parachevé l'alluvionnement. Les isobathes des graviers indiquent que l'alluvionnement du delta a été fait principalement du côté de la Plaine Roumaine par le couloir Galaţi – Tulcea et l'épaisseur des graviers mets en évidence l'intensité de l'alluvionnement.

On peut facilement observer que la sédimentation pséphitique c'est beaucoup plus réduite au nord-est (ouest Letea) du delta qu'au partie du sud (ouest Caraorman), situation dûe premièrement à l'existence du Danube et ses afluent au sud et ouest et aussi à l'existence d'une seule vallée au nord-est.

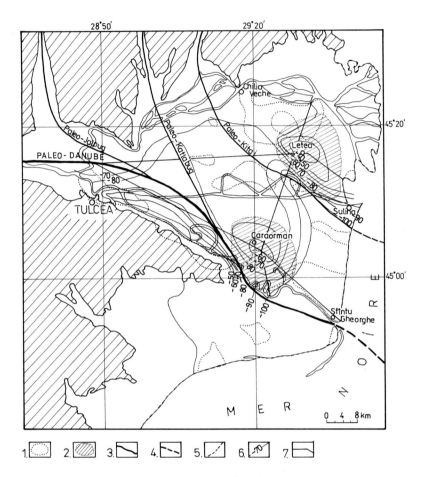

Fig. 7. Territoire du delta du Danube pendant le Villafranchienne.
1. Anomlies locales maximales correspondantes au relief prédeltaïque (îles, péninsules); 2. Périmètre des collines Letea et Caraorman; 3. Paléorivières; 4. Paléocours présumés; 5. Limite actuelle du delta; 6. Isobathes des dépôts préquaternaires; 7. Ligne des profils géomorphologiques.

Comme conséquence de la sédimentation de la zone deltaïque le perimètre des hauteurs Letea et Caraorman commence à se reduire graduellement, en présentant eu même temps une limite de recouvrement (de diminution) plus restreinte vers est et plus éöargie vers ouest.

Les intercalations d'argile du complèxe pséphitique indiquent l'existence à l'intérieur de la zone d'alluvionnements à des lacs assez élargis comme surface.

Phase Paléoeuxine – le complèxe psamitique inférieur (vieux)

Avec cette phase on achève en fait l'étape lacustre de la Mer Noir et une grade partie du relief prédeltaïque est couvert. C'est aussi la période quand se dépose la complèxe psa-

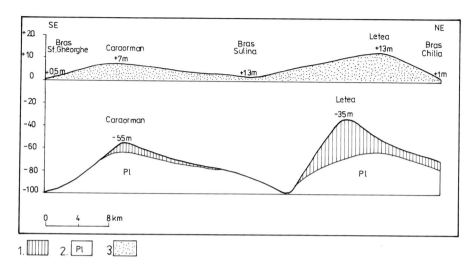

Fig. 8. Profil géomorphologique par les bourrelets Letea – Caraorman.
1. Argile rouge compacte avec des concrétions calcaires; 2. Dépôts pliocèns rélativement pas divisés; 3. Cordon littoral.

mitique inférieur (LITEANU & PRICĂJAN 1963) et qui correspond au tade glaciair Mindel du Pleistocène moyen (Fig. 14). La ligne du littoral se mentienne autour du même niveau par rapport à celui antérieur et le climat des moyennes et basses altitudes se caracterise par des températures basses avec des oscillations entre sec et humide (POSEA et collab. 1969).

Le fait que dans le processus de sédimentation se fait déjà le passage de la dimenssion pséphitique à celle psamitique grosière et moyenne nous montre un stade d'évolution plus avancé de la zone, donc une maturisation des cours d'eau, une diminution de la pente et implicitement une réduction du pouvoir erossif. La surfacetopographique qui correspond à cette période se mentienne à 40–50 m dans la partie occidentale et 50–70 m à l'est.

Les paléorivières suivent généralement le même tracé pendant que les perimètres des hauteurs Letea et Caraorman se réduisent considérable: le complèxe pséphitique inférieur recouvre déjà une grande partie de l'hauteur Caraorman qui d'ailleurs detiens aussi des altitudes plus réduites que Letea. Ce fait a déterminé aussi un déplacement de paléoDanube (paléobras Sf. Gheorghe) vers nord, situation dûe au debut de sédimentation et de subsidence au nord et aussi d'un mouvement épirogénique positif au sud. Les paléorivières bassarabiennes continuent leur déplacement lentement mais sûr vers ouest.

La sédimentation psamitique inférieure est eaucoup réduite ou même mule à l'est de l'hauteur Letea, fait qui nous indique que cette partie était pratiquement dépourvu d'un débouché ou que la plupart de la quantité d'alluvions transportés par paléoKitai était pas significative pour laisser les traces d'un alluvionnement sur une grade surface. Encore une fois ces dépôts présentent une grade épaisse au sud, autour du bras danubien et plus petite au nord et les causes ont été déjà énoncées antérieurement.

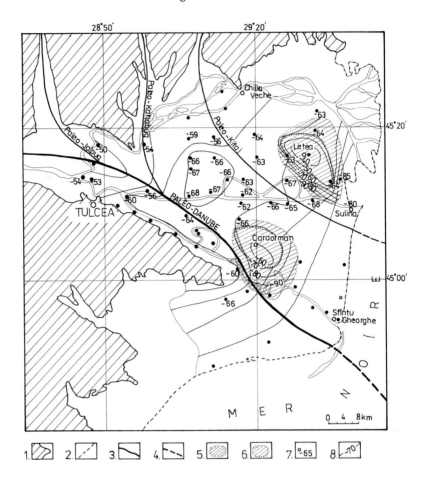

Fig. 9. Complexe pséphitique – stade Ceauda de la Mer Noire.
1. terre prédeltaïque; 2. Limite actuelle du delta; 3. Paléorivières; 4. Paléocours présumés; 5. Paléopérimètres précédents des îles Letea et Caraorman; 6. Perimètres des îles Letea et Caraorman restés découverts; 7. La profondeur du complèxe dans les forages; 8. Isobathes.

Phase Uzunlar – Karangat – le complèxe psamitique moyen

Cette phase signifie en fait le debut de l'étape fluviolacustre de la Mer Noire (ŞELARIU 1979) quand sur le territoire deltaïque se produisent des avancements et des retraites des eaux marines qui donnent naissance à des surfaces de terre ferme ou d'un golfe. On peut encadrer cette phase aux interstades Mindel-Riss et Riss-Wurm, respectivement le stade Riss, quand le climat de l'est de la Roumanie passait d'un climat doux avec nuances méditéranéennes pendant Mindel-Riss à un autre frais et humide avec variations stadiales pendant Riss et tempéré relativement doux pendant Riss-Wurm (POSEA et collab. 1973) des phases climatiques qui déterminent à leur tour une transgression en Uzunlar quand les eaux de la mer couvraient une partie du sud de la Moldavie et aussi toute la région deltaïque et quand on paraît que le niveau de la mer était situé à + 35 m (Géographie de

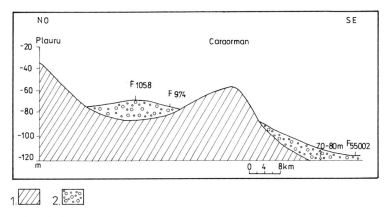

Fig. 10. Aspect du relief deltaïque pendant la phase Ceauda (Saint Prestian).
1. Relief prédeltaïque; 2. Complexe préphitique.

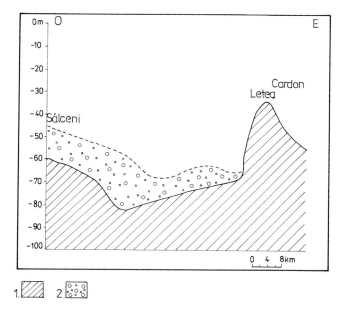

Fig. 11. Section paléogéomorphologique à travers le complexe des dépôts pséphitique.
1. Relief prédeltaïque; 2. Complexe pséphitique.

la Roumanie, vol. I, 1983), en même temps s'est produit une regression en Riss qui correspondrait à la phase Euxine Moyen, quand la ligne du littoral se trouvait quelque part à l'est par rapport à la position actuelle à −15 m (Géographie de la Roumanie, vol. I, 1983), suivit après d'une autre regression avec un retour de la ligne du littoral tout près du niveau actuel ou même à +5 m. Ces deux phases correspondraient selon les études efectués par LITEANU & PRICĂJAN (1963) sur le complèxe de sédimentation psamitique moyen (Fig. 15).

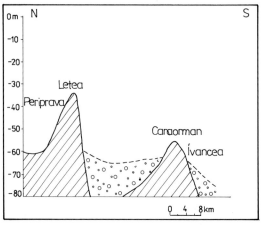

Fig. 12. Section paléogéomorphologique à travers le complexe des dépôts pséphitique.
1. Relief prédeltaïque;
2. Complexe pséphitique.

Comme conséquence à ces alternances de golfe et de terre fermé caracterisées par des sédimentations pendant les regressions, ce complèxe (Fig. 20, 21, 22) présente de temps en temps des interruptions qui peuvent indiquer la période et la dimension de regression de la zone respective et qui correspondent en fait aux tracés des vieux cours d'eaux.

Pendant cette phase tout le perimètre de l'hauteur Caraorman a été couvert avec dépôts psamitiques en retant découverts à caractère isolé, seulement quelques surfaces plus élevées pendant qu'une sédimentation plus active commence aussi au nord du delta, en face de la localité Periprava fait qui indique le changement du sens de subsidence du sud au nord.

Le fait que la transgression Karangat s'est superposée dans la plupart seulement sur la surface du bourrelet Sărăturile (Fig. 6) signifie le fait que dans cette zone existait une dépression et pas des heuteurs prédeltaïques semblables à celles de Letea et Caraorman.

Pendent que la surface topographique de l'ouest du delta se mentient entre −30 m et −40 m par rapport à celle actuelle, la partie d'est a entre −40 m et −50 m ce qui nous indique en fait le rôle de plus en plus diminué joué par les hauteurs Letea et Caraorman concernant le sédimentation et ça c'est comme conséquence du remplissement des éspaces intercollinaires par une sédimentation active.

Le rôle joué par les hauteurs Letea a permis une sédimentation plus active à l'ouest où les hauteurs ont baissé, pendant que dans la partie orientale l'épaisseur des sédiments est plus grande et présente des altitudes relativement plus prononcées.

Les cours des paléorivières, inclusivement celui du Danube, commence à former des méandres très prononcées en même temps avec leur déplacement occidental vers la position conue actuellement. La situation commence à se compliquer par une accumulation plus intense dans la zone centrale du secteur deltaïque occidental, accumulation déterminée par la torrentialité prononcée des rivières bassarabiennes et aux torrents moins importants du nord de la Dobroudja qui déposent les matériaux sous la fome des petites cônes de déjection.

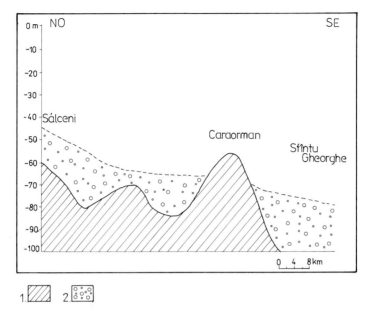

Fig. 13. Section paléogéomorphologique à travers le complexe des dépôts pséphitique.
1. Relief prédeltaïque; 2. Complexe pséphitique.

Le cours du Danube sous le poussé du matériel provenu du nord s'approche considérable de la partie continentale élevé de Dobroudja du Nord.

Le plus important évenement de cette période est que par suite de la réduction de la pente D'écoulement par l'accumulation des sédiments la vallée de Danube par surtout s'élargi que s'approfondi en se créant en même temps une vaste plaine alluviale sur laquelle coulent les eaux pendant les débordements, fait confirmé et reconfirmé par les dépôts de tourbe (Fig. 16) qui accompagnent d'une part et d'autre part son cours, en s'élargissant sur une large superficie qui ont des différentes épaisseurs ce qui nous indique d'une manière indubitable leur origine dans des cuvettes dépressionnaires situées le long de la plaine. En même temps on trouve des dépôts de tourbe qui présentent des grandes épaisseurs aussi sur le littoral actuel de la mer, entre Sulina et Sf. Gheorghe, fait ce que nous détermine à penser que par cette zone soit se dirigeait le cours de paléoKitai, soit du Danube et qui a leur tour avaient un bras secondaire par l'est de Caraorman vers Împuţita ou simplement existait un système de marécage provenu d'une surface lacustre.

Ces dépôts de tourbe sont placéc entre les stades Karangat et Néoeuxine, fait qui confirme la regression qui a eu lieu dans cet intervalle de temps et que le bras Sf. Gheorghe connaït une phase de maturation.

Phase Néoeuxine – le complèxe psamo-pélitique ou psamitique supérieur

Elle présente une importance considérable par le fait que ses dépôts ont des épaisseurs plus grandes en constituant en même temps l'objet des controversées discutions polémi-

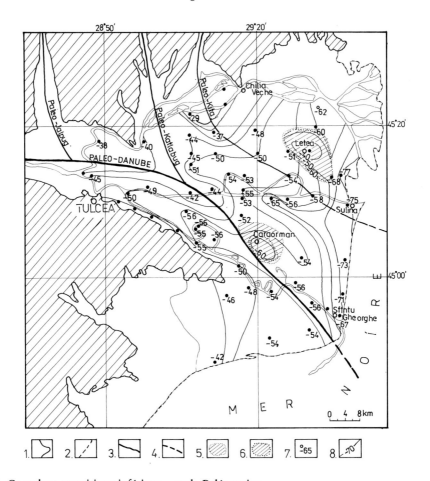

Fig. 14. Complexe psamitique inférieur – stade Paléoeuxine.
1. Terre prédeltaïque; 2. Limite actuelle du delta; 3. Paléorivières; 4. Paléocours présumés; 5. Paléopérimètres précédents des îles Letea et Caraorman; 6. Perimètres des îles Letea et Caraorman restés découverts; 7. La profondeur du complexe dans les forages; 8. Isobathes.

ques. Selon quelques auteurs cette phase correspondrait au stade Wurm I (POSEA et collab. 1973), de nature regressive avec le niveau de la mer jusqu'à –35 m (ou même à –80 m; Géographie de la Roumanie, vol. I, –40 m) mais, tout ce qui est possible qu'elle appartient en fait à toute une gamme des stades et interstades Wurmiens de nature transgressive (LITEANU & PRICĂJAN 1969).

Cette fois les grandes épaisseurs des ces dépôts se trouvent à l'ouest et aussi au centre du delta, ça signifie les hauteurs Letea et Caraorman, fait qui nous détermine à affirmer qu'un nouveau bras commence à s'organiser en suivant une direction générale d'écoulement ouest-est (Fig. 17, 18).

Pratiquement le Caraorman et Letea sont complétement couverts des sédiments Néoeuxiniques pendant que la surface topographique de l'ouest du delta se trouve seulement à une profondeur de 3–4 m par rapport au niveau actuel de la mer, et celle orientale

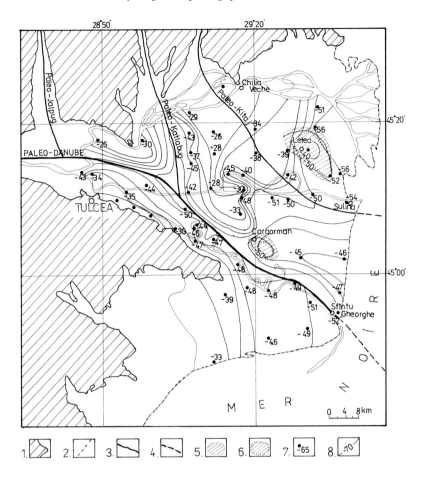

Fig. 15. Complexe psammitique moyen et supérieur – les stades Uzunlar – Karangat.
1. Terre prédeltaïque; 2. Limite actuelle du delta; 3. Paléorivières; 4. Paléocours présumés; 5. Paléopérimetres précédents des ïles Letea et Caraorman; 6. Périmètres des ïles Letea et Caraorman restés decouverts; 7. La profondeur du complexe dans les forages; 8. Isobathes.

présente encore des profondeurs jusqu'à 30–40 m et encore plus. On peut mettre en évidence le rôle joué par les hauteurs Letea et Caraorman qui ont déterminé par l'alignement formé une sédimentation favorisée à l'ouest est plus diminuée à l'est, qui en fait finit dans cette phase, en les corroborant aussi avec les transgressions stadiales locaux qui ont avancé et se sont retirées sur le littoral en fonction de ces hauteurs prédeltaïques. La prémisse d'apparition du bourrelet (du cordon) initiale de plage le long de l'alignement Jibrieni–Letea–Caraorman a été justement cette limite "d'abrupt" existente pratiquement entre les deux milieux totalment différents; ce bourrelet s'est surtout formé par l'accumulation des dépôts de plage et dérive littorale prédominante nord-sud et secondaire sud-nord.

Entre ce stade et celui Vieux de la Mer Noire il y a une deuxième couche de tourbe (Fig. 18) qui présente des caractéristiques deltaïques et épaisseurs différentes mais avec une continuité sur l'alignement ouest-est, mais cette fois prolongées sur deux secteurs longitudinals: d'une part et d'autre part du bras Sf. Gheorghe, connu comme plus vieux, déjà le classique bras et le deuxième le long de l'actuel bras Sulina, selon la constatation et les interprétations faites en lisant les forages éfectués dans le delta du Danube.

Les pentes de la partie d'ouest du delta sont déjà très réduites, de type deltaïque, de sort que dans une période très courte, ce jeune bras Sulina a réussi se ranger une vallée et se créer aussi en temps une plaine alluviale élargie qui était envahie par les eaux pendant les débordements. Les dépôts de tourbe situés sur ce dernier bras présentent une continuité et ils suivent en grande partie son vieux tracé sinueux (Marele "M" et Micul "M"). Pendant toute cette période, le secteur du nord de la Dépression Pardina était occupé par un grand lac intense colmaté.

Si on regarde avec beaucoup d'attention la carte présenté à la Fig. 18 on peut observer assez facilement que les derniers dépôts de tourbe (qui présentent des épaisseurs plus petites par comparaison aux autres) se trouvent approximativement à la limite de l'alignement Letea – Caraorman, en indiquant précisément que le littoral à cette époque-là s'y trouvait pas loin de lui et la partie d'est en étant sans doute un milieu marin et celle occidentale à ce moment-là était fluviale de type deltaïque.

En faisant une correlation entre les dépôts de tourbe qui se trouvent en fait entre les deux stades Néoeuxine et Vieux de la Mer Noire, mais en faisant aussi une comparaison entre les opinions de la plupart des chercheurs qui se sont occupés de la genèse et de l'âge du delta du Danube, on peut observer que la separation du milieu marin de celui fluvial s'est produit pendant la période ou a suivi immédiatement après le stade Néoeuxine et que ce très discuté "cordon initial" est de nature marine formé par l'accumulation des plages et aussi par la contribution du matériel porté par le courant à direction nord-sud, mais surélévé aussi pendant les périodes qui vont suivre jusqu'à l'altitude actuelle.

Cette supposition est consolidée et en quelque sorte confirmée aussi par les études de PANIN (1973, 1983), des données que l'auteur ne les considère pas édifiantes même qu'elles se fondaient sur des études réalisées entre les années 1973 et 1983 en suivant un programme de collaboration entre l'Institut de Géologie et Géophysique Bucarest et l'Université de Georgia – Étas Unis. L'hypothèse emis par l'auteur est intéressante et constitue un point de référence dans la connaissance des étapes d'évolution deltaïque pour la période Holocène et peut-être acceptées dans sa plupart. Selon N. Panin, cette hypothèse a été établi à la suite des déterminations faites avec radiocarbone (C_{14}) sur 130 échantillons de sol deltaïque avec leur entier contenu paléontologique recueillis des points considérés les plus caractéristiques. On a tenu compte de la variété des espèces des mollusques qui ont été datées séparément pour chaque échantillon, aussi de leur degré de débris dû aux conditions hydrochimiques dans lesquelles se sont déposées et encore des corrélations stratigraphiques. Dans quelques cas les analyses radiocarbones ont daté des âges très grands ou contradictoires (à l'avis de l'auteur), plus de 40000 ans (c'est pour ça qu'on analyse ici cette hypothèse au cadre de la période Pléistocène) qui font impossible l'explication de la genèse du delta pendant Holocène comme y est décrite. Ces résultats ont été exclus par l'auteur en précisant que le respectif matériel fossile est provenu à la suite des forts rémaniements et ça, selon les données déjà montrées elles peuvent être mises sous le point d'interrogation.

En réalité nous nous occupons de l'évolution Pleistocène-Holocène du delta du Danube, qui a eu quelque sorte une évolution différente par rapport à celle Holocène-Historique comme on peut voir par le fait que le delta est paru et disparu en fonction de l'amplitude des mouvements transgressifs qui l'ont couvert dans cette première période, en étant en réalité des paléodeltas, aujourd'hui avec un caractère fossile.

Ces derniers dépôts deltaïques fossiles de tourbe ont en fait l'âge de 30000–40000 ans. Généralement les plus avancés âges paraît dans la zone du bourrelet initial Letea – Caraorman où l'étude paléontologique de la faune de cette zone indique une forte mixture des espèces de Holocène avec les autres plus vieilles (probablement d'âge Karangatien – Riss-Würm). Ca serait une preuve que l'alignement du bourrelet initial a eu un rôle déterminant dans l'évolution des ces deux parties distinctes du delta, à l'ouest et respectivement à l'est de lui. On retient comme intéressante l'afirmation que les espèces des molusques à des âges avancés ont été certainement remodelés pendant les stades plus jeunes et les vrais âges du sol ont été le mieux estimé par les espèces qui avaient les âges moins avancés.

Le fait que la plupart des chercheurs ont déterminé l'apparition du bourrelet initial vers 10000 ans ne nous empêche pas à affirmer que le paléodeltaen soi qui s'est formé en Néoeuxine, à son tour, par conséquence aux autres transgressions et regressions qui ont eu lieu postérieur à cellui-ci, a pu être encore couvert par les eaux et transformé dans un golfe de type lagune à des eaux peu profondes, en le rajeunir et le faire de nouveau actif, ce qui a donné naissance à un dernier début du delta, actuellement en maturation, notamment l'étape Holocène-tard. La ligne du littoral était déjà tracée, haussée et remodelée par le vent, en pouvant séparer en même temps les deux milieux: fluvial et marin, en étant à son tour aprovisionné avec le matériel sabloneux transporté par le courant littoral à direction nord-sud. Vraiment presque tous les chercheurs scientifiques considèrent que les bourrelets Letea et Caraorman sont une création mixte de la mer et du Danube, la direction recourbée des fascicules des bourrelets et dunes indique le sens dans lequel s'est déposé le matériel alluvionnaire.

MUNTEANU-MURGOCI (1957) considère à son tour que ces bourrelets appartiennent (similaire au bourrelet Chilia) au continent préloessian et qui à son tour a été couvert par le sable soufflé par le vent et enmené par les eaux marins des plages d'autre fois et aussi de la base de loess, ce dernier fait en étant confirmé par le présent étude.

Phase La Vieille Mer Noire – le complèxe psamito-aleuritique

Correspond à l'entière période Wurmienne quand le climat a alterné entre celui tempéré froid et humide et celui sec ou tempéré chaud où le niveau de la mer a subi une transgression générale pendant laquelle se sont manifestées des stagnations ou des regressions moins accentuées pendant Wurm II et III. Le niveau de la mer était situé à –5 m (BANU 1961), niveau qui correspondrait à la ligne du bourrelet initial de limite entre les deux milieux de sédimentation, occidental et oriental, données qui soutiennent les affirmations antérieurs et selon les autres auteurs le niveau se trouvait à –35 m (POSEA et collab. 1969).

Au cadre de cette phase de sédimentation a été déposé le complèxe psamito-aleuritique (LITEANU & PRICĂJANU 1963) qui, selon les Fig. 19, 20, 21, 22, présente des petites épaisseurs à l'ouest du delta et grandes à est. Pratiquement dans la zone fluviale du delta ces dépôts de trouvent à des profondeurs entre quelques mètres jusqu'à quelques centimètres et en même temps, dans la zone orientale peuvent atteindre des épaisseurs de plus 30 m.

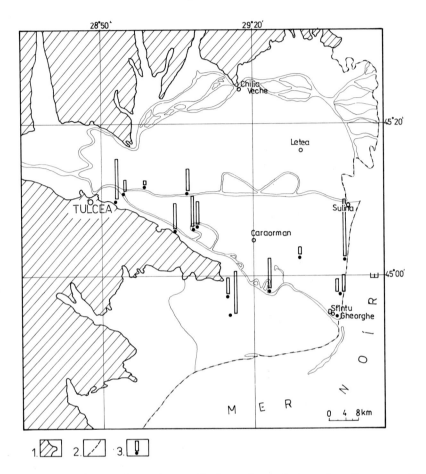

Fig. 16. Dépôts des tourbe specifique a l'intervalle d'entre les stades Uzunlar – Karangat – Neoeuxin. 1. Terre prédeltaïque; 2. Limite actuelle du delta; 3. Dépôts de tourbe (1 cm = 1 m).

En réalité la ligne du littoral était située sur l'alignement Periprava – Dunavăț où en temps s'est mis en évidence le cordon initial par des accumulations des plages.

Cette fois la sédimentation différenciée s'est produite toujours à la cause de l'existence de la bande littorale des plages, sur le fond d'un mouvement transgresif, mais au sens inverse et plus intense à l'est et ça, après que le bourrelet littoral a été percé par le bras Sulina et le niveau de la mer de transgression s'y mentenait ou paraît à la limite de celle-ci, à des petites inflexions en face de la vielle embouchures Sulina (Fig. 6).

Même après le parachèvement de cette pénultième sédimentation de la partie orientale du delta le niveau topographique des ces dépôts se mentenait autour du profondeur de 3–4 m et même –5 m par rapport au niveau actuel. On confirme en fait l'hypothèse de BANU (1961) selon laquelle le niveau de la mer était situé à –5 m par rapport à celui actuel et qui couvrait comme surface toute la partie orientale du delta.

Par les processus de sédimentation qui se manifestaient pratiquement dans la même mésure sur toute la surface du delta paraît les prémisses de la genèse de tous les bras

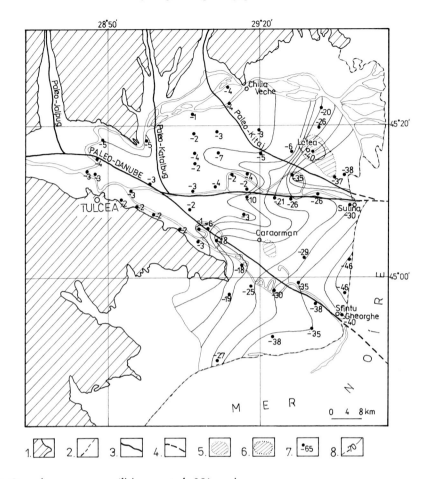

Fig. 17. Complexe psammo-pélitique – stade Néoeuxine.
1. Terre prédeltaïque; 2. Limite actuelle du delta; 3. Paléorivières; 4. Paléocours présumés; 5. Paléopérimètres précédents des îles Letea et Caraorman; 6. Périmètres des îles Letea et Caraorman restés découverts; 7. La profondeur du complexe dans les forages; 8. Isobathes.

deltaïques avec une esquise incipiente d'après toutes les possibilités aussi pour Chilia (le secteur Sireasa). On pourrait dire en fait que le paléodelta fluviale dont la surface s'est installée sur celle actuelle dans sa plupart était mise au point. En réalité les actuels dépôts aleuritiques sont une continuation avec des petites différentiations par comparaison à ceux psamito-aleuritique.

La phase La Vieille Mer Noire achève son cycle en Boreal, donc il y a aproximatif 6000 ans avant J.C. L'hypothèse de la genèse du cordon initial dans une période plus vieille que 10000 ans n'est pas nouvelle car elle a été emise par PFANNENSTIEL (1947) qui a affirmé que pendant l'intervale 65000–40000 ans qui appartient au stade La Vieille Mer Noire, le Danube submergé est étouffé par la transgression marine au niveau de l'isobathe de −21 m quand en fait commence la formation du delta du Danube à un contour initial d'estuaire en même temps avec la naissance du cordon Jibrieni–Dobroudja.

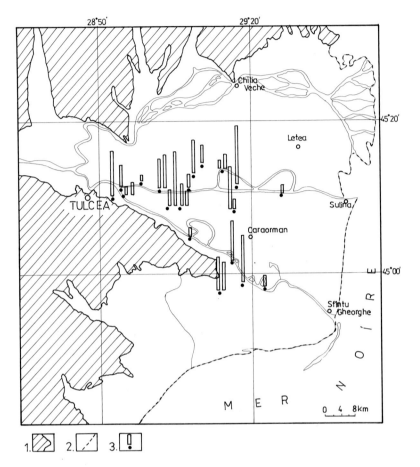

Fig. 18. Dépôts de tourbe spécifiques à l'intervalle d'entre le stade Néoeuxine supérieur et le stade Vieux de la Mer Noire.
1. Terre prédeltaïque; 2. Limite actuelle du delta; 3. Dépôts de tourbe (1 cm = 1 m).

Donc la naissance du bourrelet initial ne commence pas il y a 10000 ans, on paraît qu'il y a des antécédentes plus vieux que ce moment-là.

Un argument éloquent pour l'âge avancé de ces sédiments qui se trouvent en corrélation avec ceux déjà existents et déposés dans la même période sur tout le littoral roumain est souligné par CARAIVAN (1982) qui, en analysant une colonne des sédiments du cordon littoral Mamaia dans un niveau lumachellique du profondeur de 23–22 m formé de *Donax* sp., *Abra* sp., *Cardium* sp., *Spisula subtruncata triangula* et autres encore, a daté comme âge absolu ce matériel organogène à 26925 ± 690 ans avant J.C. ce qui confirme aussi l'âge du complèxe psamito-aleuritique de la partie orientale du delta et encore autres corrélations. Ce qui fait retenir c'est le fait que l'image actuelle de surface du delta c'est une continuation à ceux commencées dans une période plus vieille. Les derniers dépôts, donc ceux aleuritieues, à des épaisseurs réduites ne peuvent pas appartenir

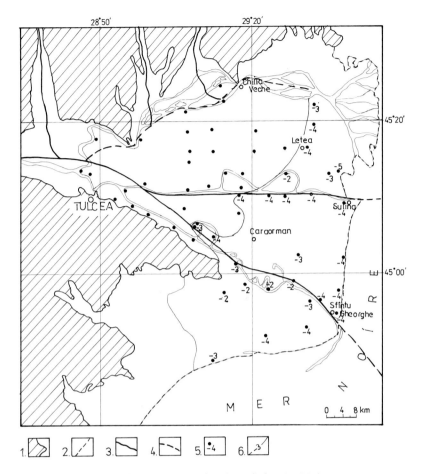

Fig. 19. Complexe psammitique-aleuritique – stade Vieux de la Mer Noire.
1. Terre prédeltaïque; 2. Limite actuelle du delta; 3. Paléorivières; 4. Paléocours présumés; 5. La profondeur du complexe dans les forages; 6. Isobathes.

qu'à la transgression actuelle et qu'ils représentent le delta présente, d'âge historique, moulé sur des caractères vieux, d'âge Holocène et même plus vieux.

Le delta, dans le vrai sens du mot, a existé aussi avant la période Holocène ou de la formation du cordon littorale classique qui séoparait le domaine occidental de celui oriental, la preuve consiste en existence de deux couches de tourbe qui se trouvent à des différents profondeurs des surfaces qu'on pourrait les dénommer *paléodeltas*.

Le parachèvement du cordon des plages et son élévation n'a pas fait autre chose que séparer les deux milieux deltaïques: l'un accidental (fluvial) et un autre oriental (fluviomarin).

Avec la phase décrite ci-dessus s'achève l'évolution Pléistocène-Holocène du Delta du Danube et celle Holocène et commence celle actuelle quand sur son territoire se sont succédées autres transgressions et regressions à des réduites amplitudes d'une grande im-

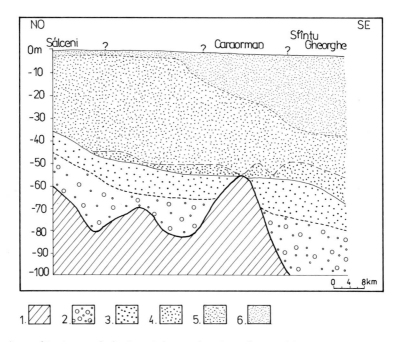

Fig. 20. Section paléogéomorphologique à travers les vieux dépôts deltaïques.
1. Relief prédeltaïque; 2. Complexe pséphitique; 3. Complexe psammitique inférieur; 4. Complexe psammitique moyen et supérieur; 5. Complexe psammo-pélitique; 6. Complexe psammito-aleuritique.

portance pour l'actuel contour du delta qui ont changé en fait, souvent partiellement, autrefois totalement l'aspect du delta ce qui sera l'objet d'étude d'un autre chapitre qui va contourer surtout l'imagé des deltas secondaires Sf. Gheorghe I, II, Sulina, Chilia etc.

Les nouvelles scientifiques à caractère personnel mentionées dans cet étude on peut les résumer de la manière suivante:

– on a mis en évidence et on a argumenté, sur la base de toute une série de forage géologiques, le rôle du fondement deltaïque dans la distribution de la couverture deltaïque et implicitement de la morphologie actuelle;

– par une analyse en détail de plus de 100 forages avec des profondeurs différentes (les uns en dépassant 500 m de profondeurs) on a reconstitué la distribution de la couverture deltaïque en espace et en temps pour chaque période de sédimentation qui s'est manifestée sur le territoire du delta;

– pour la première étape (Pléistocène) on apporte des arguments qui aident à la reconstitution de l'image des paléodeltas danubiens pour des étapes caractéristiques: Villafranchienne (Gurianā), Ceauda (Saint Prestian), Paléoeuxine, Uzunlar-Karangat, Néoeuxine et la Vieille Mer Noiree;

– on souligne aussi la présence des témoins d'érosion avec des altitudes relatives de +65 m pour Letea et +45 m pour Caraorman et en même temps on fait preuve de l'inexistence d'une telle hauteur sur le bourrelet Sărăturile;

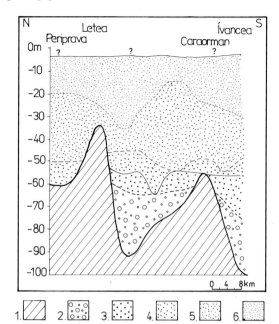

Fig. 21. Section paléogéomorphologique à travers les vieux dépôts deltaïques.
1. Relief prédeltaïque; 2. Complexe pséphitique; 3. Complexe psammitique inférieur; 4. Complexe psammitique moyen et supérieur; 5. Complexe psammo-pélitique; 6. Complexe psammito-aleuritique.

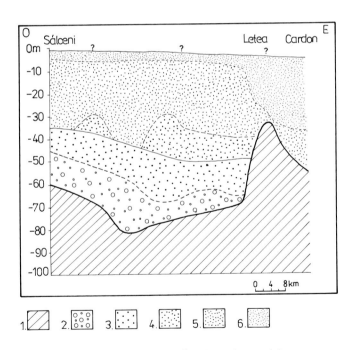

Fig. 22. Section paléogéomorphologique à travers les vieux dépôts deltaïques.
1. Relief prédeltaïque; 2. Complexe pséphitique; 3. Complexe psammitique inférieur; 4. Complexe psammitique moyen et supérieur; 5. Complexe psammo-pélitique; 6. Complexe psammito-aleuritique.

– la présence des alignements des dépôts de tourbe observés au cadre des forages constitue aussi des arguments concernant l'ancienneté et l'évolution des bras;

– la morphologie de chaque dépôt soutient et met en évidence l'inexistence d'un présumé "cordon littoral initial" et souligne l'existence d'une plage marine sur l'alignement Jibrieni – Letea – Caraorman – Dunavăț, formée à la limite d'un golfe qui avait la ligne du littoral dans cette partie du delta;

– le delta du Danube a commencé se former en saint Prestian en même temps avec les sédiments des premiers dépôts à caractère deltaïque. Les dépôts accumulés dans cette période et jusqu'en Holocène représentent des paléodeltas. Les derniers sédiments, ceux Holocènes qui appartient au complèxe aleuritique, représentent en fait l'actuel delta du Danube et qui est d'âge plus récent qu'on pensait.

Références

AIRINEI, ȘT. (1968): Măsurători gravimetrice-magnetometrice în Delta Dunării. pentru hărțile geofizice ale României. – D.S. Com. Geol., vol. LII/3 (1965–1966), Buc.

– (1969): Măsurători gravimetrice pe teritoriul Deltei Dunării. – Rev. Peuce, vol. I, Muz. Deltei Dunării-Tulcea.

ALEXANDRU, M. (1980): Câteva date privind spectrele sporopolinice ale unor depozite caracteristice din Delta Dunării. – Hidrobiologia 16, Ed. Acad. R.S. România.

ANTIPA, GR. (1913–1914): Câteva probleme științifice și economice privitoare la Delta Dunării. – Anal. Acad. fom., Mem. Secț. Șt., Seria II, t. XXXVI, Buc.

BALTAC, G. (1964): Fauna de moluște corespunzătoare stadiului Vechi al Mării Negre din depozitele Deltei Dunării. – Hidrobiologia, vol. V, Ed. acad. R.P. România, Buc.

BANU, A. (1961): Observații și măsurători asupra oscilațiilor de nivel, actuale și seculare, ale apelor Mării Negre la țărmul romanesc. – Hidrobiologia, vol. II, Ed. Acad. R.P. România, Buc.

BLEAHU, M. (1963): Observații asupra zonei Histria în ultimele trei milenii. – Probl. de geogr., vol. IX, Buc.

BRĂTESCU, C. (1923): Delta Dunării. – Imprimeria Fundației Culturale "Principele Carol", Buc.

– (1943): Oscilațiile de nivel ale apelor și bazinului Mării Negre în Quaternar. – M.Q., Imprimeria Națională, Buc.

CARAIVAN, G. (1982): Evoluția zonei Mamaia în Cuaternarul târziu. – Pontica, nr. XV, Muz. de Istorie Națională și Arheologie, Constanța.

COTET, P. (1969): Delta Dunării – geneză și evoluție. – Rev. Peuce, vol. I, Muz. Deltei Dunării-Tulcea.

FEODOROV, P.V. & L.A. SKIBA (1961): Oscilațiile Mării Negre și Mării Caspice în Holocen. – Anal. Rom.-Sov., Geol.-Geogr., t. XV, nr. 1, Cuc.

GÂȘTESCU, P. & B. DRIGA (1981): Evoluția țărmului Mării Negre între brațele Sulina și Sf. Gheorghe. – Delta Dunării-studii și comunicări de entomologie, vol. 2, Tulcea.

GÂȘTESCU, P. și colab. (1986): Modificările țărmului românesc al Mării Negre. – Univ. din București.

IONESCU-DOBROGEANU, M. (1921): Delta Dunării. – Bul. Soc. Regale Rom. de Geogr., 40, Buc.

LITEANU, E., A. PRICAJAN & G. BALTAC (1961): Transgresiunile cuaternare ale Mării Negre pe teritoriul Deltei Dunării. – St. si Cerc. Geol., vol. VI, nr. 4, Buc.

LITEANU, E. & A. PRICĂJAN (1963): Alcătuirea geologică a Deltei Dunării. – Hidrobiologia, vol. IV, Ed. Acad. R.P. România, Buc.

MIHĂILESCU, N. (1989): The evolution of the fluviatile network of the Danube delta in the Pleistocene and Holocene. – Travaux du Museum d'Historie Naturell Gr. Antipa, 30, Buc.

MURATOV, M.V. (1961): Istoricul cuaternar al Mării Negre si compararea acestuia cu istoricul Mării Mediterane. – Anal. Rom.-Sov., Seria Geol.-Geogr., 3, Buc.

MURGOCI, GH.M. (1957): Opere alese. – Ed. Acad. R.P. România, Buc.

PANIN, N. (1974): Evoluția Deltei Dunării în timpul Holocenuliu. – Inst. geol., Geofiz., stud. de geol. cuaternarului, seria H., nr. 5, Buc.
– (1983): Black Sea coast line changes in the last 10000 years a new attempt at identfying the Danube mouths as described by ancients. – Dacia, N.S., t. XXVII, nr. 1–2.
PASKOFF, R. (1985): Les deltas. – "Les Litoraux", Masson.
PETRESCU, I.GH. (1957): Delta Dunării – geneză și evoluție. – Ed. Științifică, Buc.
PFANNENSTIEL, M. (1950): Die Quartärgeschichte des Donaudeltas. – Selbstverlag des Geographischen Instituts der Universität Bonn.
POPP, N. (1958): Foraje în Delta Dunării. Interpretare geomorfologică și hidrogeomorfologică. – Hidrobiologia, vol. I, Ed. Acad. R.P. România, Buc.
POSEA, GR. (1983): Pedimentele din Dobrogea. – Sinteze geografice. Ed. Did. și Ped., Buc.
POSEA, GR. și colab. (1974): Relieful României. – Ed. Șt., Buc.
ROMANESCU, GH. (1992): Noi interpretări ce privesc factorii genetici care au favorizat apariția și dezvoltarea Deltei Dunării. – Stud. și Com. "Geographica Timisiensis", Univ. Timișoara.
– (1992): Date noi cu privire la controversata insulă Peuce. – Lucr. sem. Geogr. "Dimitrie Cantemir", Iași.
ȘELARIU, O. (1979): Studiu morfohidrografic al platformei continentale din sectorul românesc al Mării Negre (teză de doctorat). – Inst. de Geogr., Buc.
VÂLSAN, G. (1935): Remarques complementaires a propos de la nouvelle hypothese sur le delta du Danube. – Bul. Soc. Rom. Geogr., Buc.
ZENKOVICI, V.P. (1957): Enigma Deltei Dunării. – Anal. Rom.-Sov., Geol.-Geogr., nr. 1, Buc.
– (1963): Zona de vărsare a Dunării. – Monografie hidrologică, Buc.
– (1967): Dunărea între Bazias și Ceatalul Ismail. – Monografie hidrologică, Inst. Stud. Hidro., Buc.

Adresse de l'auteur: Cercet. dr. GH. ROMANESCU, Academia Română – filiala Iași, Bulevardul Copou 8A, 6600 Iași, România.